21

A Specialist Periodical Report

Colloid Science
Volume 1

A Review of the Literature Published during
1970 and 1971

Senior Reporter
D. H. Everett, M.B.E., *Department of Physical Chemistry,
University of Bristol*

Reporters
S. G. Ash, *Thornton Research Centre, Chester*
J. M. Haynes, *University of Bristol*
R. H. Ottewill, *University of Bristol*
K. S.W. Sing, *Brunel University*
B. Vincent, *University of Bristol*

The Chemical Society
Burlington House, London W1V 0BN

ISBN: 0 85186 508 9

Library of Congress Catalog Card No. 72–95096

Printed in Great Britain by
Adlard & Son Ltd.
Bartholomew Press, Dorking

Preface

The initiation of any new series of progress reports in science involves a number of policy decisions. The first concerns the choice of a convenient date from which to commence a detailed literature survey, and the second, the extent to which earlier work should be incorporated. Most of the chapters in this first volume cover the literature from late 1969 to the end of 1971. However, it is not feasible to give a useful account of recent progress without assessing the state of development of the subject at the start of the period under review. This has been achieved in some cases by reference to earlier review articles, but in several instances it has been considered desirable to give a much fuller introductory exposition. In part, at least, the need to prepare a number of basic reviews before describing recent work has made the interval between the end of the period covered and publication longer than desirable; it is hoped to rectify this in subsequent volumes.

Additional problems arise in a field as diverse as that of Colloid Science, where it is difficult, and in some ways undesirable, to define the boundaries of the subject too sharply. The number and range of journals in which relevant work is published then makes it difficult to ensure complete coverage. The primary bibliography for the present volume has been drawn from *Chemical Abstracts*, despite certain shortcomings in the coverage of fringe areas, and the inevitable delay which reliance on an abstract journal involves.

The scientific principles underlying Colloid Science are in some cases not firmly established and an important function of reports such as this is to attempt a critical assessment of new ideas so that readers can make a considered judgment of their relevance and importance. At the same time an essentially complete bibliography of related but less fundamental papers is often of considerable value. Some balance of treatment has been aimed at: the reader may judge the adequacy of the resultant text for his particular needs.

It is clearly impossible to review the whole of Colloid Science in one volume. The first volume, therefore, deals with a selected group of topics which will be supplemented in the second and subsequent volumes to cover most of the important areas of Colloid Science. Certain of the major fields of endeavour will be reviewed in each volume, while others will be dealt with at longer intervals.

To assist the reporters in ensuring the widest possible coverage, and to avoid delays arising from the need to search abstracts, authors are invited to forward reprints of their relevant papers to the Senior Reporter, Specialist Periodical Report on Colloid Science, School of Chemistry, Bristol BS8 1TS, England.

Bristol, June 1973   D. H. Everett

Contents

Chapter 3 Polymer Adsorption at the Solid/Liquid Interface
           By S. G. Ash

Chapter 4 Capillarity and Porous Materials: Equilibrium
           Properties
           By D. H. Everett and J. M. Haynes

Chapter 5 Particulate Dispersions
By R. H. Ottewill

1
Adsorption at the Gas/Solid Interface

BY K. S. W. SING

1 Introduction

Numerous studies of gas/solid interfacial systems were reported in the years 1970 and 1971, with much attention given to the physisorption of gases on carbons, oxides, clays, and zeolites and to chemisorption on metal and oxide catalysts. The main objective of much of this work has been to elucidate the surface structure and texture of porous and finely divided materials of industrial importance (*e.g.* catalysts, desiccants, and pigments) and various experimental techniques were developed for this purpose. Advances were also made in the application of the principles of statistical thermodynamics and quantum mechanics to both chemisorption and physisorption.

It would be impossible to deal adequately in this Report with all of these areas of research. By restricting the scope of this chapter to physisorption and its role in the characterization of solid surfaces, an introduction is provided to certain of the later chapters in this first volume and also to the more specialized aspects of the gas/solid interface to be surveyed in subsequent volumes. This approach should avoid any appreciable overlap of subject matter with the series of Specialist Periodical Reports on 'Surface and Defect Properties of Solids', which deals *inter alia* with various aspects of chemisorption and catalysis.

Physisorption occurs whenever a gas (the *adsorptive*) is brought into contact with an evacuated solid (the *adsorbent*). The phenomenon is thus a general one and is dependent on those intermolecular attractive and repulsive forces which are responsible for the condensation of vapours and the deviations from ideality of real gases. In physisorption (as distinct from chemisorption) there is no electron exchange between the adsorbed species (the *adsorbate*) and the adsorbent.

In the least complicated cases (the adsorption of a non-polar molecule on a homopolar surface), dispersion forces provide the main attraction between the adsorbate molecule and the assembly of force centres in the adsorbent. In other cases (the adsorption of a polar molecule on a heteropolar surface), various types of specific adsorbent–adsorbate interactions may contribute to the adsorption energy.[1,2] The importance of specificity within the context of

[1] A. V. Kiselev, *Discuss. Faraday Soc.*, 1965, No. 40, 205.
[2] R. M. Barrer, *J. Colloid. Interface Sci.*, 1966, **21**, 415.

1

physisorption has only been fully appreciated in recent years and increasing attention is being given to its assessment. It seems appropriate therefore to discuss this aspect of physisorption in some detail.

Most adsorbents of high surface area are porous; to discuss the effect of porosity on physisorption it is helpful to classify pores into three groups on the basis of their effective width.[3] The narrowest pores, of width not exceeding about 2.0 nm (20 Å) are called *micropores*; the widest pores, of width exceeding about 50 nm (0.05 μm or 500 Å) are called *macropores*. The pores of intermediate width, which were for a time termed intermediate or transitional pores,[4] are now referred to as *mesopores*.

The whole of the accessible *micropore volume* may be pictured as adsorption space,[4] since in pores of these dimensions the adsorption fields of opposite walls overlap. The micropore volume is thus filled by adsorbate molecules at fairly low relative pressure (*i.e.* within the region of the adsorption isotherm below the conventional 'monolayer capacity'). The filling of micropores may therefore be regarded as a primary physisorption process. On the other hand, capillary condensation in mesopores is always preceded by the formation of an adsorbed layer on the pore walls and is consequently a secondary process; this aspect of the problem is dealt with in Chapter 4.

Although a clear distinction may be drawn in principle between the processes occurring in micropores and mesopores, in practice it is difficult to specify this difference in terms of the characteristic features of a real system. There are two underlying problems: first, the adsorbent properties of a solid are determined both by its texture (area and porosity) and by the adsorbent–adsorbate and adsorbate–adsorbate interactions, which occur in a unique way in each adsorption system; secondly, the pores in a real solid are generally distributed over a wide range of both size and shape. In view of these difficulties, it is hardly surprising that the interpretation of isotherms, in terms of monolayer–multilayer adsorption and micropore filling, has been the subject of much debate over the past few years.

With the growing awareness of this complexity of physisorption has come the appreciation of the need for the determination of standard adsorption data on carefully prepared and well-characterized solids. Graphitized carbon blacks probably represent the best examples of adsorbents with uniform (homotattic) surfaces. Certain low-temperature isotherms (*e.g.* of Ar or Kr) on graphitized carbon blacks exhibit a stepwise character indicating well-defined layer-by-layer adsorption. Oxide surfaces are generally energetically heterogeneous; they are hydrated (hydroxylated) unless they have been heated to a high temperature. The effect of the surface dehydroxylation of silica on the physisorption of a number of vapours has been studied in great detail and the results have revealed the sensitivity of the adsorption heats

[3] I.U.P.A.C. Manual of Symbols and Terminology, Appendix 2, Part I, Colloid and Surface Chemistry, *Pure Appl. Chem.*, 1972, **31**, 578.
[4] M. M. Dubinin, *J. Colloid Interface Sci.*, 1967, **23**, 487.

and isotherms to the change in the character of the adsorbent–adsorbate interactions.

In this Report, current theories of physisorption are discussed in relation to the adsorption potential and to the adsorption isotherm, with a final section devoted to empirical methods of isotherm analysis. Emphasis is thus placed on the interpretation of adsorption data, rather than on any theoretical treatment *per se*. Reference is made to a wide range of gas–solid systems, but the surface properties of particular adsorbents are left for detailed consideration in a subsequent Report.

2 The Adsorption Potential

Adsorbate–Adsorbent Interactions on Non-porous Solids.—The importance of *a priori* calculations of the potential energy (ϕ) of an atom or molecule in the force field of a solid surface has long been recognized, and a considerable literature dealing with this subject has accumulated. The earlier work has been discussed in some detail by Young and Crowell,[5] and more recent developments have been reviewed by Crowell,[6] Steele,[7] and Pierotti and Thomas.[8]

The potential energy depends both on the distance, z, of the atom from the surface and on the location relative to the lattice of the solid. If the position of the foot of the normal from the atom to the surface is defined in terms of a vector τ in the plane of the surface, relative to some chosen point in the lattice, then

$$\phi = \phi(z, \tau). \tag{1}$$

In the earlier work concerned mainly with adsorption by ionic crystals, the values of ϕ were calculated for certain specific values of τ, *e.g.* over the centre of the lattice cell and the midpoints of the lattice edges. Recently, various attempts have been made to express $\phi(z, \tau)$ as a periodic function of τ. Attention has been confined mainly to simple adsorptives such as the noble gases, which in the case of He, requires that account be taken of quantum mechanical effects. Studies of the adsorption of He on Kr and Xe crystals are of special interest because they may be expected to improve our understanding of energetic heterogeneity in physisorption.

The Lennard-Jones (12:6) potential* is generally used[9-15] to give the

---

* By an ($m:n$) potential is meant one which gives the potential energy of interaction between two molecular centres as the following bireciprocal function of their separation (r):
$$\varepsilon(r) = Ar^{-m} - Br^{-n}$$

[5] D. M. Young and A. D. Crowell, 'Physical Adsorption of Gases', Butterworth, London, 1962.
[6] A. D. Crowell in 'The Solid/Gas Interface', ed. E. A. Flood, Dekker, New York, 1967, vol. 1, p. 175.
[7] W. A. Steele, *Adv. Colloid Interface Sci.*, 1967, **1**, 3.
[8] R. A. Pierotti and H. E. Thomas in 'Surface and Colloid Science', ed. E. Matijevic, Wiley-Interscience, New York, 1971, vol. 4, p. 93.

single pair He–substrate atom interaction, $\varepsilon(r)$:

$$\varepsilon(r) = \varepsilon^0 \left[\left(\frac{r^0}{r} \right)^{12} - 2 \left(\frac{r^0}{r} \right)^6 \right], \tag{2}$$

where r is the distance between the two interacting particles, and r^0 is the equilibrium distance corresponding to the minimum potential energy ε^0. In the case of He–Xe, $\varepsilon^0 = 66.24 \times 10^{-23}$ J and $r^0 = 0.372$ nm (3.72 Å), as calculated by the usual combination rules for unlike pairs.[9] Ricca and his co-workers[9-11] carried out their calculations of potential energy as follows: for each point of co-ordinates x, y, z (x- and y-axis lying in the plane of surface atoms, z-axis outwards), the potential energy $\phi(x, y, z)$ was calculated as the discrete summation over all atoms of solid contained within a sphere centred at point x, y, z with radius equal to three times the edge, a^0, of the unit cell of the solid together with integration for uniform density outside the sphere.

The potential energy of adsorption is generally expressed in the topographical form, *i.e.* $\phi^0(x, y)$ at z^0 as defined by the conditions $(\partial\phi/\partial z)_{z=z^0} = 0$. For convenience the potential energies are often normalized or referred to the minimum value for single pair interaction. Maps of such potential energy surfaces have served to identify the adsorption sites.[9-12] For example, in the case of a He atom on the (110) face of solid Xe, the adsorption site is located at the centre of the elementary surface cell, directly above an atom in the underlying layer.[10] Two saddle points of different energies connect adjacent sites.

An alternative method of representing the variation of potential energy in three dimensions, adopted by Ricca and Garrone,[10, 11] is to plot the iso-potential contours on the three orthogonal planes intersecting at the point of minimum potential. Such a representation provides a clearer picture of the gradients of potential energy in various directions.

The closely spaced contour lines on the x–y co-ordinates (*i.e.* the plane parallel to the adsorbing surface) indicate the steep increase in potential as the adsorbed He atom moves across the surface towards the surface atoms of Xe. This lateral movement is equivalent to the penetration of the adsorbed atom into the crystal structure of the adsorbent since the point of minimum potential is at a distance of only 0.134 nm from the surface plane. This analysis also showed that the probability of migration from one site to another is very different in the directions parallel to the two normal edges of the surface cell.

The other maps referring to the vertical planes containing the two different saddle points revealed the importance of the role played by atoms in the

[9]   F. Ricca, C. Pisani, and E. Garrone, *J. Chem. Phys.*, 1969, **51**, 4079.
[10] F. Ricca and E. Garrone, *Trans. Faraday Soc.*, 1970, **66**, 959.
[11] C. Pisani and F. Ricca, *J. Vac. Sci. Technol.*, 1971, **9**, 926.
[12] R. J. Bacigalupi and H. E. Neustadter, *Surface Sci.*, 1970, **19**, 396.
[13] J. S. Brown, jun., *Surface Sci.*, 1970, **19**, 259.
[14] H.-W. Lai, C.-W. Woo, and F. Y. Wu, *J. Low Temp. Phys.*, 1970, **3**, 463.
[15] C. E. Campbell and M. Schick, *Phys. Rev. (A)*, 1971, **3**, 691.

second layer of the solid. Ricca and Garrone[10] concluded that the shape of the potential hole, which is particularly flat near the minimum, as well as the strong asymmetry normal to the surface, exclude any possibility that the adsorbed atom may be satisfactorily treated as a harmonic three-dimensional oscillator centred at the point of minimum potential energy.

Studies[10] of He adsorption on the (110) and (100) faces of Xe have confirmed that the minimum potential energies are close in the two cases [-349.1×10^{-23} J on the (110) face, and -339.9×10^{-23} J on the (100) face]. Quantum mechanical calculations, however, indicated that the energies for the fundamental state differ by about 20% (-252.3×10^{-23} J and -210.4×10^{-23} J, respectively), the (110) face providing the more stable adsorption. This result was taken to confirm the need for an adequate quantum mechanical treatment to provide a satisfactory evaluation of the effect of surface heterogeneity on the adsorption energy.

Brown[13] calculated the enthalpy of adsorption of He atoms on the basal plane of a graphite surface and included quantum corrections in the zero-point energy by adapting the Zucker approximation for condensed inert gases. The effect of including the quantum corrections was found to shift both the lattice spacing and the potential energy, *i.e.* to increase z^0 by about 6% and to decrease the enthalpy of adsorption by about 30—50%.

Avgul, Kiselev, Lygina, and Poskus[16-18] have calculated the interaction energy of a large number of simple and complex molecules with the basal plane of graphite. The basis for these semi-empirical calculations was the relation

$$\phi(z) = -C_{ij1} \sum r_{ij}^{-6}(z) - C_{ij2} \sum r_{ij}^{-8}(z) - C_{ij3} \sum r_{ij}^{-10}(z) \\ - A_{ij} \sum r_{ij}^{-6}(z) + B_i \sum e^{-r_{ij}/\rho_{ij}}, \quad (3)$$

where r_{ij} is the distance of the centre of either the adsorbed molecule i (or for a complex molecule, its atom or group i) from the centres of the carbon atoms in the graphite lattice j, and z is the distance of the centre i from the plane that passes through the centres of the outer layer of carbon atoms.

The ($6:8:10$) function (first three terms) arises from the induced dipole–dipole, dipole–quadrupole, and quadrupole–quadrupole interactions. The dispersion force constants C_{ij1}, C_{ij2}, and C_{ij3} were calculated with the aid of the Kirkwood–Müller equation and analogous equations derived by Kiselev and Poskus. It was estimated that the dipole–quadrupole and quadrupole–quadrupole terms contributed about 10% and 1%, respectively, to the total value of the dispersion interaction; the latter was therefore ignored in most cases.

The fourth term in the expression for $\phi(z)$ allows for the interaction between permanent dipoles of adsorbate molecules with induced dipoles in the graphite

[16] N. N. Avgul and A. V. Kiselev, *Chem. Phys. and Carbon*, 1970, **6**, 1.
[17] A. V. Kiselev, *J. Chromatog.*, 1970, **49**, 84.
[18] D. Poskus, *J. Chromatog.*, 1970, **49**, 146.

lattice. The induction interaction constant, A_{ij}, was estimated by means of the equation

$$A_{ij} = -\alpha_j \mu_i^2, \tag{4}$$

where α_j is the polarizability of the graphite and μ_i the dipole moment of the adsorbate molecule. The fifth term in equation (3) allows for the short-range repulsion, and the constant B_i was adjusted to satisfy the condition $(\partial\phi/\partial z)_{z=z_0}=0$. The summations of r_{ij}^{-6}, r_{ij}^{-8}, and r_{ij}^{-10} were made over 100—250 centres j and of $e^{-r_{ij}/\rho_{ij}}$ over 40—50 j; the remaining interaction was estimated by integration. In the case of complex molecules, the interaction energy of the whole molecule was estimated by summation over i groups, taking into account the possible spatial arrangements of the molecule. The value of the potential energy of adsorption at 0 K, ϕ_0, was given as the minimum of the curve. This corresponds to adsorption at zero coverage.

The enthalpy data obtained by Kiselev and his co-workers[17] for the adsorption of many different molecules on graphitized carbon are plotted in Figure 1 as a function of the molecular polarizability, α. The experimental values of q_0^{st} (minus the differential enthalpy of adsorption at zero surface coverage) were obtained by the calorimetric, isosteric, and gas-chromatographic methods.[19, 20] The agreement between the calculated values of $-\phi_0$ and the experimental values of q_0^{st} is remarkable, as is the linear dependence on α. This holds true not only for the noble gases and saturated hydrocarbons, but also for polar molecules which have lone electron pairs or π-bonds and those containing various functional groups (OH and NH). It is evident that the basal plane of graphite interacts in an essentially non-specific manner with all types of adsorbate molecules.

The somewhat higher values of $-\phi_0$ as compared with those of q_0^{st} (*e.g.* for n-C_6H_{14} and n-C_7H_{16} in Figure 1) are accounted for by the fact that these values of q_0^{st} were determined at higher temperatures. The direct measurement of the heat capacity of certain adsorption systems has revealed[21] that the enthalpies of adsorption generally decrease with increasing temperature (over 100 °C this may amount to a 5—10% change in q_0^{st}). This also explains the small difference in q_0^{st}, as determined by calorimetry at room temperature and gas chromatography at elevated temperatures.

It is clear that no allowance has been made by Kiselev and his co-workers for electrostatic interaction between a polar molecule and the graphite surface. Avgul and Kiselev[16] have suggested that Crowell's calculation[22] gives too high a value for the electrostatic interaction energy term for a simple dipole interacting with its image in the graphite. Crowell[23] has now extended this

[19] A. V. Kiselev, A. V. Kuznetsov, I. Yu. Filatova, and K. D. Scherbakova, *Zhur. fiz. Khim.*, 1970, **44**, 1272 (*Russ. J. Phys. Chem.*, 1970, **44**, 705).

[20] E. V. Kalaschnikova, A. V. Kiselev, R. S. Petrova, and K. D. Scherbakova, *Chromatographia*, 1971, **4**, 495.

[21] G. I. Berezin, A. V. Kiselev, and V. A. Sinitzyn, *Zhur. fiz. Khim.*, 1970, **44**, 734 (*Russ. J. Phys. Chem.*, 1970, **44**, 408).

[22] A. D. Crowell, *J. Chem. Phys.*, 1968, **49**, 892.

[23] A. D. Crowell, *Surface Sci.*, 1971, **24**, 651.

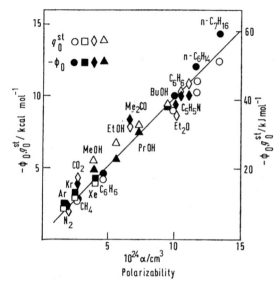

Figure 1 *The calculated potential energy of adsorption,* $- \phi_0$, *and the initial differential enthalpy of adsorbtion,* q_0^{st}, *plotted as a function of the molecular polarizability,* α, *for different substances adsorbed on graphitized carbon black.*

(Reproduced by permission from *J. Chromatog.*, 1970, **49**, 84)

calculation to the case of a nearly spherical molecule possessing a significant classical quadrupole moment, Q. It was assumed that interaction with graphite can be represented as the sum of the pairwise (12:6) potential and the interaction of Q with its image formed by the graphite conduction electrons, *i.e.*

$$\phi(z, \theta) = \phi_{12:6}(r_i) + \phi_Q(z, \theta). \tag{5}$$

According to Crowell,[23] the image interaction energy $\phi_Q(z, \theta)$ is given by

$$\phi_Q(z, \theta) = -(3Q^2/256)\,(z-\delta)^{-5}[3 + 2\cos^2\theta + 3\cos^4\theta], \tag{6}$$

where δ is the distance of the image plane from the basal plane and θ is the angle of the linear quadrupole with the normal to this plane. The maximum contribution was assessed for nitrogen by assuming $\theta = 0$ and taking $\delta = b/2$ (b, the interlaminar spacing, is 0.335 nm). The inclusion of the quadrupole term had the effect of displacing the minimum potential from $z_0 = 0.348$ nm to $z_0 = 0.341$ nm. At the uncorrected minimum, $\phi_{12:6} = 8.8$ kJ mol^{-1} compared with $\phi(z_0) = -9.5$ J mol^{-1}.

The significance of this calculation is still uncertain, but it does serve to emphasize the need for an improvement in the accuracy of the experimental data. The empirical values of q_0^{st} reported in the literature[16, 24] for the adsorp-

[24] S. Ross and J. P. Olivier, 'On Physical Adsorption', Interscience, New York, 1964.

tion of nitrogen on graphitized carbon black are spread over the range 9.2—11.3 kJ mol^{-1}.

Heteropolar or ionic solids generally exhibit energetic heterogeneity with respect to the physisorption of the noble gases and saturated hydrocarbons. A marked decrease in the magnitude of the differential enthalpy of adsorption with increase in surface coverage is due in part to the presence of growth steps and dislocations and to the difference in density of the various crystallographic planes.[25-28] Adsorption potentials have been calculated for Ar adsorption on KCl[25,26] and MgO;[27,28] in the latter case, although account was taken[27] of the effects of surface heterogeneity it was not possible to explain the high experimental values of the enthalpy of adsorption other than by assuming that the MgO contained micropores (see later).

The interaction energy for a polar adsorbate molecule on a heteropolar or ionic surface may be expressed[2] as the sum

$$\phi = \phi_D + \phi_P + \phi_{F\mu} + \phi_{\dot{F}Q} + \phi_R \qquad (7)$$

in which the dispersion energy, ϕ_D, and the short-range repulsion energy, ϕ_R, may be regarded as originating from non-specific interactions, whereas the polarization, ϕ_P, the field-dipole, $\phi_{F\mu}$, and the field gradient-quadrupole, $\phi_{\dot{F}Q}$, terms represent the specific interaction energy contributions. It appears that the contribution of ϕ_P is usually small, but that $\phi_{F\mu}$ or $\phi_{\dot{F}Q}$ may be very important with adsorbate molecules possessing permanent dipole or quadrupole moments, especially if the concentration of charge density is located on the periphery of the molecule.

Kiselev and his co-workers have studied in great detail the adsorbent properties of silica in relation to its specific and non-specific interactions with many different adsorbate molecules.[29] Their earlier investigations[1] revealed that the differential enthalpy curves for the adsorption of molecules with π-bonds (*e.g.* benzene, ethylene, and nitrogen) or peripheral lone electron pairs (*e.g.* diethyl ether, pyridine, and tetrahydrofuran) were changed drastically by a decrease in the number of hydroxy-groups per unit area of silica surface (dehydroxylation by heat treatment). On the other hand, the differential enthalpies of adsorption for comparable non-polar molecules (*e.g.* n-hexane, ethane, and argon) were not affected to any significant extent. This demonstrated very clearly that the ϕ_D, ϕ_P, and ϕ_R contributions are insensitive to the degree of surface hydroxylation, whereas $\phi_{F\mu}$ and $\phi_{\dot{F}Q}$ are dependent on the concentration of the surface hydroxy-groups.

Recent studies[17,19] have been made of the effect of the modification of the silica surface by reacting the OH groups with ClSiMe$_3$ and by preadsorbing dense monolayers. Such treatment generally leads to an overall reduction in

[25] A. V. Kiselev, A. A. Lopatkin, and E. R. Razumova, *Zhur. fiz. Khim.*, 1970, **44**, 150 (*Russ. J. Phys. Chem.*, 1970, **44**, 82).
[26] M. Leard and A. Mellier, *Compt. rend.*, 1971, **272**, B, 1477.
[27] P. J. Anderson and R. F. Horlock, *Trans. Faraday Soc.*, 1969, **65**, 251.
[28] P. R. Anderson, *Surface Sci.*, 1971, **27**, 60.
[29] A. V. Kiselev, N. V. Kovaleva, and Yu. S. Nikitin, *J. Chromatog.*, 1971, **58**, 19.

ϕ by decreasing the concentration of force centres in the outer surface and also removing (or shielding) the specifically active OH groups. It has been found[30-32] that the presence of small amounts of aluminium or boron has a profound effect on the adsorbent properties of silica. On dehydroxylation (by outgassing at temperatures around $1000\,^{\circ}C$) such impurity centres give rise to highly energetic sites with respect to the adsorption of triethylamine and tetrahydrofuran. After evacuation of the impure silica (containing about 0.4 % Al) at $1000\,^{\circ}C$, the initial differential enthalpy of adsorption for triethylamine was about $200\,kJ\,mol^{-1}$; evacuation at $200\,^{\circ}C$ (with the surface hydroxyls remaining undisturbed) resulted in a nearly constant differential enthalpy of absorption (around $82\,kJ\,mol^{-1}$) over a wide range of surface coverage of triethylamine. These results suggest that chemisorption of triethylamine occurs on the dehydrated Al sites, whereas specific physisorption takes place on the hydroxylated silica.

The similarity of their molecular size and polarizability would suggest that argon and nitrogen should be alike in their physisorption behaviour. In fact, there is remarkably close agreement between the corresponding enthalpies of adsorption, provided that the ϕ_{FQ} contribution for nitrogen is insignificant.[2,33] This is the case with graphitized carbon black,[1,16] dehydroxylated silica,[1] and various molecular solids,[34,35] but the interaction between the nitrogen quadrupole and hydroxylated silica (and other hydrated oxides[33]) results in an appreciable ϕ_{FQ} contribution ($\sim 3\,kJ\,mol^{-1}$) with these surfaces.

Specific interactions between polar adsorbate molecules and various other heteropolar surfaces have been studied in recent years.[1,2,17,29,32,33,36] Unlike hydroxylated silica, the basal plane of phthalocyanine[37] does not appear to interact specifically with benzene or other aromatic hydrocarbons. Specific interaction is shown by phthalocyanine, however, towards polar molecules with lone electron pairs (*e.g.* acetone) or certain functional groups (*e.g.* alcohols).[17] Ionic solids generally exhibit pronounced specificity, especially if multi-charged cations protrude at the surface of the crystal and the negative charge is distributed over large complex anions.[17] In the case of barium sulphate,[38] the specific contribution to the enthalpy of adsorption of benzene amounts to about $21\,kJ\,mol^{-1}$. This effect is due to the strong π-electron interaction with the surface Ba^{2+} ions.

If the cations and anions are small with similar radii, the electrostatic field changes rapidly and uniformly in the *xy*-plane above the surface. The contribution of the specific interaction energy to the total potential energy of

[30] R. E. Day, A. V. Kiselev, and B. V. Kuznetsov, *Trans. Faraday Soc.*, 1969, **65**, 1386.
[31] S. G. Ash, A. V. Kiselev, and B. V. Kuznetsov, *Trans. Faraday Soc.*, 1971, **67**, 3118.
[32] A. V. Kiselev, *Discuss. Faraday Soc.*, 1971, No. 52, 14.
[33] K. S. W. Sing, *J. Oil Colour Chemists' Assoc.*, 1971, **54**, 731.
[34] L. M. Dormant and A. W. Adamson, *J. Colloid Interface Sci.*, 1968, **28**, 459.
[35] N. K. Nair and A. W. Adamson, *J. Phys. Chem.*, 1970, **74**, 2229.
[36] M. Sanesi and V. Wagner, *Z. Naturforsch.*, 1970, **25a**, 693.
[37] A. V. Kuznetsov, C. Vidal-Madjar, and G. Guiochon, *Bull. Soc. chim. France*, 1969, 1440.
[38] L. D. Belyakova, A. V. Kiselev, and G. A. Soloyan, *Chromatographia*, 1970, **3**, 254

B

adsorption is then small if the adsorbate molecule is relatively large. This would appear to explain the fact that the specific interaction is absent in the adsorption of benzene on cubic crystals of MgO.[17]

3 Adsorbate–Adsorbent Interactions on Microporous Solids

The enhancement of adsorption energies by the 'overlapping of force fields' in pores of molecular dimensions was recognized many years ago by de Boer and Custers.[39] Although this effect has frequently been discussed in a qualitative and semi-quantitative manner,[2,4,7,40,41] few attempts[27,42,43] have been made to undertake detailed calculations of potential energies for micropore filling. The simple estimates made by de Boer and Custers[39] and by Barrer[2,40] were for cavities and apertures just large enough to accommodate an adsorbed molecule. These calculations indicated that although in certain environments considerable enhancement of the dispersion energy would be given, the magnitude of the effect would depend on the size of the pore; in sufficiently large pores, this energy of adsorption—expressed as a function of the distance across the pore—would exhibit two minima, one near each wall, separated by a maximum at the centre of the pore.[41,44]

Dubinin[4] has adopted a similar but rather less quantitative approach to distinguish between adsorption in micropores and mesopores. In his view, a characteristic feature of micropore filling is a 'substantial increase' in the adsorption energy over that given by the adsorption of the same vapour on the internal surface of mesopores.

The calculations of the enhancement of the adsorption energy in slit-shaped and cylindrical pores made some years ago by Steele and Halsey[42] were based on an $(\infty:6)$ potential (*i.e.* assuming 'hard sphere' repulsion). This work was directed towards the calculation of Henry's law constants for adsorption and little use has been made of the results, largely because of the lack of suitable experimental data with which to compare them.

Recently, Gurfein, Dobychin, and Koplienko[43] have made more extensive calculations of the adsorption energy in cylindrical capillaries. For simplicity, the pore is pictured as being made from a sheet, of one molecule thickness, wrapped into cylindrical form; each force centre in the pore wall is assumed to interact with the adsorbate molecule according to a Lennard-Jones (12:6) potential. The summation of the pairwise interaction energies was

[39] J. H. de Boer and J. F. H. Custers, *Z. phys. Chem.* (*B*), 1934, **25**, 225; see also J. H. de Boer, 'The Dynamical Character of Adsorption', Oxford University Press, 2nd edn. 1968.

[40] R. M. Barrer, *Proc. Roy. Soc.*, 1937, **A161**, 476; see also R. M. Barrer, *Nature* 1958, **181**, 176.

[41] D. A. Cadenhead and D. H. Everett, 'Industrial Carbon and Graphite', Society of Chemical Industry, London, 1958, p. 272.

[42] W. A. Steele and G. D. Halsey, *J. Phys. Chem.*, 1955, **59**, 57.

[43] N. S. Gurfein, D. P. Dobychin and L. S. Koplienko, *Zhur. fiz. Khim.*, 1970, **44**, 741 (*Russ. J. Phys. Chem.*, 1970, **44**, 411).

[44] R. M. Barrer and W. I. Stuart, *Proc. Roy. Soc.*, 1957, **A243**, 172.

replaced by integration over the continuous cylindrical surface and the potential energy, ϕ, within the pore was computed as a function of x, the distance of the molecule from the nearest point on the pore wall. A similar method was used to obtain the potential energy, ϕ^0, for the molecule at a distance x from an infinite, planar, monomolecular layer of the same density as the layer forming the walls of the pore. The ratio ϕ/ϕ^0 was expressed as a function of $2x/d$, where d, the 'diameter of the adsorbate molecule', is apparently defined as twice the equilibrium distance of the molecular centre to the wall. There is some ambiguity in the paper as to how the position of the molecule is defined in relation to the wall. Thus, in using the (12:6) equation, x is defined as the distance between two interacting molecules and must therefore refer to the separation of molecular centres; however, when x is used as the 'distance from the wall to the centre of the molecule' it appears to be measured from the *external surface* of the molecules comprising the wall. Similarly, although the pore diameter, D, is not specifically defined, it appears again to be measured from the 'surface' of the pore wall; however, since the repulsive energies are not hard-sphere repulsions this 'surface' is somewhat indefinite. Accepting these limitations, the conclusions of this work are that when $D/d > 1.5$, ϕ/ϕ^0 has two minima, each at a distance of approximately $d/2$ from the wall. For smaller pores, these merge into a single deep minimum whose depth increases as D/d decreases to 1.1. At this point the adsorption potential has been enhanced by a factor of about 3.37 over that for a plane surface.* As the pore diameter is reduced, repulsive forces begin to play a significant role and the depth of the minimum decreases rapidly until at $D/d \sim 0.93$ the adsorption potential is zero. It is clear that when $D/d \leqslant 2$ the pore can only accommodate one molecule in its cross-section so that in the range $2 \geqslant D/d \geqslant 1.5$ the molecule will move in a double minimum (better described as a ring minimum); the apparent packing density of the adsorbate will then be lower than that of the pure liquid adsorptive. Consequently, it is suggested that the decrease in apparent pore volume often observed[45] when probe molecules of increasing size are used to measure pore volumes by displacement methods may arise from a change in the packing density of the adsorbate in small pores rather than from a simple molecular sieve effect. The conclusions drawn from the results of 'molecular probe' techniques must therefore be treated with caution, as was noted in earlier studies.[45,46]

Calculations were also made by Gurfein *et al.*[43] of the effect of a distribution of pore sizes on the mean adsorption energy ϕ (obtained by weighting each energy with the appropriate Boltzmann factor). If D is taken to be a gaussian

---

* This differs from the ratio of about 4.8 obtained by Steele and Halsey,[42] who considered a cylindrical pore in a solid matrix rather than the space inside a cylindrical sheet.

[45] S. J. Gregg and K. S. W. Sing, 'Adsorption, Surface Area, and Porosity', Academic Press, London, 1967.

[46] D. H. Everett, in 'Structure and Properties of Porous Materials', ed. D. H. Everett and F. S. Stone, Butterworth, London, 1958, p. 95.

function, then ϕ decreases from its maximum value of $3.37\phi^0$ for a homogeneous assembly of pores of relative size $D/d = 1.1$, to a value of $1.4\phi^0$ when the variance about this mean reaches 0.5 (this conclusion must presumably depend on the ratio of ϕ^0 to T, but this is not stated). With the advent of computers of increasing power one may now expect calculations of this kind to be extended to pores of different shapes and to assumed intermolecular potentials of different forms: when they are, however, it is to be hoped that full numerical data will be made available, rather than only a few illustrative examples.

As noted above, experimental evidence for the enhancement of adsorption potentials in micropores in MgO has been reported by Anderson and Horlock.[27] They calculated the adsorption energy for isolated Ar atoms in slit-shaped pores of MgO, assuming them to be bounded by parallel (100) faces, and estimated that for a slit-width of 0.4—0.5 nm the initial enthalpy of adsorption, q_0^{st}, would be about 13.5 kJ mol^{-1}. The values of q_0^{st} calculated for adsorption at a surface step and at a re-entrant corner were 12.7 and 13.7 kJ mol^{-1}, respectively. The experimental values of q_0^{st} for Ar adsorption on microporous MgO were found to be above 13 kJ mol^{-1}, appreciably greater than those obtained on non-porous MgO smoke (~ 9 kJ mol^{-1}). The calculations of Anderson and Horlock[27] indicated that widening of the pores to around 1.0 nm should cause a decrease in q_0^{st} to ~ 9 kJ mol^{-1}. It was concluded that high initial enthalpies of physisorption are not generally associated with the coverage of high-energy crystallographic faces or surface defects, but are more likely to be due to the presence of geometric features (*e.g.* cracks, steps, or re-entrant corners) or micropores. Other systems in which high values of q_0^{st} have been attributed to the effect of micropore filling are argon[2] and benzene[4] adsorption on porous carbons, and benzene,[47] carbon tetrachloride,[48] and nitrogen[49] adsorption on porous silicas.

A serious difficulty which has impeded progress in the study of adsorption in micropores is the lack of independent knowledge of the pore shape and size distribution in materials such as amorphous carbons and silicas. Fortunately, more success has been achieved in recent years in the study of the adsorptive properties of porous crystals, such as zeolites, whose structures can be examined by other methods.

The dimensions of the zeolite cavities and windows are determined by the structure of the aluminosilicate framework and the nature and distribution of the exchangeable cations. Extensive studies have been made—notably by Barrer[2,40,44,50-53] and Kiselev[1,17,54,55]—of the interactions of zeolites with

[47] G. D. Cancela, F. Rouquerol, and J. Rouquerol, *J. Chim. Phys.*, 1970, **67**, 609.
[48] K. S. W. Sing, in Proceedings of the International Symposium on Surface Area Determination, 1969, ed. D. H. Everett and R. H. Ottewill, Butterworth, London 1970, p. 25.
[49] D. Dollimore and T. Shingles, *J. Chem. Soc.* (*A*), 1971, 872.
[50] R. M. Barrer and B. Coughlan, 'Molecular Sieves', Society of Chemical Industry, London, 1968, pp. 141 and 233.
[51] R. M. Barrer and J. A. Davies, *Proc. Roy. Soc.*, 1970, A320, 289.
[52] R. M. Barrer and J. A. Davies, *Proc. Roy. Soc.*, 1971, A322, 1.
[53] R. M. Barrer and E. V. T. Murphy, *J. Chem. Soc.* (*A*), 1970, 2506.

both polar and non-polar adsorbate molecules: this work has led to a broad understanding of the various specific and non-specific contributions to the interaction energy. Certain zeolites (*e.g.* decationated chabazite and NaX) have been found to give a nearly constant differential enthalpy of adsorption for the noble gases and n-paraffins over an appreciable range of uptake.[51, 52, 55] In other cases[52, 53, 56] (*e.g.* for Ar and Kr adsorption on mordenites, Ca-rich chabazite, and CaA) a greater degree of energetic heterogeneity (a decrease in q^{st} with coverage) is revealed. Energetic heterogeneity of this type appears to be a direct consequence of the preferential filling of the narrowest of the micropores, *e.g.* the side pockets lining the wide channels of mordenite. The most numerous and polarizable lattice atoms are oxygen, and their interactions with the adsorbate molecule provide the main contribution to ϕ_D. The ϕ_P contributions cannot be assessed accurately because of the rapid variation of electrical field near the wall of the cavity and in the vicinity of the cation.[51, 57, 58] However, it seems likely that the constancy of q^{st} with coverage is due in part to the compensating effect of a decrease in the magnitude of ϕ_P and an increase in that of the adsorbate–adsorbate interaction.

The importance of the ϕ_P term has been indicated by the work of Barrer and Davies[51] who found a difference of 2.9 kJ mol^{-1} between the values of q_0^{st} for Ar adsorption in Ca-chabazite and H-chabazite. This difference was attributed to the much greater local electrostatic field in the vicinity of the Ca^{2+}, which gives rise to a higher value of ϕ_P for Ca-chabazite than for H-chabazite. (Since the aluminosilicate framework remains unchanged, the difference in ϕ_D should be small.) According to this view, the difference in the ϕ_P contribution results in a displacement of the q_0^{st}–polarizability curve, as indicated in Figure 2. It will be of interest to ascertain whether the initial isosteric heats for the adsorption of Ne, Kr, and Xe on Ca-chabazite all lie on the smooth curve drawn through the single point for Ar.

Barrer and Davies[52] have also determined the values of q_0^{st} for C_2H_6, C_3H_8, and n-C_4H_{10} adsorption on H-chabazite. In these cases, there is a linear dependence of q_0^{st} on adsorbate polarizability, as was found[17] with n-paraffins on zeolite NaX. The results of André and Fripiat[59] on decationated zeolite Y are therefore surprising: they found that the values of q_0^{st} for O_2, Ar, Kr, and CH_4 adsorption appeared to *decrease* with increase in the polarizability of the adsorbate. This behaviour was tentatively explained by the assumption that the interaction energy involved an additional term to

[54] A. V. Kiselev, *Adv. Chem. Ser.*, 1971, No. 102, p. 37.
[55] A. G. Bezus, A. V. Kiselev, Z. Sedlacek, and Q. D. Pham, *Trans. Faraday Soc.*, 1971, **67**, 468.
[56] P. Braeuer, A. A. Lopatkin, and G. F. Stepanets, *Adv. Chem. Ser.*, 1971, No. 102, p. 97.
[57] A. K. K. Lee and D. Basmadjian, *Canad. J. Chem. Eng.*, 1970, **48**, 682.
[58] O. M. Dzhigit, K. Karpinskii, A. V. Kiselev, K. N. Mikos, and T. A. Rakhmanova, *Zhur. fiz. Khim.*, 1971, **45**, 1504 (*Russ. J. Phys. Chem.*, 1971, **45**, 848).
[59] J. M. André and J. J. Fripiat, *Trans. Faraday Soc.*, 1971, **67**, 1821.

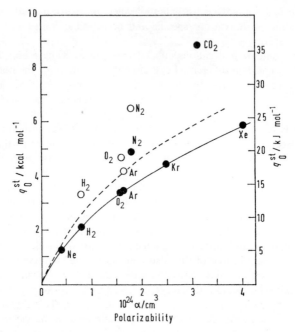

Figure 2 *The initial differential enthalpy of adsorption, q_0^{st}, plotted against the polarizability of molecules adsorbed on* H-chabazite *(●) and* Ca-chabazite *(○).*

(Reproduced by permission from *Proc. Roy. Soc.*, 1970, **A320**, 289)

allow for the change in the dipole moment of the surface hydroxy-groups. It should be noted, however, that the isotherms were only determined at two temperatures and the lack of data at low surface coverage made it difficult to estimate q_0^{st} with accuracy; more work on decationated X- and Y-zeolites is therefore required to confirm these findings.

As would be expected from earlier studies of specificity in physisorption,[1,2] zeolites have been found to exhibit pronounced energetic heterogeneity with respect to the adsorption of molecules possessing permanent dipole or quadrupole moments. The essentially new feature of the recent work on the decationation of zeolites is its revelation of the importance of the specific cation–adsorbate interaction.

The relative magnitude of the ϕ_{FQ} contribution was first assessed by Kington and MacLeod[60] for the adsorption of N_2, CO, and CO_2 on chabazite by taking for each system the difference in isosteric heat, $q_1^{st}-q_2^{st}$, at two coverages. A similar procedure was adopted by Barrer and Coughlan[50] for the adsorption of N_2 and CO_2 by NaX and other zeolites and it was concluded that the quadrupole energy provided the major contribution to q^{st}. Sargent

[60] G. L. Kington and A. C. MacLeod, *Trans. Faraday Soc.*, 1959, **55**, 1799.

and Whitfield[61] have estimated that $\phi_{\dot{F}Q} > 2\phi_D$ at the maximum interaction between CO_2 and the cation in CaA zeolite, and have suggested that this results in strongly preferred orientations of the adsorbate molecule at low surface coverage.

In their recent studies, Barrer and his co-workers[51, 53] have made use of their standard q_0^{st}–polarizability curve[2] (see Figure 2) to assess the non-specific energy contributions to the isosteric enthalpies for CO_2 adsorption on various zeolites. Thus, the difference between the experimental q_0^{st} and the interpolated value of q_0^{st} on the appropriate reference curve for non-polar adsorbate molecules appeared to provide a simple and effective method of estimating the specific energy contribution. Decationation of Ca-chabazite[51] (and to a lesser extent Na-mordenite[53]) was found to result in a marked reduction in the $\phi_{\dot{F}Q}$ contribution. Similar results have been reported[62] for the decationation of the K and Na–K forms of zeolite L, and the significance of the ion–quadrupole interaction has been confirmed with other zeolites.[63, 64]

The role of the field–dipole interaction energy, $\phi_{F\mu}$, has been investigated, especially in relation to the adsorption of ammonia and water vapour. Khvoshchev, Zhdanov, and Shubaeva[62] have compared the enthalpies of adsorption of NH_3 and CO_2 on L-type zeolites and have attributed the higher values of q^{st} obtained with NH_3 to the larger contribution of the ion–dipole interaction energy compared with the quadrupole term (*i.e.* $\phi_{F\mu} > \phi_{\dot{F}Q}$). Schirmer and his co-workers[65-70] have investigated in some detail the adsorption of NH_3 by A-type zeolites and have found the enthalpy to depend on both the nature of the cation (its charge and radius) and its location.[65, 70] Steps observed in the q^{st} *versus* coverage curves for NH_3 on NaCaA zeolites were assigned to the localization of adsorbate molecules around the cations in particular regions in the zeolite cages.[65, 69, 70] In the case of NaMgA zeolite, it was postulated[66-68] that NH_3 is initially adsorbed at Mg^{2+} sites located at the six-membered ring of O atoms (*i.e.* S_{II} sites). The second step in the q^{st} curve was attributed[68] to adsorption at Na^+ sites associated with four-membered O rings (S_{III} sites). Further studies[68, 69] on bi- and multi-valent cations in A-type zeolites revealed that the adsorption energy of NH_3 increases with increasing cationic charge.

Perhaps the most important, and certainly the most complex, of all zeolite

[61] R. W. H. Sargent and C. J. Whitford, *Adv. Chem. Ser.*, 1971, No. 102, p. 144.
[62] S. S. Khvoshchev, S. P. Zhdanov, and M. A. Shubaeva, *Doklady Akad. Nauk S.S.S.R.*, 1971, **196**, 1391 (*Doklady Phys. Chem.*, 1971, **196**, 198).
[63] R. J. Harper, G. R. Stifel, and R. B. Anderson, *Canad. J. Chem.*, 1969, **47**, 4661.
[64] N. Dupont-Pavlovsky and J. Bastick, *Bull. Soc. chim. France*, 1970, 24.
[65] W. Schirmer, A. Grossmann, K. Fiedler, K. H. Sichhart, and M. Buelow, *Chem. Tech. (Leipzig)*, 1970, **22**, 405.
[66] T. Peinze, K. Fiedler, H. Stach, and W. Schirmer, *Monatsber. Deut. Akad. Wiss. Berlin*, 1970, **12**, 855.
[67] T. Peinze, K. Fiedler, H. Stach, and W. Schirmer, *Monatsber. Deut. Akad. Wiss. Berlin*, 1970, **12**, 867.
[68] H. Stach, T. Peinze, K. Fiedler, and W. Schirmer, *Z. Chem.*, 1970, **10**, 229.
[69] H. Stach, K. Fiedler, T. Tilmann, and W. Schirmer, *Z. Chem.*, 1970, **10**, 473.
[70] K. H. Sichhart, P. Kolsch, and W. Schirmer, *Adv. Chem. Ser.*, 1971, No. 102, p. 132.

adsorption systems are those in which water is the adsorbate. The application of Barrer's method[2, 51] of assessing the non-specific contribution to q_0^{st} for the adsorption of water in zeolites yields approximate values of 11.3 kJ mol^{-1} for NaX and NaY, 16.3 kJ mol^{-1} for H-mordenite, and 24.3 kJ mol^{-1} for (Ca,Na)-chabazite.[2] The fact that the experimental values[51, 71-73] of q_0^{st} for these systems are all in excess of 80 kJ mol^{-1} demonstrates the high degree of specificity of water adsorption. In addition, the sharp decrease in q^{st} with increase in coverage at low coverage[71, 73] confirms the strong heterogeneity of the electrostatic field.

Following the suggestion of Habgood[74] that there may be some abnormally exposed cations, Barrer and Cram[73] have calculated the interaction energy between a water molecule and an *isolated* ion (Na$^+$ and Ca^{2+}) as the sum $(\phi_D + \phi_P + \phi_{F\mu} + \phi_{FQ} + \phi_R)$ and have obtained the values 105 kJ mol^{-1} for Na$^+$-H$_2$O and 256 kJ mol^{-1} for Ca^{2+}-H$_2$O. Because of the strong dipolar nature of the H$_2$O molecule, the contribution of $\phi_{F\mu}$ is considerably larger than that of ϕ_{FQ} and the molecular orientation of the molecule is determined by the dipole, so that the small ϕ_{FQ} is effectively endothermic. The experimental values of q_0^{st} (84—125 kJ mol^{-1}) for the Na-forms of zeolites X, Y, and A were found[73] to be comparable to the calculated value for the isolated cation–H$_2$O interaction, but the enthalpy of adsorption at low coverage (around $\theta = 0.15$) for (Ca,Na)-chabazite was only 134 kJ mol^{-1}, *i.e.* lower than anticipated for Ca^{2+}-H$_2$O interaction. It is possible that the Na$^+$ ions in natural chabazite may occupy the more exposed positions than the more numerous Ca^{2+} ions. The differential enthalpy is known to increase very steeply as the coverage tends to zero and therefore the true value of q_0^{st} may be considerably higher than 134 kJ mol^{-1} and may reach even 250 kJ mol^{-1}.

Enthalpies of adsorption of water vapour on X-zeolites containing Li$^+$, Na$^+$, K$^+$, Rb$^+$, and Cs$^+$ cations were determined calorimetrically by Dzhigit *et al.*[71] The highest enthalpies at low coverage (~ 80 kJ mol^{-1}) were given by the LiNaX and NaX zeolites. The extent of the initial region of rapid decrease in q^{st} was found to diminish with increase in cation size until with the CsNaX zeolite q^{st} remained nearly constant (at 63—67 kJ mol^{-1}) over an appreciable range of coverage. The relation between the initial value of q^{st} and the cation radius is similar to that shown by the enthalpies of hydration of the cations and it appeared that the first water molecules are adsorbed predominantly on cations located at S$_{III}$ sites.

At higher coverage (4—8 molecules per large cavity) the adsorption was thought to occur mainly at cations situated on S$_{II}$ sites and in this region the lowest enthalpy was observed with the LiNaX zeolite. It was suggested that

[71] O. M. Dzhigit, A. V. Kiselev, K. N. Mikos, G. G. Muttik, and T. A. Rakhmanova, *Trans. Faraday Soc.*, 1971, **67**, 458.
[72] M. M. Dubinin, A. A. Isirikyan, A. I. Sarakhov, and V. V. Serpinskii, *Izvest. Akad. Nauk S.S.S.R., Ser. khim.*, 1969, 2355 (*Bull. Acad. Sci. U.S.S.R., Chem. Sci.*, 1969, 2209).
[73] R. M. Barrer and P. J. Cram, *Adv. Chem. Ser.*, 1971, No. 102, p. 105.
[74] H. W. Habgood, *Chem. Eng. Progr. Symp.*, 1967, **63**, 45.

the small Li^+ cation, located on an S_{II} site, is embedded in the six-membered ring of O atoms so that a molecule of water cannot be oriented in a favourable position with respect to both the Li^+ cation and the negatively charged oxygen ring. As the size of the cation is increased so it protrudes above the oxygen ring and is displaced into a large cavity. The least difference in the interaction energies between H_2O molecules and the cations on the S_{III} and S_{II} sites was obtained with the CsNaX zeolite and therefore q^{st} remains nearly constant in this case. After the adsorption of about $8H_2O$ per large cavity (*i.e.* one molecule on each cation), the value of q^{st} on KNaX zeolite was found to decrease sharply but then to increase again as the cavity is filled because more favourable H_2O-H_2O interaction is then made possible. For the other zeolites, the decrease in q^{st} at uptake greater than one water molecule:one cation was not observed. It was suggested[71] that with the large Rb^+ and Cs^+ cations, additional molecules can interact not only with the framework but also with cations. On the other hand, with LiNaX and NaX zeolites, hydrogen bonding is possible between the water molecules directly attached to the cations and those subsequently adsorbed.

It is clear from this work that the influence of the cation is not always strong; it depends not only on the size and charge of the cation, but also on the degree of screening by the oxygen atoms of the lattice. By comparing the energies of adsorption of ethylene and ethane (two molecules of similar size and polarizability) Kiselev and his co-workers[55] have shown that there is a clear inverse relationship between the energy of specific interaction, *i.e.* $q_0^{st}(C_2H_4)-q_0^{st}(C_2H_6)$, and the cation radius for the family LiX, NaX, KX, RbX, and CsX. On the other hand, Tsitsishvili and Andronikashvili[75] have found that the partial replacement of Na^+ by Li^+ in NaX zeolite results in a slight decrease in the specific interaction energy. The expected increase was only found when the cation exchange was taken over 80%. Similar results were obtained[75] with other unsaturated hydrocarbons and carbon monoxide. These results indicate that the cations which are first exchanged are those which interact the most weakly with molecules specifically adsorbed. This type of behaviour goes some way at least to explain the discrepancies among enthalpy data in the literature and underlines the importance of using well-defined zeolites for future studies.

4 Adsorbate–Adsorbate Interactions

As we have seen, an initial decrease in the magnitude of the enthalpy of adsorption with coverage is always attributed to heterogeneity in the adsorbent–adsorbate interactions;[5] conversely, a constant enthalpy of adsorption is usually taken to indicate energetic homogeneity of the adsorbent surface.[7,16] With some systems, however, adsorbate–adsorbate interactions appear to become significant, even at low coverage, and constancy of the enthalpy

[75] G. V. Tsitsishvili and T. G. Andronikashvili, *Adv. Chem. Ser.*, 1971, No. 102, p. 217.

(*e.g.* for the adsorption of n-butyl alcohol on NaX zeolite[76]) may then be somewhat misleading. The adsorbate–adsorbate interactions are usually manifested as a maximum in q^{st} at coverages approaching monolayer completion[5,7] (or in the final stages of micropore filling[1,52]). Well-defined maxima in the $q^{st}-\theta$ curves have been reported recently for the adsorption of nitrogen on silica,[49] carbon dioxide, hydrocarbons, carbon tetrachloride, alcohols, and the noble gases on graphitized carbon blacks,[16,77,78] hydrocarbons and argon on active carbon,[79] and hydrocarbons, ethers, alcohols, and water on zeolites.[71,76,80] In view of the wide variety of these systems, it appears that the enthalpy increase may be associated with either specific or non-specific interactions between the adsorbed molecules.

Calculations have been made[16,81] of the potential energy change brought about by the lateral adsorbate–adsorbate interactions in close-packed monolayers of polar and non-polar molecules adsorbed on the basal plane of graphite. The approximate values of the adsorbate–adsorbate interaction potential, $-\phi_{ii}$, obtained by summing the ϕ_D, $\phi_{F\mu}$, ϕ_{FQ}, and ϕ_R terms for the most favourable packing of the monolayer, gave remarkably good agreement with $(q^{st}_{max}-q^{st}_0)$, *i.e.* the difference between the maximum enthalpy and that at zero coverage. Individual values of $-\phi_{ii}$ ranged between 4.2 kJ mol^{-1} for diethyl ether and 8.4 kJ mol^{-1} for carbon tetrachloride.[16]

There seems little doubt that the enthalpy of adsorption maxima obtained with homogeneous surfaces such as graphite are primarily due to the attractive forces between the adsorbate molecules. With other adsorbents (*e.g.* zeolites) the situation is perhaps more complicated, and Parsonage[82] has shown that repulsive forces may also play a part. The principle was illustrated by reference to a simple model adsorbent which provides two types of adsorption site such that the simultaneous occupancy of the lower state (α) and its nearest-neighbour upper state (β) leads to a repulsive contribution to ϕ. At low coverage, the number of β-sites occupied will be determined by the Boltzmann factor, but as the α-sites approach saturation this number will be reduced—to minimize the repulsion between near neighbours—with the net result of a reduction in energy of the adsorbed phase. This treatment was extended by Parsonage[83] to the case of a plane surface consisting of a square lattice. In this case, the enthalpy maxima are expected to be located near half-coverage of the α-sites.

[76] N. N. Avgul, A. G. Bezus, and O. M. Dzhigit, *Adv. Chem. Ser.*, 1971, No. 102, p. 184.
[77] R. J. Tyler and H. J. Wouterlood, *Carbon*, 1971, **9**, 467.
[78] G. I. Berezin, A. V. Kiselev, and V. A. Sinitsyn, *Zhur. fiz. Khim.*, 1970, **44**, 734 (*Russ. J. Phys. Chem.*, 1970, **44**, 408).
[79] R. Ruckh and D. Schuller, *Z. phys. Chem. (Frankfurt)*, 1969, **68**, 194.
[80] O. M. Dzhigit, A. V. Kiselev, and L. G. Ryabukhina, *Zhur. fiz. Khim.*, 1970, **44**, 1790 (*Russ. J. Phys. Chem.*, 1970, **44**, 1007).
[81] B. G. Aristov, A. G. Bezus, G. I. Berezin, and V. A. Sinitsyn, in '*Osn. Probl. Teor. Fiz. Adsorbtsii, Tr. Vses. Konf. Teor. Vop. Adsorbtsii, 1st 1968*', ed. M. M. Dubinin, 'Nauka', Moscow, 1970, p. 367 (*Chem. Abs.*, 1971, **74**, 68041).
[82] N. G. Parsonage, *Trans. Faraday Soc.*, 1970, **66**, 723.
[83] N. G. Parsonage, *J. Chem. Soc. (A)*, 1970, 2859.

Few systematic studies appear to have been made of the variation of the enthalpy of adsorption with temperature. Calorimetric measurements over a range of temperature are laborious and changes in slope of the $\log p$ versus $1/T$ plots (used in the calculation of q^{st}) are difficult to interpret. That such studies are rewarding is, however, illustrated by the results obtained by Berezin, Kiselev, and their co-workers[84] for carbon tetrachloride on graphitized carbon black over the temperature range -48 to $+20\,^\circ\text{C}$. The differential enthalpy (calorimetrically determined) was plotted against surface coverage at constant temperature and the progressive change in the enthalpy curve with temperature was then revealed. A region of constant enthalpy, corresponding to two-dimensional condensation, was obtained at temperatures below the two-dimensional critical temperature T_{c2}, but the position of maximum enthalpy was not very sensitive to temperature.

Berezin and Kiselev[78, 81, 85] have shown that the dependence of the heat capacity on coverage can be employed to compare the real adsorption systems with particular models, which do not give appreciably different enthalpy curves. Thus, the heat capacity of a two-dimensional imperfect gas on a homogeneous surface undergoes little change with coverage, whereas an adsorbed phase made up of two-dimensional complexes would give a maximum in the heat capacity–coverage curve. A comparison of the variation of the molar heat capacity, C_m, of the adsorbate with coverage has been made[78, 86] for benzene, n-hexane, and ethanol on graphitized carbon black. The values of C_m for adsorbed benzene remain lower than the heat capacity of the liquid, even at uptakes exceeding two molecular layers; the adsorbate–adsorbate interactions are weak because the aromatic ring is lying parallel to the basal plane. In the case of n-hexane, C_m rapidly approaches the value for the liquid at very low coverage and it seems clear that the adsorbate–adsorbate interactions (essentially $\phi_D + \phi_R$) are similar to those in the liquid hydrocarbon. The pronounced maximum in C_m for ethanol is especially distinctive and probably originates as the energy required to bring about partial dissociation of hydrogen-bonded molecular complexes. Similar results have been obtained[85] for n-propyl and t-butyl alcohols. The formation of these complexes allows strong co-operative adsorption of the alcohol molecules to take place at low p/p^0.

Over the past few years a number of investigations have been made of the heat capacity of adsorbed helium.[87–93] At first it appeared that helium films

[84] G. I. Berezin, A. V. Kiselev, R. T. Sagatelyan, and M. V. Serdobov, *Zhur. fiz. Khim.*, 1969, **43**, 224 (*Russ. J. Phys. Chem.*, 1969, **43**, 118).
[85] G. I. Berezin, A. V. Kiselev, I. V. Kleshnina, and V. A. Sinitsyn, *Zhur. fiz. Khim.*, 1970, **44**, 523 (*Russ. J. Phys. Chem.*, 1970, **44**, 292).
[86] A. V. Kiselev, *J. Colloid Interface Sci.*, 1968, **28**, 430.
[87] W. A. Steele, *J. Low Temp. Phys.*, 1970, **3**, 257.
[88] J. L. Wallace and D. L. Goodstein, *J. Low Temp. Phys.*, 1970, **3**, 283.
[89] J. G. Dash, R. E. Peierls, and G. A. Stewart, *Phys. Rev. (A)*, 1970, **2**, 932.
[90] M. Bretz and J. G. Dash, *Phys. Rev. Letters*, 1971, **26**, 963.

on different substrates all had low-temperature heat capacities characteristic of two-dimensional solids.[87,88] It is now clear, however, that there are appreciable differences in the thermal properties of ^{3}He and ^{4}He monolayers on various substrates. Adsorbed ^{4}He on Ar-coated Cu was found[88] to behave as a two-dimensional solid at fractional coverages as low as 0.1, but other studies[90] on graphite revealed that there was no indication of two-dimensional condensation at ~ 1 K. Bretz and Dash[90] found that ^{3}He and ^{4}He films on graphite exhibited quasiclassical two-dimensional gas properties at low coverage and $T \geqslant 2$ K. This model was unsatisfactory for ^{4}He at lower temperatures since the heat capacity curves had broad peaks between 1 and 2 K. Roy and Halsey[94] have suggested that substrate inhomogeneities tend to force adatoms into dense surface aggregates, but it seems unlikely that this would explain the formation of the two-dimensional solid ^{4}He on Ar–Cu unless the substrate was porous.

Brewer, Evenson, and Thomson[91–93] have studied the adsorption of ^{3}He, ^{4}He, and ^{3}He–^{4}He mixtures on Vycor porous glass over the temperature range 1—4 K. At fractional coverages of 0.9—1.1 the heat capacities showed T^2 temperature-dependence, characteristic of collective excitations of a two-dimensional array of harmonic oscillators, but the values of the two-dimensional Debye temperature, θ_{D2}, were considerably less than would be expected for a two-dimensional He lattice. At coverages above the monolayer, the specific heat against T^2 plots gave positive intercepts; this is probably due to the translational and additional vibrational contributions from the second layer. In the case of mixtures,[91] it appears that preferential adsorption of ^{4}He takes place first with the enrichment of ^{3}He in the outer layers. An important observation is that even when there is insufficient helium to form the complete monolayer, the ^{3}He atoms tend to form a second layer on top of the ^{4}He.

The heat capacity of Ne adsorbed on graphitized carbon black has been determined[95] over the temperature range 1.5—30 K for two coverages, $\theta = 0.5$ and $\theta = 1.5$. The heat capacities for both coverages indicated that the state of the adsorbate is liquid-like at temperatures close to the boiling point of the adsorptive. At lower temperatures, the peak in the heat capacity was found to depend on the coverage, *i.e.* it becomes more pronounced at the higher coverage. It was suggested[95] that the localized adatoms may be regarded as asymmetric three-dimensional oscillators with the two-dimensional component contributing a larger share to the heat capacity. It is of interest to note that the observed heat capacities for $\theta = 1.5$ confirmed that the state of the adsorbate differs markedly from that of the bulk Ne.

[91] A. Evenson, D. F. Brewer, and A. L. Thomson, in 'Proceedings of the 11th International Conference on Low Temperature Physics, 1968', ed. J. F. Allen, 1969, vol. 1, p. 125.
[92] D. F. Brewer, A. Evenson, and A. L. Thomson, *J. Low Temp. Phys.*, 1970, **3**, 603.
[93] D. F. Brewer, A. Evenson, and A. L. Thomson, *Phys. Letters (A)*, 1971, **35**, 307.
[94] N. N. Roy and G. D. Halsey, *J. Chem. Phys.*, 1970, **53**, 798.
[95] A. A. Antoniou, P. H. Scaife, and J. M. Peacock, *J. Chem. Phys.*, 1971, **54**, 5403.

5 The Adsorption Isotherm

Experimental Methods.—Developments in apparatus and technique have enabled adsorption measurements to be made over wide ranges of pressure and temperature.[96] High-pressure adsorption studies have been conducted with helium,[97] nitrogen,[97, 98] carbon dioxide,[99] and various hydrocarbons;[99, 100] measurements at low pressure have been made with the noble gases,[101] nitrogen,[102] ethylene,[103] and ethylene glycol.[104] Gas-chromatographic techniques have been developed[105-118] for the measurement of the uptake of gas at low surface coverage, especially at temperatures higher than those which can be used with conventional static methods.

Volumetric methods are still widely used for BET surface area determination.[45, 119-121] Procedures have been described for handling condensable vapours,[122-124] correcting for adsorption on the walls of the apparatus,[124, 125] and for calculating the non-ideality correction factors.[126]

[96] P. G. Menon, *Adv. High Pressure Res.*, 1969, **3**, 313.
[97] E. Papirer, J. B. Donnet, and P. Badie, *Compt. rend.*, 1970, **271**, *C*, 969.
[98] C. Gachet and Y. Trambouze, *J. Chim. phys.*, 1970, **67**, 380.
[99] A. J. Gonzalez and C. D. Holland, *Amer. Inst. Chem. Engineers J.*, 1971, **17**, 470.
[100] Y. Hori and R. Kobayashi, *J. Chem. Phys.*, 1971, **54**, 1226.
[101] J. P. Hobson, *J. Phys. Chem.*, 1969, **73**, 2720.
[102] J. W. Coleman and F. A. Inkley, Proceedings of the 4th International Vacuum Congress, 1968, Pt. 1, p. 159.
[103] M. Troy and J. P. Wightman, *J. Vac. Sci. Technol.*, 1971, **8**, 515.
[104] G. T. Elsmere and R. A. G. Rawson, *J. Phys.* (*E*), 1970, **3**, 1013.
[105] T. N. Gvosdovich, A. V. Kiselev, and Ya. I. Yashin, *Chromatographia*, 1969, **6**, 234.
[106] D. Dollimore, G. R. Heal, and D. R. Martin, *J. Chromatog.*, 1970, **50**, 209.
[107] G. Edel, B. Chabert, and J. Chauchard, *Bull. Inst. Textile France*, 1970, **24**, 13.
[108] Y. Shigehara and A. Ozaki, *Nippon Kagaku Zasshi*, 1970, **91**, 940.
[109] B. Dvorak and J. Pasek, *J. Catalysis*, 1970, **18**, 108.
[110] G. Racz, G. Szekely, K. Huszar, and K. Olah, *Period. Polytech., Chem. Eng.*, 1971, **15**, 111 (*Chem. Abs.*, 1971, **75**, 40 783).
[111] I. L. Mar'yasin, S. L. Pishchulina, I. S. Rafal'kes, T. D. Snegireva, T. V. Tekunova, and L. M. Borodina, *Zavodskaya Lab.*, 1971, **37**, 41 (*Chem. Abs.*, 1971, **74**, 80 156).
[112] I. S. Krasotkin and R. L. Dubrovinskii, *Obogashch. Rud.*, 1970, **15**, 62 (*Chem. Abs.*, 1971, **74**, 144 779).
[113] A. Imre and I. Baan, *Banyasz. Kohasz. Lapok. Kohasz.*, 1970, **103**, 137 (*Chem. Abs.*, 1970, **73**, 48 833).
[114] R. L. Dubrovinskii, I. S. Krasotkin, and A. S. Kuz'menko, *Nov. Issled. Tsvet. Met. Obogashch.*, 1969, 171 (*Chem. Abs.*, 1971, **75**, 80 566).
[115] M. F. Burke and D. G. Ackerman, jun., *Analyt. Chem.*, 1971, **43**, 573.
[116] H. Engelhardt and B. P. Engelbrecht, *Chromatographia*, 1971, **4**, 66.
[117] V. K. Solyakov, M. E. Magdasieva, and A. G. Sokker, *Konstr. Mater. Osn. Grafita*, 1970, No. 5, 222 (*Chem. Abs.*, 1971, **75**, 10 530).
[118] M. Krejci and D. Kourilova, *Chromatographia*, 1971, **4**, 48.
[119] 'Methods for the Determination of Specific Surface of Powders, Pt. I, Nitrogen Adsorption (BET Method)', B.S. 4359, Part 1, 1969.
[120] L. G. Berg and R. A. Abdurakhmanov, *Zhur. fiz. Khim.*, 1970, **44**, 527.
[121] J. B. Donnet and B. Lespinasse, ref. 48, p. 211.
[122] M. Lason and J. Zolcinska, *Roczniki Chem.*, 1970, **44**, 2199.
[123] B. V. Zheleznyi, *Zhur. fiz. Khim.*, 1971, **45**, 1258 (*Russ. J. Phys. Chem.*, 1971, **45**, 711).
[124] V. R. Deitz and N. H. Turner, *Rep. Nav. Res. Lab. Progr.*, 1970, July, p. 1 (*Chem. Abs.*, 1970, **73**, 123 849).
[125] V. R. Deitz and N. H. Turner, *J. Vac. Sci. Technol.*, 1970, **7**, 577.
[126] W. V. Loebenstein, *J. Colloid Interface Sci.*, 1971, **36**, 397.

Recording microbalances[127-130] and other new devices[131,132] have been adapted to provide a continuous record of adsorbent mass and adsorptive pressure and programmed to give an incremental or continuous measurement, both up and down the isotherm. Various sources of error (*e.g.* slow thermal equilibration,[133-135] displacement of impurity from the apparatus walls,[136] the adsorption of mercury vapour,[137] and the effect of light on certain systems[138]) have been overcome by improvement in design and operational procedure. Mass spectrometry[139] and radiochemical techniques[138,140,141] have been employed to measure low uptakes of gas and to investigate isotope exchange in physical adsorption.

The widespread use of gas adsorption in many areas of science and technology has led to the development of routine methods and simplified apparatus for surface area determination.[142-152] Although such techniques may be satisfactory for the routine comparison of adsorbent activity, comparative studies[153,154] have indicated that they cannot replace the more conventional methods for the accurate determination of adsorption isotherms.[153]

The Isotherm at Low Surface Coverage.—*Henry's Law.* The linear adsorption isotherm,

$$x = k_1 p, \qquad (8)$$

[127] E. Robens, G. Sandstede, and G. Walter, *Vac. Microbalance Tech.*, 1971, **8**, 111.
[128] H. Drexler and K. Gierschner, *Chem.-Ing.-Tech.*, 1971, **43**, 691.
[129] C. H. Garski and L. E. Stettler, *J. Colloid Interface Sci.*, 1971, **37**, 918.
[130] S. Tilenschi, ref. 48, p. 249.
[131] M. R. St. John, G. Constabaris, and J. F. Johnson, *Amer. Chem. Soc., Div. Petrol. Chem., Prepr.*, 1969, **14**, 17 (*Chem. Abs.*, 1971, **74**, 6725).
[132] G. R. Landolt, *Analyt. Chem.*, 1971, **43**, 613.
[133] C. J. Williams, *Amer. Lab.*, 1969, June, 40.
[134] P. A. Cutting, *Vac. Microbalance Tech.*, 1970, **7**, 71.
[135] G. Beurton and P. Bussiere, *J. Phys. (E)*, 1970, **3**, 875.
[136] M. M. Dubinin, K. M. Nikolaev, N. S. Polyakov, and N. I. Seregina, *Izvest. Akad. Nauk S.S.S.R., Ser. khim.*, 1970, 761 (*Bull. Acad. Sci. U.S.S.R., Chem. Sci.*, 1970, 717).
[137] F. S. Baker and K. S. W. Sing, *Nature Phys. Sci.*, 1971, **229**, 27.
[138] J. B. Donnet, H. Dauksch, and R. Battistella, *Compt. rend.*, 1971, **272**, C, 576.
[139] E. M. Trukhanenko and L. I. Panfilova, *Zhur. fiz. Khim.*, 1970, **44**, 2954 (*Russ. J. Phys. Chem.*, 1970, **44**, 1687).
[140] P. C. Rankin, A. T. Wilson, and I. D. Beatson, *J. Colloid Interface Sci.*, 1971, **36**, 340.
[141] S. Thibault and J. Talbot, *Meas. Methods Corros. Prot., Event Eur. Fed. Corros., 42nd 1968*, 1969, p. 1 (*Chem. Abs.*, 1971, **74**, 15 940).
[142] R. A. G. Rawson, *J. Soil Sci.*, 1969, **20**, 325.
[143] E. Wirsing, jun., L. P. Hatch, and B. F. Dodge, *U.S. At. Energy Comm.*, 1970, BNL-50-254 (*Nuclear Sci. Abs.*, 1971, **25**, 23 644).
[144] D. P. Roelofsen, *Analyt. Chem.*, 1971, **43**, 631.
[145] A. P. Dzisyak and V. P. Kostyukov, *Zavodskaya Lab.*, 1971, **37**, 702.
[146] G. Bliznakov, I. Bakurdzhiev, and E. Gocheva, *Izvest. Otd. Khim. Nauki, Bulg. Akad. Nauk*, 1971, **4**, 11 (*Chem. Abs.*, 1971, **75**, 91 483).
[147] J. E. Benson and R. L. Garten, *J. Catalysis*, 1971, **20**, 416.
[148] I. V. Uvarova and V. V. Panichkina, *Zavodskaya Lab.*, 1970, **36**, 306.
[149] Z. Spitzer, *Sbornik Pred U.V.P.*, 1968, **12**, 236 (*Chem. Abs.*, 1970, **72**, 104 176).
[150] K. Watanabe and T. Yamashina, *J. Catalysis*, 1970, **17**, 272.
[151] M. G. Farey and B. G. Tucker, *Analyt. Chem.*, 1971, **43**, 1307.
[152] G. Bliznakov, I. Bakardjiev, and E. M. Gocheva, *J. Catalysis*, 1970, **18**, 260.
[153] A. S. Joy, ref. 48, p. 391.
[154] J. F. Padday, ref. 48, p. 401.

where k_1 is the Henry's law constant and x is the amount adsorbed, is generally assumed to hold at low pressures.[5,7] The application of gas–solid chromatography (g.s.c.) has confirmed that many high-temperature isotherms do exhibit a limited range of linearity at low surface coverage.[17] Symmetrical g.s.c. peaks have been reported recently for the adsorption of C_6—C_{10} alkenes on graphitized carbon black[155] over the temperature range 348—498 K and for various polar and non-polar vapours on a macroporous silica[156] over the range 373—473 K.

Curvature in the isotherm (convex with respect to the adsorption axis) at very low pressure may be due to the effect of either substrate heterogeneity[157,158] or microporosity.[4] We have seen already that specific adsorbent–adsorbate interactions usually give rise to energetic heterogeneity and it is not surprising to find that the corresponding isotherms tend to be non-linear at low pressure. Coleman and Inkley,[102] investigating the adsorption of nitrogen at room temperature on Pt–alumina and Pt–silica catalysts, obtained curved isotherms at pressures as low as 10^{-8} Torr. Troy and Wightman[159] found that nitrogen on stainless steel gave non-linear isotherms down to $p < 10^{-9}$ Torr (at 77—90 K), whereas an argon isotherm[160] (at 90 K) on the same material obeyed Henry's law below about 10^{-7} Torr. The effect of porosity may explain the results of Hobson,[101] who detected a deviation from Henry's law at p/p^0 as low as 10^{-13} with the isotherms of Ar, Kr, and Xe on a porous Ag catalyst.

Some low-energy surfaces give linear isotherms over a wide range of pressure. For example, the isotherm of n-butane on polyvinylidene chloride[161] at 294 K was approximately linear up to a pressure of about 200 Torr. Isotherms of Ar and O_2 at 77 K on samples of lunar materials[162] were remarkably linear up to $p/p^0 \sim 0.8$, whereas CO and N_2 isotherms showed marked curvature at $p/p^0 < 0.2$. The isotherms of hexane, octane, and other alkanes on polytetrafluoroethylene[163] and alcohol-covered aluminium[164] exhibited a small, but distinct knee at low p/p^0 and were then nearly linear over an appreciable range of higher p/p^0. The knee was probably associated with the adsorption on a small high-energy fraction of the surface, and the linear part of the isotherm with adsorbate–adsorbate interactions which extend into the multilayer.

[155] O. G. Eisen, A. V. Kiselev, A. E. Pilt, S. A. Raig, and K. D. Scherbakova, *Chromatographia*, 1971, **4**, 448.
[156] N. K. Bebris, A. V. Kiselev, B. Ya. Mokeev, Yu. S. Nikitin, Ya. I. Yashin, and G. E. Zaitseva, *Chromatographia*, 1971, **4**, 93.
[157] G. D. Halsey, jun. and C. M. Greenlief, *Ann. Rev. Phys. Chem.*, 1970, **21**, 129.
[158] S. Ross, ref. 48, p. 143; see also p. 205.
[159] M. Troy and J. P. Wightman, *J. Vac. Sci. Technol.*, 1970, **7**, 429.
[160] M. Troy and J. P. Wightman, *J. Vac. Sci. Technol.*, 1971, **8**, 743.
[161] P. G. Hall and H. F. Stoeckli, *Trans. Faraday Soc.*, 1969, **65**, 3334.
[162] E. L. Fuller, H. F. Holmes, R. B. Gammage, and K. Becker, Proceedings of the Second Lunar Science Conference, M.I.T. Press, 1971, vol. 3, p. 2009.
[163] J. W. Whalen, *Vac. Microbalance Tech.*, 1971, **8**, 121.
[164] T. D. Blake, J. L. Cayias, W. H. Wade, and J. A. Zerdecki, *J. Colloid Interface Sci.*, 1971, **37**, 678.

The low-pressure linearity of isotherms on zeolites, aluminas, and hydroxylated silicas may be improved by the pre-adsorption of polar molecules on the most active sites. For example, the nitrogen isotherm at 323 K on NaX zeolite was found[165] to be linear up to at least 120 Torr after the presorption of 4.3 mmol H_2O (g NaX)$^{-1}$, but a lower residual quantity of H_2O was insufficient to block all the active sites and remove the isotherm curvature.

It seems clear that the low-pressure linearity shown by some Type II and stepwise isotherms is directly associated with the energetic homogeneity of the adsorbent–adsorbate interactions. The situation is more complex, however, with Type III (or poorly defined Type II) isotherms, which may show extensive linearity; in these cases, the adsorbate–adsorbate interactions must be taken into account, even at low coverage. To distinguish between these effects and to explore the Henry's law region in terms of adsorbent–adsorbate interactions it is therefore important to undertake adsorption studies at high temperature and very low coverage. For this purpose, gas-chromatographic techniques appear to have great practical advantages, but it is important to assess their limitations.

The computation of Henry's law constants and thermodynamic quantities ($\Delta\mu$, q_0^{st}, $\Delta\bar{S}$, and ΔC_p) from g.s.c. data has been discussed in some detail.[166-168] Boucher and Everett,[166] using hydrogen as a carrier gas and helium as a marker, investigated the adsorption of noble gases, CH_4, and N_2 on an active carbon. They assessed the various sources of error and concluded that k_1 could be determined to within about $\pm 3\%$.

If Henry's law holds at low coverage, the isosteric enthalpy at 'zero' coverage is given by

$$q_0^{st} = -RT^2 \, \partial \ln k_1/\partial T. \tag{9}$$

In the calculation of q_0^{st} from the retention volume or isotherm data at different temperatures, it is generally assumed that q_0^{st} does not itself vary over the temperature range studied. Kiselev and his co-workers[167] have adopted an alternative approach: they assume a temperature dependence for q_0^{st} and $\Delta\bar{S}$ and attempt to calculate the heat capacity change ΔC_p accompanying adsorption. This approach appears to be of doubtful value at the present time, however, because of the limited accuracy of the g.s.c. data.

The theoretical basis for Henry's law behaviour within the context of high-temperature adsorption has been reviewed by Everett.[169] In the special case of an energetically uniform surface, ϕ, the potential energy with respect to a single gas atom, may be assumed independent of x and y (i.e. in the plane parallel to the adsorbent surface) and the application of the Boltzmann

[165] V. Patzelova, *Chromatographia*, 1970, **3**, 170.
[166] E. A. Boucher and D. H. Everett, *Trans. Faraday Soc.*, 1971, **67**, 2720.
[167] T. I. Bertush, A. V. Kiselev, A. A. Lopatkin, and R. S. Petrova, *Chromatographia*, 1970, **3**, 369.
[168] J. P. Okamura and D. T. Sawyer, *Analyt. Chem.*, 1971, **43**, 1730.
[169] D. H. Everett, ref. 48, p. 181.

distribution law is therefore considerably simplified. If the gas phase behaves ideally, the surface excess, n^σ, is given by

$$n^\sigma = \frac{Ap}{RT} \int_{z_0}^{\infty} [e^{-\phi(z)/kT} - 1]\, dz, \qquad (10)$$

where A is the surface area and z the distance normal to the surface. If ϕ may be expressed as a function of z,

$$\phi/\phi^* = f(z/z_0), \qquad (11)$$

where ϕ^* is the depth of the potential energy well at z_0, it follows that

$$\int_{z_0}^{\infty} [e^{-\phi(z)/kT} - 1]\, dz = z_0 F(\phi^*/kT). \qquad (12)$$

Equation (10) predicts that the isotherm of n^σ against p is linear and the slope k_1 is thus given by

$$k_1 = \frac{Az_0}{RT} F\left(\frac{\phi^*}{kT}\right). \qquad (13)$$

Kiselev[17, 54, 170] and Poskus[18] and their co-workers[56] have provided a more explicit statement of equations (10) and (13). They and others[171] have discussed the effects of molecular rotation and libration on the separation of various isotopes and isomers. The differences between the statistical functions of hydrocarbons containing H and D atoms have been calculated[18] and related to the differences in the values of k_1. The results of these calculations indicate that the isotopic separational effect is due mainly to differences in the potential functions rather than to quantum effects.

If an adequate theoretical treatment is available to enable $z_0 F(\phi^*/kT)$ to be calculated, experimental measurement of k_1 enables equation (13) to be used to find the surface area A. The use of this procedure for determination of surface areas has been reviewed critically by Everett[169] who concludes that, in view of the uncertainties involved in the calculation of ϕ^* and z_0, high accuracy cannot be expected in the calculation of values of A, even for high-temperature adsorption on plane, homogeneous surfaces. If the surface is heterogeneous or non-planar the difficulties are considerably greater, especially if both heterogeneity and porosity are present together.

The case of adsorption in the Henry's law region by a microporous solid can be analysed in a straightforward manner provided that the adsorption potential is essentially constant within each pore.[169] We then have the simple relationship

$$k_1 = V^P/RT[e^{-\phi/kT} - 1], \qquad (14)$$

[170] A. V. Kiselev, D. Poskus, and A. Ya. Afreimovich, *Zhur. fiz. Khim.*, 1970, **44**, 981 (*Russ. J. Phys. Chem.*, 1970, **44**, 545).
[171] P. L. Gant, K. Yang, M. S. Goldstein, M. P. Freeman, and A. I. Weiss, *J. Phys. Chem.*, 1970, **74**, 1985.

where V^P is the pore volume accessible to the centres of mass of the adsorbed molecules. We should therefore expect Henry's law to be approached if the enthalpy is constant at low uptake, as with non-porous adsorbents.

The elaborate calculations carried out by Kiselev and his co-workers[16,17,54,56,170] of the temperature dependence of the Henry's law constant have been remarkably successful when applied to the adsorption of the noble gases, hydrogen, and nitrogen on graphitized carbon,[16,17] but rather less so in the case of the adsorption of hydrocarbons on graphitized carbon[170] and the noble gases on zeolites.[54,56] The hydrocarbons were treated as quasi-rigid molecules[16] and it was found necessary to introduce empirical correction factors into the potential functions. As mentioned already, computations on zeolites are at present hampered by the lack of precise information on the orientation of the adsorbate molecule—especially in relation to the cation—and on the various interaction energy contributions.

The Virial Equation. We have seen the need for a general molecular-statistical theory of adsorption which takes into account adsorbate–adsorbate interactions and in addition allows for heterogeneity of the substrate. The theory should provide the foundation for the mathematical description of adsorption isotherms over an appreciable range of coverage and at different temperatures. Although much remains to be done before this difficult problem can be solved, encouraging progress has been made in recent years[7,16,50,51,55] by the application of the virial equation to isotherms obtained on certain well-defined adsorbents—notably graphitized carbon black and zeolites. The approach has been made at two levels: first, to provide a mathematical analysis of adsorption data over a range of pressure and temperature and secondly, to enable a statistical thermodynamic interpretation to be made of the adsorption mechanism.[172] The treatment is analogous in many respects to the analysis of solution thermodynamics and its strength lies in the fact that no model has to be assumed before the virial equation is applied.

The two-dimensional equation of state of an adsorbate may be expressed in the virial form

$$\pi = RTx + b_1 x^2 + b_2 x^3 + \ldots, \tag{15}$$

where π may be regarded as the two-dimensional pressure of the adsorbate and $b_1, b_2 \ldots$ are constants. Equation (15) when combined with the Gibbs adsorption equation gives

$$p = x \exp(C_1 + C_2 x + C_3 x^2 + \ldots), \tag{16}$$

where

$$C_1 = \ln(1/k_1), \quad C_2 = 2b_1/RT, \quad \text{and} \quad C_3 = 3b_2/2RT. \tag{17}$$

Equation (16) can be applied to both Type I isotherms and isotherms with a point of inflection at low pressure. It is clear that the number of coefficients

[172] D. Poskus, ref. 81, p. 9.

required will tend to increase as the shape of the isotherm becomes more rectangular.

The application of equation (16) will be illustrated by reference to the investigation of Bezus *et al.*[55] in which adsorption isotherms of ethane and ethylene were measured on X-zeolites containing various cations. Values of the coefficients were obtained to provide the best fit for equation (16) to each isotherm over selected ranges of x and with different numbers of terms in the series. At low coverage, C_1 was found to be nearly independent of the number of terms in the exponent and its value could be determined to within 1—2%. The accuracy in the determination of C_2 was lower (especially with ethylene) and only approximate values of C_3 and C_4 could be obtained.

The fact that the values of the coefficients C_1 and C_2 were independent of the method of calculation confirmed that they could be regarded as physicochemical constants for the given adsorbent–adsorbate system at a specified temperature. The inclusion of the additional terms can only be justified at present on an empirical basis to provide better agreement with data over a wider range of surface coverage.

The general p–x–T relation is obtained from equation (16) as follows. Differentiation with respect to temperature gives

$$(\partial \ln p/\partial T)_x = (\mathrm{d}C_1/\mathrm{d}T) + (\mathrm{d}C_2/\mathrm{d}T)\,x + (\mathrm{d}C_3/\mathrm{d}T)\,x^2 + \dots \quad (18)$$

Since $(\partial \ln p/\partial T)_x = q^{st}/RT^2$, the isosteric enthalpy, q^{st}, is given by

$$q^{st} = (\mathrm{d}C_1/\mathrm{d}T)\,RT^2 + (\mathrm{d}C_2/\mathrm{d}T)\,RT^2 x + (\mathrm{d}C_3/\mathrm{d}T)\,RT^2 x^2 \dots \quad (19)$$

If we now put

$$(\mathrm{d}C_i/\mathrm{d}T)\,RT^2 = q^{st}_{i-1}, \quad (20)$$

the dependence of q^{st} on x is obtained in the form of the series

$$q^{st} = q^{st}_0 + q^{st}_1 x + q^{st}_2 x^2 + \dots, \quad (21)$$

where q^{st}_0 is the differential enthalpy of adsorption at $x = 0$ and q^{st}_1, q^{st}_2 ... are constants. Over a small interval of T, q^{st}_{i-1} may be assumed independent of temperature and therefore

$$C_i = B_i - q^{st}_{i-1}/RT, \quad (22)$$

where B_i is a constant of integration of equation (20). Substituting for the coefficients in equation (16) now yields the expression

$$p = x \exp\left(\sum_i B_i x^{i-1}\right) \exp\left(-\sum_i q^{st}_{i-1} x^{i-1}/RT\right). \quad (23)$$

It will be noted that q^{st}_0, which is characteristic of the adsorbent–adsorbate interactions, may be calculated according to equation (22) from the temperature dependence of C_1 and that the constant q^{st}_1, characteristic of the adsorbate–adsorbate interactions, is similarly calculated from the temperature dependence of C_2. The constants B_i and q^{st}_{i-1} in equation (23) can be determined either in the manner already described, *i.e.* adsorption measurements over a range of p and T, or from one adsorption isotherm together with calorimetric enthalpy data. This method of isotherm analysis has been applied so far to only a limited number of systems, *e.g.* the adsorption of inert gases[51] and hydrocarbons[52] on decationated zeolites, and carbon dioxide on zeolites[50] and graphitized carbon black.[173] It is expected that the use of the virial equation will increase as more adsorption data are obtained on other well-defined adsorbents.

Besides providing a useful general method of analysing isotherms, the use of virial type equations is important since (just as with three-dimensional virial equations) the successive terms have well-defined statistical mechanical interpretations.[7] Thus, as indicated above, C_1 is equal to the logarithm of the reciprocal of the Henry's law constant, and is determined solely by interactions between the adsorbent and the adsorbate molecule, the second coefficient C_2 is related to interactions between pairs of molecules which are simultaneously under the influence of the adsorbent, C_3 is concerned with simultaneous interaction of triplets of adsorbed molecules and so on.

The extensive theoretical calculations of Henry's law constants by Kiselev and his co-workers referred to above depend for their verification on reliable methods of finding C_1 from experimental data, which in many cases the virial method of analysis provides.

Theoretical studies of C_2 have so far dealt successfully only with adsorption on a plane homogeneous surface, in which case C_2 is related to the two-dimensional virial coefficient B^* by

$$C_2 = -2B^*/A, \qquad (24)$$

where A is the surface area. The coefficient B^* is defined in terms of the interaction potential $u^*(r)$ of a pair of adsorbed molecules a distance r apart by

$$B^* = -N_A\pi \int_0^\infty [e^{-u^*(r)/kT} - 1]\, r\, dr, \qquad (25)$$

where N_A is Avogadro's constant. In principle, when $u^*(r)$ is known, B^* can be calculated. If C_2 is then measured experimentally, it is possible to find A, the surface area of the solid from equation (24). This method of determining surface areas depends on one's ability to establish the function $u^*(r)$. The simplest assumption, that $u^*(r) = u(r)$, the interaction potential in the bulk gas state, is found to be invalid and it is necessary to employ measurements of the temperature coefficient of C_2 to find the parameters of a chosen form for $u^*(r)$ [*e.g.* a (12:6) or a (12:6:3) potential] which best fits the data. The present status of this method of surface area determination has recently been reviewed critically by Everett:[169] provided measurements can be made over a wide range of temperature approaching the Boyle temperature (where

[173] N. N. Avgul, A. S. Guzenberg, A. V. Kiselev, L. Ya. Kurdyukova, and A. M. Ryabkin, *Zhur. fiz. Khim.*, 1971, **45**, 442 (*Russ. J. Phys. Chem.*, 1971, **44**, 243).

$B^* = 0$) of the two-dimensional gas, reliable estimates of the areas of homogeneous surfaces can be made. Its applicability to heterogeneous surfaces is less certain. Everett suggests that surface heterogeneity has a relatively small effect on the method, but this is contested by Ross.[158]

The problem of the interpretation of C_2 in the case of porous adsorbents is more complicated. If the adsorption potential within the pores is more or less uniform, C_2 will be related to a quasi-three-dimensional second virial coefficient, but defined by an integral taken only within the available pore space. However, no detailed theoretical work has been carried out; nor are there many reliable experimental data relating to C_2.

The Dubinin–Radushkevich–Kaganer Equation. In his treatment of micropore filling, Dubinin[4] utilized the 'characteristic curve' principle of the potential theory and on this basis Dubinin and Radushkevich (DR) derived their isotherm equation. In recent years the DR equation has been applied to many systems as it appears to provide a simple method for the determination of micropore volume. A number of attempts[174-183] have been made to improve and extend the application of the characteristic curve, but it is recognized that it must remain an empirical relationship.

It is against this background that Kaganer's modification of the DR equation has been used for the determination of surface area. Kaganer[184] retained the principle of the characteristic curve and the gaussian distribution of potential, but replaced the concept of micropore filling by that of surface coverage; this gives the Dubinin–Radushkevich–Kaganer (DRK) equation

$$\ln \frac{x}{x_m} = -D \left(RT \ln \frac{p^0}{p} \right)^2, \tag{26}$$

where x_m is the monolayer capacity and D is a constant, which characterizes the gaussian distribution. Equation (26) has the same mathematical form as the original DR equation, but x/x_m now replaces the fractional filling of the pore volume. According to equation (26) the plot of $\ln x$ against $(RT \ln p^0/p)^2$ should be linear with an intercept equal to $\ln x_m$.

[174] M. M. Dubinin and V. A. Astakhov, *Adv. Chem. Ser.*, 1971, No. 102, p. 69.
[175] A. Cointot, J. Cruchaudet, and M. H. Simonot-Grange, *Bull. Soc. chim. France*, 1970, 497.
[176] J. L. Ginoux and L. Bonnetain, *Compt. rend.*, 1970, **270**, C, 1484.
[177] M. Roques and M. Bastick, *Compt. rend.*, 1971, **273**, C, 609.
[178] B. P. Bering and V. V. Serpinskii, *Izvest. Akad. Nauk S.S.S.R., Ser. khim.*, 1971, 847 (*Bull. Acad. Sci. U.S.S.R., Chem. Sci.*, 1971, 760).
[179] K. Kawazoe and V. A. Astakhov, *Seisan-Kenkyu*, 1970, **22**, 373 (*Chem. Abs.*, 1970, **73**, 102 328).
[180] W. D. McCain, jun. and T. D. Stacy, *Soc. Petrol. Eng. J.*, 1971, **11**, 4.
[181] J. L. Ginoux and L. Bonnetain, *Compt. rend.*, 1971, **272**, C, 879.
[182] P. J. Reucroft, W. H. Simpson, and L. A. Jonas, *J. Phys. Chem.*, 1971, **75**, 3526.
[183] B. Kindl, E. Negri, and G. F. Cerofolini, *Surface Sci.*, 1970, **23**, 299.
[184] M. G. Kaganer, *Zhur. fiz. Khim.*, 1959, **33**, 2202 (*Russ. J. Phys. Chem.*, 1959, **33**, 352).

In a number of recent investigations,[160,185-189] linear DRK plots have been obtained in the region of low surface coverage. For example, Troy and Wightman,[159,160] studying the physisorption of several gases on non-porous stainless steel at low pressure and fractional coverages below 0.01, obtained a common linear DRK plot for nitrogen isotherms[159] over the temperature range 77—90 K. Argon and krypton isotherms[160] also gave nearly linear DRK plots, but in these cases the temperature dependence was not described adequately by equation (26).

The results of comparisons[190-194] between the values of x_m estimated by the DRK and BET methods present a somewhat confused picture. Granville *et al.*[190] measured adsorption isotherms of krypton on evaporated films of Ni, Fe, and Ti at various stages of sintering, and on catalyst sponges of Fe and Ti; they found that the overall average for the ratio of the monolayer capacities (DRK/BET) was 1.02 with a standard deviation of 0.22. Klemperer[191] reported the results of a similar comparison made with Xe on various metal films; the average ratio was again 1.02, but there was a tendency for this to decrease to around 0.70 with a decrease in the BET C value. By comparing the mathematical form of the DRK and the BET isotherm equations, Gottwald[192] concluded that close correspondence between the two values of monolayer capacity is a mathematical requirement if both equations hold over the different ranges of relative pressure. In practice, of course, this is rarely the case and differences between the two values of x_m are then inevitable.

We are forced to the conclusion that at best the DRK plot may provide a useful empirical method of analysing isotherm data at relative pressures below the normal BET range. In principle, however, the DRK equation appears to have very little to contribute in any attempt to elucidate the nature of the adsorption process.

The Isotherm at High Surface Coverage.—*The Hill–de Boer Equation and Two-dimensional Condensation.* Renewed interest has been shown[16,195] in recent years in the Hill–de Boer equation for the adsorption of a mobile monolayer on a homogeneous surface,

[185] B. A. Gottwald and R. Haul, Proceedings of the 4th International Vacuum Congress, 1968, Pt. 1, p. 96.
[186] K. E. Tempelmeyer, *U.S. Clearinghouse, Fed. Sci. Tech. Inform.*, AD 1970, No. 707 844 [from *U.S. Govt. Res. Develop. Rep.*, 1970, **70** (16), 218].
[187] K. E. Tempelmeyer, *Cryogenics*, 1971, **11**, 120.
[188] K E. Tempelmeyer, *U.S. Clearinghouse Fed Sci. Tech. Inform.*, AD 1970, No. 700 981 [from *U.S. Govt. Res. Develop. Rep.*, 1970, **70** (7), 70].
[189] R. A. Outlaw, N.A.S.A. Technical Note, 1970 NASA TN D-5777 (*Chem. Abs.*, 1970, **73**, 29 252).
[190] A. Granville, P. G. Hall, and C. J. Hope, *Chem. and Ind.*, 1970, 435.
[191] D. F. Klemperer, ref. 48, p. 55.
[192] B. A. Gottwald, ref. 48, p. 59.
[193] P. L. Walker, jun. and R. L. Patel, *Fuel*, 1970, **49**, 91.
[194] V. D. Pomeshchikov and V. V. Pozdeev, *Kinetika i Kataliz.*, 1971, **12**, 794.
[195] J. C. P. Broekhoff and R. H. van Dongen, in 'Physical and Chemical Aspects of Adsorbents and Catalysts,' ed. B. G. Linsen, Academic Press, London, 1970, p. 63.

$$p = \frac{\theta}{k_1(1-\theta)} \exp\left(\frac{\theta}{1-\theta} - k_2\theta\right). \tag{27}$$

Here k_1 is the Henry's law constant and $k_2 = 2a_2/b_2kT$. Equation (27) may be derived from the two-dimensional analogue of the van der Waals equation; a_2 and b_2 are the two-dimensional constants, which are approximately related to a and b for the bulk vapours. The two-dimensional critical temperature, T_{c2}, is also related to the three-dimensional critical temperature ($T_{c2} \sim 0.5T_c$), if the molecules in the adsorbed state have lost only one degree of translational freedom.

Equation (27) may be re-arranged to give the linear form

$$k_2\theta + \ln k_1 = \frac{\theta}{1-\theta} + \ln \frac{\theta}{1-\theta} - \ln p \tag{28}$$
$$= W(p, \theta),$$

and the plot of W against θ should give k_1 and k_2. The linearity of equation (28) has been tested by application to isotherms of various vapours on graphitized carbon blacks.[16,195,196] Carbon tetrachloride isotherms[196] at temperatures between 231 and 273 K gave linear and nearly parallel W *versus* θ plots at $\theta < 0.5$. The value of the enthalpy of adsorption at zero coverage calculated from the slope of the $\ln k_1$ against $1/T$ plot was 37.6 kJ mol^{-1}, in reasonably good agreement with $q_0^{st} = 35.6$ kJ mol^{-1} (obtained by extrapolation of the isosteric enthalpy curve to zero coverage).

Hall and Stoeckli[161] found that their isotherms for n-butane on polyvinylidene chloride obeyed equation (27) over the range $\theta = 0.2 - 0.5$. The integral entropy change ($\Delta S_s^a = -27$ J mol^{-1} K^{-1}), calculated from the isosteric enthalpy, was, however, found to be appreciably less than that calculated for the formation of a mobile, two-dimensional film with the loss of one degree of translational freedom ($\Delta S_s^{tr} = -49.7$ J mol^{-1} K^{-1}).

Whalen[197] detected a small knee in the isotherms of hexane on polytetrafluoroethylene and employed the Hill–de Boer equation to estimate the extent of heterogeneity. He calculated that the high-energy sites accounted for about 3% of the total area.

Broekhoff and van Dongen[195] have discussed in some detail the validity of equation (27). They suggest that the BET monolayer capacity should not be used to calculate θ for a mobile film since the localized monolayer (implied in the BET model) has a different structure. Instead, they select the 'best' values of the three adjustable parameters (θ, k_1, and k_2) to match experimental isotherms with equation (27). This approach, of course, presents difficulties: it demands high accuracy in the experimental data and cannot allow for the onset of multilayer formation or the effect of heterogeneity.

In spite of these difficulties, Broekhoff and van Dongen[195] found that equation (27) is remarkably successful as an empirical relation in fitting the

[196] C. Pierce, *J. Phys. Chem.*, 1968, **72**, 1955.
[197] J. W. Whalen, *J. Colloid Interface Sci.*, 1968, **28**, 443.

isotherms of Ar, N_2, Kr, Xe, and C_6H_6 on graphitized carbon blacks. They conclude[195] that the simple van der Waals equation of state gives in general a satisfactory description of most of the characteristics of adsorption in the sub-monolayer region on uniform surfaces.

Equation (27) has been applied to isotherms measured at temperatures both above and below the two-dimensional critical temperature, T_{c2}. At $T < T_{c2}$, a marked first-order transition as revealed in the sub-monolayer region of the isotherm indicates the occurrence of two-dimensional condensation.[5, 7] As might be expected, many of the recently reported systems showing two-dimensional condensation feature graphitized carbon as the adsorbent.[16,195,196,198-201] Well-defined sub-monolayer steps were also registered with Kr adsorption isotherms on copper,[202] nickel,[203] layer-type halides[204, 205] (*e.g.* $NiCl_2$, $CoCl_2$, and $CdCl_2$), Ar, Kr, and Xe isotherms on alkali-metal halides,[204-207] and Ar on a layer of pre-adsorbed ethylene.[208]

It is important to recognize that not all stepwise isotherms reveal first-order transitions in the sub-monolayer region. In some cases,[209-211] the step is only evident in the multilayer range, and at lower coverage the isotherm is Type II in character. This behaviour is presumably due to preferential coverage of high-energy sites or to the participation of micropore filling. If the heterogeneity is confined to a small area, the presence of a shallow low-pressure knee may not obscure the first step (*e.g.* with Kr on $CdCl_2$).[205]

Thomy and Duval[198-200] have reported a remarkable set of stepwise isotherms for Kr, Xe, and CH_4 on exfoliated graphite which had been carefully prepared and was apparently of exceptional homogeneity. The sub-monolayer step was found to be made up to several parts and it was suggested that the first condensed phase is a two-dimensional liquid which undergoes 'solidification' as the pressure is increased. A similar feature was noted by Prenzlow[208] with Ar on pre-adsorbed ethylene. The small vertical discontinuity near monolayer completion was again attributed to a phase transition between the two-dimensional condensed phases.

There is an alternative explanation for the formation of secondary steps.

[198] A. Thomy and X. Duval, *J. Chim. phys.*, 1969, **66**, 1966.
[199] A. Thomy and X. Duval, *J. Chim. phys.*, 1970, **67**, 286.
[200] A. Thomy and X. Duval, *J. Chim. phys.*, 1970, **67**, 1101.
[201] C. E. Brown and P. G. Hall, *Trans. Faraday Soc.*, 1971, **67**, 3558.
[202] J. H. de Boer and J. H. Kaspersma, *Proc. k. ned. Akad. Wetenschap., Ser. B*, 1969, **72**, 289.
[203] J. M. Delolme and L. Bonnetain, *Compt. rend.*, 1970, **270**, C, 123.
[204] Y. Larher, *J. Colloid Interface Sci.*, 1971, **37**, 836.
[205] Y. Larher, *J. Chim. phys.*, 1971, **68**, 796.
[206] S. Ross and J. J. Hinchen, in 'Clean Surfaces: Their Preparation, Characterisation and Interfacial Study', ed. G. Goldfinger, Marcel Dekker, New York 1970, p. 115.
[207] A. V. Kiselev, A. A. Lopatkin, B. I. Lourie, and S. Shpigel, *Zhur. fiz. Khim.*, 1969, **43**, 2660 (*Russ. J. Phys. Chem.*, 1969, **43**, 1498).
[208] C. F. Prenzlow, *J. Colloid Interface Sci.*, 1971, **37**, 849.
[209] R. B. Gammage, E. L. Fuller, jun., and H. F. Holmes, *J. Colloid Interface Sci.*, 1970, **34**, 428.
[210] S. A. Selim, R. S. Mikhail, and R. I. Razouk, *J. Phys. Chem.*, 1970, **74**, 2944.
[211] P. G. Hall, V. M. Lovell, and N. P. Finkelstein, *Trans. Faraday Soc.*, 1970, **66**, 2629.

Delolme and Bonnetain[203] found that whereas powdered Ni, obtained by careful reduction of pure NiO, gave a single monolayer step in the Kr isotherm, Ni obtained from carbonyl and heated *in vacuo* at temperatures > 400 °C gave two steps in the sub-monolayer range. Other discontinuities were apparent in the multilayer region. It was suggested that the second phase change in the monolayer was due to the presence of islands of graphite which occupied about 20 % of the total Ni surface. If the Ni from the carbonyl was slightly oxidized and then reduced by hydrogen the resulting isotherm was similar in shape to that on the Ni from NiO. The decomposition of a small amount of CO on the clean Ni surface also gave rise to a Kr isotherm with secondary steps.

Rather similar results were reported by de Boer and Kaspersma[202] who studied Kr adsorption on Cu, obtained by the reduction of pure CuO. In this work, the secondary steps were steeper in the second-layer region of the isotherm than in the monolayer part. Evidence was given to show that the small steps were associated with the presence of three exposed crystal faces of Cu, *i.e.* the (111), (100), and (110) planes. Stepwise isotherms were also found[212] for the adsorption of Ar, Xe, He, and N_2 on freshly cleaved Iceland Spar crystals. The steps in these isotherms were thought to be due to heterogeneities in the surface.

Reference must be made to the detailed studies of van Olphen[213-215] on interlayer adsorption of water in clays. Well-defined stepwise isotherms were obtained for water vapour on sodium and magnesium vermiculite and these were related to X-ray studies of the *c*-spacings of the expanding clay mineral at various stages of hydration. In these systems, it was clear that the pronounced step in the isotherm does not simply involve either the superposition of a second water layer on the first or a sub-monolayer phase transformation. It was suggested that the interlayer cations change their configuration from the original association with $2H_2O$ to the octahedral environment of $6H_2O$.

The work of Brewer *et al.*[216] has shown that the nature of the substrate cannot be ignored in any analysis of phase transitions in the sub-monolayer range, even at high temperature. Taken together, all these results appear to present a confused picture. It is evident that stepwise isotherms may originate in a number of different ways. It is impossible to draw firm conclusions about the mechanism of the adsorption process on the basis of an isotherm at one temperature, no matter how carefully the measurements have been made. We may expect much more attention to be given to these interesting systems and that an effort will be made to correlate isotherms with calorimetric and structural studies.

The Ross–Olivier method[24] for isotherm analysis provides an elegant

[212] G. M. Yaryshev and P. E. Suetin, *Zhur. fiz. Khim.*, 1970, **44**, 2028 (*Russ. J. Phys. Chem.*, 1970, **44**, 1147).
[213] H. van Olphen, *J. Colloid Sci.*, 1965, **20**, 822.
[214] H. van Olphen, Proceedings of the International Clay Conference, 1969, vol. 1, p. 649.
[215] H. van Olphen, ref. 48, p. 255.
[216] D. F. Brewer, A. Evenson, and A. L. Thomson, *J. Low Temp. Phys.*, 1970, **2**, 137.

attempt to extend the application of Hill–de Boer equation to heterogeneous substrates. The surface is pictured as an assembly of uniform, or 'homotattic', patches distributed in gaussian manner over a range of adsorption potentials. Ross[158] has explained the origin of the term *homotattic*, which is not synonymous with iso-energetic. The Hill–de Boer equation is used to describe the form of the isotherm on each homotattic patch and the weighted sum of the coverages of all patches is employed to give the composite isotherm for the heterogeneous surface. A set of standard isotherms have been calculated[24] for different degrees of heterogeneity and the comparison with an experimental isotherm then provides a means of estimating the heterogeneity (width of the distribution function), the mean adsorption potential, and the monolayer capacity.

Jaycock and Waldsax[217] have discussed the solution of the Hill–de Boer equation (27) in an attempt to improve on the original treatment of Ross and Olivier. They confirm that steps are to be expected if the number of homotattic patches (*N* homogeneous areas) is small.[218] Increasing the value of *N* increases the degree of smoothness of the isotherm, but the value $N = 50$, used by Ross and Olivier,[24] is still too small to eliminate the irregularity. By increasing *N* to 200—250, Jaycock and Waldsax[217] obtain smooth isotherms, even in cases where the distribution function is very wide.

Localized Adsorption and the Langmuir Equation. In its simplest form localized adsorption is represented by the Langmuir equation

$$\theta = \frac{bp}{1 + bp}. \tag{29}$$

The constant *b* is a function of the enthalpy and the entropy of adsorption. Equation (29) assumes the surface to be uniform (sites all of equivalent energy) with no interactions between the adsorbed molecules.

Many attempts[5] have been made to modify the Langmuir model to bring it closer to real systems. Adsorbate–adsorbate interactions are allowed for in the treatment of Fowler and Guggenheim;[219] the random distribution of the adsorbed molecules is retained in the statistical thermodynamic derivation of the equation

$$p = \frac{\theta}{k_1(1 - \theta)} \exp\left(2\theta w/kT\right), \tag{30}$$

where *w* is the mutual interaction energy of a pair of neighbouring molecules in the localized monolayer on a uniform surface. When $w = 0$ equation (30) reduces to the Langmuir equation (29), which in turn reduces to Henry's law at very low coverage. The point of inflection lies at $\theta = 0.5$ and the

[217] M. J. Jaycock, and J. C. R. Waldsax, *J. Colloid Interface Sci.*, 1971, **37**, 462.
[218] B. G. Baker and L. A. Bruce, *Trans. Faraday Soc.*, 1968, **64**, 2533.
[219] R. H. Fowler and E. A. Guggenheim, 'Statistical Mechanics', Cambridge University Press, Cambridge, 1949.

conditions for two-dimensional condensation are also determined by equation (30).

The Fowler–Guggenheim equation has been found[16] to describe the CO_2 isotherm on graphitized carbon black at 193.5 K for surface coverage up to $\theta \sim 0.7$, but it is not so successful with ethane at 173.2 K on the same surface. As an empirical relation, equation (30) appears to be less satisfactory than equation (29) for isotherms of a number of vapours on graphitized carbons.

The formidable problem of handling localized adsorption on a heterogeneous surface has been approached in different ways.[220–225] Misra[220] and Cerofolini[221] have extended the treatment of Sips[226] and have supposed that the Langmuir equation can be adopted as a 'local' isotherm, *i.e.* valid for a set of homotattic sites. If the sites are non-interacting, an overall isotherm can be established to correspond with a particular energy distribution function. Misra[220] obtained a new isotherm equation, which corresponds with a simple exponential function and is perhaps useful for a surface with a very limited distribution of energy. Cerofolini,[221] on the other hand, imposed the condition of an overall DRK isotherm in order to compute the energy distribution function. This turns out to be somewhat different to the gaussian which is normally assumed as a pre-requisite for the DRK equation.

D'Arcy and Watt[222] adopted the summation procedure to obtain a composite isotherm. Langmuir-type processes were assumed to occur independently on sites with high affinity and sites which have much weaker affinities, such that the monolayer on the weak sites is only partially completed at saturation pressure. The multilayer also forms at higher pressure, but in this region the high-energy sites are completely covered. The composite equation appears to have general applicability to a wide range of water isotherms, but it must be regarded as an empirical equation.

We return now to the ideal localized monolayer model and the application of equation (29), which is usually rearranged to the linear form

$$p/x = 1/x_m b + p/x_m, \qquad (31)$$

where x is the amount of gas adsorbed at pressure p and x_m is the monolayer capacity.

It is now recognized that a linear plot of p/x against p does not in itself imply adsorption in accordance with the Langmuir model: Everett[227] and others[5,228,229] have established that the isosteric enthalpy must be

[220] D. N. Misra, *J. Chem. Phys.*, 1970, **52**, 5499.
[221] G. F. Cerofolini, *Surface Sci.*, 1971, **24**, 391.
[222] R. L. D'Arcy and I. C. Watt, *Trans. Faraday Soc.*, 1970, **66**, 1236.
[223] Yu. S. Khodakov, A. A. Berlin, G. I. Kalyaev, and Kh. M. Minachev, *Teor. i eksp. Khim.*, 1969, **5**, 631.
[224] C. V. Heer, *J. Chem. Phys.*, 1971, **55**, 4066.
[225] D. S. Jovanovic, *Kolloid-Z.*, 1969, **235**, 1203.
[226] R. Sips, *J. Chem. Phys.*, 1948, **16**, 490.
[227] D. H. Everett, *Proc. Chem. Soc.*, 1957, 38.
[228] B. G. Baker, *J. Chem. Phys.*, 1966, **45**, 2694.
[229] M. J. Sparnaay, *Surface Sci.*, 1968, **9**, 100.

independent of coverage and that $\Delta \bar{S}$, the differential entropy of adsorption, must take the form

$$\Delta \bar{S} = \Delta \bar{S}^* - R \ln \left[\frac{\theta}{(1-\theta)} \right], \qquad (32)$$

with $\Delta \bar{S}^*$, the standard entropy of adsorption (at $\theta = 0.5$), in principle being amenable to theoretical calculation in accordance with the model. Barrer and his co-workers[52, 53] have shown that the variation of $\Delta \bar{S}$ with θ provides a sensitive measure of the energetic heterogeneity of an adsorbent. That no system has been found to satisfy all the requirements for ideal localized adsorption over the complete range of surface coverage is not surprising in view of the complexities already noted, even with uniform substrates. It is interesting to note that linear Langmuir plots (p/x against p) have been obtained for the adsorption of certain gases by zeolites[51, 230-232] and other microporous solids.[233, 234] In only a few cases has the variation of q^{st} with coverage been followed, but, as mentioned previously, generally micropore filling and specific adsorbent–adsorbate interactions are known to give rise to energetic heterogeneity and this in itself indicates that the ideal localized monolayer model cannot strictly hold.

The work of Egerton and Stone[230] is of particular interest because it illustrates how the Langmuir isotherm may be used to analyse adsorption data for a complex system. Langmuir plots were constructed for the adsorption of CO on Ca-exchanged zeolite (i) by taking the adsorption x in equation (31) as the total amount adsorbed and (ii) by first subtracting a certain fraction corresponding to the non-specific adsorption (as given by the original Na Y zeolite). On making this correction, it was found[230] that the linearity of the Langmuir plots was considerably improved. Such a procedure was justified by the fact that q^{st} remained nearly constant with coverage and the $\Delta \bar{S}$ against θ curve was consistent with equation (32). It was therefore concluded that for CO adsorption on Ca Y, the Langmuir model gives at least a self-consistent description of the *specific* contribution to the adsorption process.

Barrer[50, 51, 233] has employed the Langmuir equation in the analysis of adsorption data obtained with many zeolites and other microporous systems and has shown that it is convenient to study the variation of the Langmuir quotient, $K_L = p(1-\theta)/\theta$, with θ. It should be noted that this is a more searching test of mathematical conformity of the isotherm to equation (29) than is the more usual plot of p/x against p. In some cases (*e.g.* Ar, O_2, and Kr in H-chabazite[51]) both K_L and q^{st} remain nearly constant over a wide range of

[230] T. A. Egerton and F. S. Stone, *Trans. Faraday Soc.*, 1970, **66**, 2364.
[231] G. Gnauck, E. Rosner, and E. Eichhorst, *Chem. Tech. (Leipzig)*, 1970, **22**, 680.
[232] F. Wolf and R. Reisdorf, *Z. Chem.*, 1971, **11**, 265.
[233] R. M. Barrer and D. L. Jones, *J. Chem. Soc. (A)*, 1971, 2594.
[234] A. D. Zorin, V. Ya. Dudorov, T. S. Rogoshnikova, and E. A. Ryabenko, *Zhur. fiz. Khim.*, 1970, **44**, 717 (*Russ. J. Phys. Chem.*, 1970, **44**, 398).

coverage; in others (*e.g.* H$_2$, N$_2$, and Xe in H-chabazite[51]) q^{st} varies with coverage and in some cases the plots of log K_L tend to be linear functions of θ. The latter behaviour corresponds with the requirements of the Fowler–Guggenheim equation, but the picture of localized adsorption with nearest-neighbour interaction is again too simple because the variation of the slope with temperature is not in accordance with equation (30). Barrer and Davies[51] have suggested that increase in θ may lead to a squeezing together of the molecules with a resultant change in thermal entropy. They conclude that no general interpretation of adsorption by zeolites in terms of localized adsorption is feasible.

The Type I isotherm obtained by Wade and his co-workers[235] for the adsorption of n-aliphatic alcohols on oxidized aluminium foil provides excellent examples of gas-phase 'autophobicity', in which the high-energy heterogeneous surface of Al$_2$O$_3$ has attached to it a strongly held oriented monolayer. The alkyl groups of the alcohol molecules present a low-energy surface for second- and higher-layer adsorption so that the multilayer formation is delayed until a high p/p^0 is reached. Lateral interactions undoubtedly play some part in the monolayer formation, and the linearity of Langmuir plots, especially evident with the increase of chain length, is probably due to compensation between the effects of heterogeneity of the substrate and lateral interactions in the monolayer.

The question of the validity of the Langmuir x_m value is still under discussion. There are two opposing views: that of Brunauer,[236] who maintains that the plateau of the Type I isotherm corresponds to the completion of the monolayer; and that of Dubinin,[4, 237] who reasons that the Type I isotherm (for physical adsorption) is necessarily associated with micropore filling and that the plateau therefore provides a measure of the pore volume. This question is mentioned again later, because a similar problem arises in the application of the BET method for the determination of the surface area of porous solids. It may be noted, however, that, in view of the uncertainty in interpreting x_m for microporous solids, Barrer[238] has suggested that the micropore adsorption capacity should be expressed either as a volume or as a 'monolayer equivalent area', *i.e.* the area obtained by removing the adsorbate from the micropore and spreading it as a close-packed monolayer on a molecularly smooth surface.

It is evident that the linearity of the p/x against p plot by itself tells us nothing about the mechanism of the adsorption process. The fact that an isotherm appears to approach at some stage a limiting value of uptake may be due to an appreciable difference in energy requirements for second-layer formation, as in the case of 'autophobicity', or it may be associated with the filling of micropores, as with zeolites. In the latter case, multilayer formation

[235] W. H. Wade and T. D. Blake, *J. Phys. Chem.*, 1971, **75**, 1887.
[236] S. Brunauer, ref. 48, p. 63.
[237] M. M. Dubinin, ref. 48, p. 123.
[238] R. M. Barrer, ref. 48, p. 89.

occurs without difficulty in the usual range of p/p^0, but the uptake is comparatively very low because it is restricted to a small *external* surface.

6 The Brunauer–Emmett–Teller Equation and the Determination of Surface Area

There has been a general awareness for many years of the short-comings of the Brunauer–Emmett–Teller (BET) theory. Opinion is still divided,[5,33,45,158,236,237,239] however, on the range of applicability of the BET method for surface area determination and the validity of the BET monolayer capacity.

The importance of surface area determination is firmly established in many areas of science and technology.[119,153,215,240] In spite of the weakness in its theoretical foundations, the BET method continues to overshadow all other available methods for the determination of the surface area of porous and finely divided materials. The conditions required for the successful application of the BET method have been discussed at length,[5,35] and amongst the large number of recent publications on the subject it is difficult to find any original approach which casts new light on the assessment of the use of gas adsorption for surface area determination.

It is customary to apply the BET equation in the linear form

$$\frac{p}{x(p^0-p)} = \frac{1}{x_m C} + \frac{(C-1)}{x_m C}\frac{p}{p^0}. \tag{33}$$

According to the theory, C is a constant related to the net enthalpy of adsorption, but in fact it is a free-energy term which is dependent on both the enthalpy and entropy of adsorption.[236] It is well known that the BET plot, $p/x(p^0-p)$ against p/p^0, is generally linear in the approximate p/p^0 range of 0.05—0.30. With uniform substrates, however, the range of linearity is often much more restricted[241,242] and with graphitized carbon[45] it may not extend beyond $p/p^0 = 0.1$.

Various attempts have been made to modify the BET equation to improve the agreement with experimental isotherm data over a wide range of p/p^0. One of the most recent was that of Brunauer and his co-workers,[243] who derived a modified form of BET equation,

$$\frac{kp}{x(p^0-kp)} = \frac{1}{x_m C} + \frac{C-1}{x_m C}\cdot\frac{kp}{p^0}, \tag{34}$$

where k is an additional parameter (<1) to allow for the finite number of layers deposited on a non-porous solid as $p \to p^0$.

[239] J. H. de Boer, ref. 48, p. 7.
[240] G. Beurton and P. Bussiere, ref. 48, p. 217.
[241] Yu. F. Berezkina, M. M. Dubinin, and A. I. Sarakhov, *Izvest. Akad. Nauk S.S.S.R., Ser. khim.*, 1969, 2653 (*Bull. Acad. Sci. U.S.S.R., Chem. Sci.*, 1969, 2495).
[242] B. W. Davis, *J. Colloid Interface Sci.*, 1969, **31**, 353.
[243] S. Brunauer, J. Skalny, and E. E. Bodor, *J. Colloid Interface Sci.*, 1969, **30**, 546.

Equation (34) had been proposed many years before by Anderson,[244] but had received little attention.* In fact, as an empirical relation with an additional parameter it appears to be more successful than equation (33) for isotherms of nitrogen on non-porous oxide.[243] From a theoretical standpoint it takes some account of the shortcomings of the BET assumption that the second and higher adsorbed layers have the properties of the bulk liquid: the deviation of k from unity is a crude measure of the inadequacy of this assumption.

Takizawa,[245] in constructing BET plots from composite isotherms on multicomponent surfaces, has revealed the insensitivity of the BET plot to a change in the number of sites of different energy. A rather similar type of investigation has been made by Beruto and Merli,[246] who introduce a parameter for energetic heterogeneity in the BET equation. Such an approach could perhaps be justified if the localized monolayer model (on which the BET theory is essentially based) were found to hold for uniform surfaces. Unfortunately, the range of linearity of both the Langmuir and the BET plots is generally better with heterogeneous surfaces than with carefully prepared uniform substrates.[45,158] De Boer[239] and others[5,45] have stressed the fundamental significance of point B (*i.e.* the view originally expressed by Halsey[247]) and have suggested that the true role of the BET method is to provide a simple graphical procedure for locating point B. This reinforces the need for a critical assessment of x_m values before the BET surface areas are accepted as valid. It also directs attention to the opportunity for devising other mathematical procedures for the analysis of the isotherm.[248-250]

The BET surface area, A_{BET}, can only be calculated from x_m if a value is assumed for the average area, a_m, occupied by the adsorbate molecule in the completed monolayer. For a number of reasons[5,45] (theoretical and practical) nitrogen is generally employed as a standard adsorbate, with $a_m(N_2) = 0.162$ nm^2 (the value first calculated by Brunauer and Emmett from the density of liquid nitrogen at 77 K, assuming hexagonal close-packing). Comparisons made with the geometric areas of glass fibre[251] and glass spheres[127,252] appear to confirm the correctness of A_{BET} and $a_m(N_2)$ to within about 5%. The uncertainty is rather greater with aluminium foil[127] and titanium carbide,[253] and it is possible that in these cases surface roughness

---

* Guggenheim ('Applications of Statistical Mechanics', Oxford University Press, Oxford 1966, Chapter 11, p. 186) has pointed out that an equation of this form was in fact derived by Langmuir (*J. Amer. Chem. Soc.*, 1918, **40**, 1375).

[244] R. B. Anderson, *J. Amer. Chem. Soc.*, 1946, **68**, 686.
[245] A. Takizawa, *Kolloid-Z.*, 1968, **222**, 143.
[246] D. Beruto and C. Merli, *Ann. Chim. (Italy)*, 1970, **60**, 630.
[247] G. D. Halsey, *Discuss. Faraday Soc.*, 1950, No. 8, 54.
[248] V. T. Belov, *Zhur. fiz. Khim.*, 1970, **44**, 2388 (*Russ. J. Phys. Chem.*, 1970, **44**, 1352).
[249] P. T. John and J. N. Bohra, *Indian J. Technol.*, 1970, **8**, 34.
[250] R. E. Mardaleishvili and S. N. Khadzhiev, *J. Res. Inst. Catal., Hokkaido Univ.*, 1968, **16**, 581.
[251] V. R. Deitz and N. H. Turner, ref. 48, p. 43.
[252] E. Robens, ref. 48, p. 51.
[253] B. W. Davis and R. G. Varsanik, *J. Colloid Interface Sci.*, 1971, **37**, 870.

accounts for the difference of about 30%. The electron-microscopic examination of graphitized carbon black (*e.g.* Sterling FT), on the other hand, has given[253] a significantly *higher* value of the surface area than that calculated from x_m with $a_m(N_2)$ taken as 0.162 nm^2. This result indicates that the nitrogen monolayer may be slightly more open [with $a_m(N_2) \sim 0.19 \text{ nm}^2$] on graphitized carbon than it is on a hydroxylated silica surface. We shall return to this point later.

Only a few other comparative studies appear to have been made with well-characterized low-area adsorbents. Krypton adsorption measurements on a pristine glass fibre[251] (of geometric area 4.6 m^2) indicated a value for $a_m(Kr)$ of 0.204 nm^2, in excellent agreement with the value of 0.202 nm^2 reported[160] for a stainless steel (geometric area, 47.5 cm^2) and close to that recommended for Kr on boron phosphate[254] (0.215 nm^2). The position appears to be more complicated with argon (at 77 K), and comparisons with geometric areas[160,251,253] have yielded values of $a_m(Ar)$ of 0.134, 0.147, and 0.182 nm^2. Studies of the adsorption of ethylene on stainless steel[103] and of tetramethylsilane and other vapours on aluminium foil[235] have also provided useful information on the effective cross-sectional areas of adsorbates on known geometric areas.

The many attempts[5,45,255] made in recent years to improve the accuracy of BET areas by the somewhat arbitrary adjustment of the adsorbate cross-sectional area cannot be discussed in any detail. Unless it is based on independent evidence, this type of approach is unsatisfactory for two reasons: (i) it necessarily assumes the validity of the BET monolayer capacity, irrespective of the true nature of the adsorption process; (ii) it assumes the constancy of the value of a_m for the 'primary standard' adsorbate (usually nitrogen) on all substrates.

At the risk of repetition, it must be emphasized that the BET analysis takes no account of the possibility of micropore filling[48,256] or interlamellar penetration[233] occurring in the BET range of the isotherm and thereby distorting the value of x_m. There is also the difficulty of interpretation when x_m is obtained from an isotherm with a low C value: it seems generally agreed[45] that the BET monolayer capacity should *not* be calculated from a Type III isotherm, but the significance of x_m when derived from a Type II isotherm with $C < 50$ is still uncertain. Related to this is the problem of selecting for a particular system the appropriate range of p/p^0 for the BET plot, especially with systems giving a small knee at low pressure.[163,164] High C values present a different problem, because it seems likely that they are associated with localized monolayer formation;[250,256,257] the exact value of a_m will then depend on the substrate structure. In view of these considerations, it is

[254] A. J. Knowles and J. B. Moffat, *J. Colloid Interface Sci.*, 1971, **37**, 860.
[255] A. L. McLellan and H. F. Harnsberger, *J. Colloid Interface Sci.*, 1967, **23**, 577.
[256] C. Pierce, *J. Phys. Chem.*, 1968, **72**, 3673.
[257] J. D. Carruthers, D. A. Payne, K. S. W. Sing, and L. J. Stryker, *J. Colloid Interface Sci.*, 1971, **36**, 205.

strongly recommended that C values be recorded along with BET surface areas.

To gain a clearer understanding of the limitations of the BET method it is rewarding to examine closely the adsorption data obtained after the modification of a well-defined non-porous surface. A number of such studies have been made recently with such adsorbents as graphitized carbon,[241] rutile,[258] thorium oxide,[259] and calcium fluoride;[260] adsorption isotherms were determined both before and after the pre-adsorption of molecules which remained relatively strongly attached to the surface (either by chemisorption or because at the low temperature their p^0 was very small). Investigations have been conducted with pre-adsorbed layers of ethylene,[261] xenon, water, and methanol[241] on graphitized carbon, water on thorium oxide,[259] water, ethanol, hexan-1-ol, and hexane-1,6-diol on rutile,[258] and sodium oleate on calcium fluoride.[260] In every case so far studied, the presence of the pre-adsorbed layer produces a decrease in the adsorbent–adsorbate interaction energy (weakening of the dispersion forces through a decrease in the surface density) and a progressive change in the isotherm character. An increase in the fraction of surface covered by the pre-adsorbed species causes the isotherm under investigation (Ar, N_2, *etc.*) to change towards Type III with a gradual decrease in the value of the parameter C and also in the BET area. Day, Parfitt, and Peacock[258] found that the presorption of alcohol reduced the surface area of rutile, as assessed by BET–nitrogen, from 10.2 to 7.2 $m^2 g^{-1}$ (the corresponding C values were changed from 450 to 39). Similar results for the adsorption of nitrogen and argon on methanol-coated graphitized carbon were obtained by Dubinin and his co-workers,[241] as indicated in the Table.

Table *Effect of pre-adsorption of methanol on BET parameters for adsorption of nitrogen and argon on graphitized carbon blacks*

Fractional coverage with MeOH	$A_{BET}/m^2 g^{-1}$	C
	Nitrogen at -195.6 °C on sample CT [$a_m(N_2)=0.162$ nm²]	
0	39.1	150
0.50	31.4	90
0.80	30.0	40
1.3	29.9	32
1.6	28.1	27
2.9	25.4	25
	Argon at -183 °C on sample CG [$a_m(Ar)=0.137$ nm²]	
0	28.8	700
0.54	24.6	112
0.76	24.4	43
1.2	24.9	15
2.0	29.9	6

[258] R. E. Day, G. D. Parfitt, and J. Peacock, *Discuss. Faraday Soc.*, 1971, No. 52, 215.
[259] R. B. Gammage, E. L. Fuller, and H. F. Holmes, ref. 48, p. 161.
[260] P. G. Hall, V. M. Lovell, and N. P. Finkelstein, *Trans. Faraday Soc.*, 1970, **66**, 1520.
[261] C. F. Prenzlow, H. R. Beard, and R. S. Brundage, *J. Phys. Chem.*, 1969, **73**, 969.

C

It seems reasonable to suppose that the precoating of a *non-porous* solid up to—and probably beyond—monolayer coverage can be achieved with only negligible change in the real surface area. If this is the case, it implies that the BET areas calculated from isotherms with low C values may be over 20% in error, even on a comparative basis. It is interesting to note that the BET–nitrogen area is more seriously affected than the BET–argon area, although since they both change in the same direction the ratio is not changed as drastically as are the individual values.

The situation with some microporous solids is different since the BET area is appreciably reduced by the pre-adsorption of even a small amount of adsorbate[165] which is retained in the pores at low p/p^0. A method of micropore filling by pre-adsorption has been devised by Gregg and Langford[262] for the assessment of microporosity. The results of these studies on the effect of the pre-adsorption of nonane support the Dubinin concept[4] of micropore filling but further work is now required on a range of well-characterized porous materials.

The adsorption of water vapour has been investigated[45,162,165,175,210,213, 255,257,258,260,263-269] in some detail, but comparisons with the BET–nitrogen areas present a very confused picture. McLellan and Harnsberger[255] have recommended a value of 0.125 nm² for the molecular area of water, but in fact the BET monolayer capacity for water may be considerably smaller (*e.g.* on graphitized carbon[7,16]) or much larger (*e.g.* in chromium oxide gel[265,266] or hardened Portland cement[236,269]) than would be expected on the basis of the BET–nitrogen area. This anomalous behaviour is associated with the high degree of specificity of the interactions of water with solid surfaces.[1,32] Although the adsorption of water vapour cannot be safely used for the determination of the *total* area, it appears that a careful analysis of water isotherms can yield valuable information concerning the chemical structure of the surface.[257]

7 Empirical Methods for Isotherm Analysis

We have seen that none of the current theories of adsorption is capable of describing an experimental isotherm over its entire range of relative pressure. In the sub-monolayer region, the virial equation[16] can provide a satisfactory mathematical description of an adsorption isotherm, with additional coefficients introduced as required. The Hill–de Boer[195] and the

[161] S. J. Gregg and J. F. Langford, *Trans. Faraday Soc.*, 1969, **65**, 1394.
[162] R. B. Gammage, W. S. Brey, and B. H. Davis, *J. Colloid Interface Sci.*, 1970, **32**, 256.
[164] J. B. Moffat and K. F. Tang, *Ind. and Eng. Chem.* (*Product Res. and Development*), 1970, **9**, 570.
[165] F. S. Baker, K. S. W. Sing, and L. J. Stryker, *Chem. and Ind.*, 1970, 718.
[166] F. S. Baker, J. D. Carruthers, R. E. Day, K. S. W. Sing, and L. J. Stryker, *Discuss. Faraday Soc.*, 1971, No. 52, 173.
[167] O. P. Mahajan, *Amer. Chem. Soc., Div. Fuel Chem., Prepr.*, 1969, **13**, 4.
[168] Y. Iguchi and M. Inouye, *Tetsu to Hagane*, 1971, **57**, 15 (*Chem. Abs.*, 1971, **74**, 78, 551).
[169] O. P. Mchedlov-Petrosyan and D. A. Uginchus, *Silikattechnik*, 1969, **20**, 416.

Dubinin–Radushkevich–Kaganer[184] equations are also applicable, on an empirical basis, to certain systems over restricted ranges of low coverage. Other equations are required for the multilayer section of the isotherm. Although the BET equation does not itself provide a useful empirical relation,[45] the modified form proposed by Anderson[244] and by Brunauer[243] is more successful. The Harkins–Jura equation may also be fitted to some isotherms on non-porous solids[5,45] in the intermediate range of p/p^0. Except with microporous solids, the isotherm curvature is not usually pronounced in the range of $p/p^0 = 0.1$—0.6, so that the test of a mathematical relation is not severe. At high p/p^0, the isotherm on a non-porous solid usually becomes steep and may appear to approach the $p/p^0 = 1$ axis asymptotically; under these conditions, the Frenkel–Halsey–Hill[270] (FHH) equation is generally obeyed,[5,271,272]

$$\ln(p/p^0) = -k/\theta^s, \tag{35}$$

where k and s are constants.

According to the 'slab model',[5,270,271] at an appreciable distance from the surface (*i.e.* in the multilayer range) the adsorbent–adsorbate interaction energy decays with the third power of distance and $s = 3$. Various attempts have been made to verify this cube law, but the overall picture is still far from clear, partly because of the difficulty of detecting the onset of capillary condensation.[45] In the work of Evenson *et al.*,[273] helium isotherms were determined on porous glass and nitrogen-coated glass and allowance was made for the curved shape of the adsorbed film in the pores. On this basis, the cube law appears to fit the multilayer isotherm over the range $\theta = 2$—5, above which capillary condensation occurs.

Lando and Slutsky[274] studied the adsorption of neopentane and tetramethylsilane on gold films at very high coverage ($\theta = 10$—40) and found that the values of s in equation (35) are close to 2. These data were re-examined by Roy and Halsey[94] with a view to fitting the cube law along with an extra term, containing a compatibility factor as used originally by Singleton and Halsey.[271] The approach adopted followed that of Adamson and Ling,[275] taking account of the relaxation of the adsorbate structure from layer to layer. Although it was possible in this way to fit the high-coverage data to a modified cube law, Roy and Halsey[94] were forced to the conclusion that the cube law is not strictly valid for this system over the large range of distance investigated.

Following the earlier work of Pierce,[272] Zettlemoyer and his co-wor-

[270] T. L. Hill, *Adv. Catalysis*, 1952, **4**, 211.
[271] J. H. Singleton and G. D. Halsey, *Canad. J. Chem.*, 1955, **33**, 184.
[272] C. Pierce, *J. Phys. Chem.*, 1960, **64**, 1184; *ibid.*, 1968, **72**, 1955.
[273] A. Evenson, D. F. Brewer, A. J. Symonds, and A. L. Thomson. *Phys. Letters (A)*, 1970, **33**, 35.
[274] D. Lando and L. J. Slutsky, *J. Chem. Phys.*, 1970, **52**, 1510.
[275] A. W. Adamson and I. Ling, *Adv. Chem. Ser.*, 1964, No. 43, p. 57.

kers[276-278] have tested equation (35) by plotting log θ against log (log p^0/p). With nitrogen isotherms on a range of non-porous solids it was noted that oxides and other high-energy (polar) surfaces give $s = 2.75$, whereas low-energy polymer surfaces give $s = 2.12$. The case of nitrogen on graphitized carbon black is especially interesting because the FHH plot has two linear parts, intersecting at $p/p^0 = 0.4$: the slope of the branch at higher p/p^0 is very close to that of the 'ideal isotherm' on low-energy surfaces but the slope in the range $p/p^0 = 0.2$—0.4 is steeper. Pierce[256,279] and Zettlemoyer[276] suggest that a uniform graphitized carbon surface adsorbs a nitrogen monolayer as an open array (*i.e.* a localized monolayer) with $a_m(N_2) \sim 0.20$ nm^2. Zettlemoyer[274] has also noted that water FHH plots on hydroxylated surfaces (TiO_2, Fe_2O_3, SiO_2) give s values around 2.4 whereas with water on hydrophobic surfaces (dehydroxylated SiO_2 *etc.*) the s values are appreciably lower (1.3—1.8).

It is evident that the FHH plot can provide useful comparative information[280] about the nature of the multilayer. It can also be used to separate the effects of multilayer adsorption and capillary condensation, as originally suggested by Pierce.[272] The fact that nitrogen isotherms on various high-energy surfaces give similar s values is related to the concept of the 'universal' multilayer isotherm for nitrogen.[256] Indeed, in pore-size distribution determinations, constant numerical values of k and s are generally assumed in the application of equation (35) for the calculation of the multilayer thickness correction curve.[281]

Halsey[282] has pointed out that a linear combination of a virial type of expansion and a FHH term, giving an equation of the form

$$p/p^0 = (b\theta + c\theta^2 + \ldots)\,\phi + \exp(-k/\theta^3) \qquad (36)$$

with ϕ as a 'blending function', satisfies the Henry's law requirement at low coverage and also gives the FHH multilayer at high p/p^0. Greenlief and Halsey[283] show that this form of equation is reasonably satisfactory for an argon isotherm on graphitized carbon precoated with a xenon monolayer, but that it does not hold for the isotherm of xenon on silver, although the two isotherms are similar at high p/p^0. It is clear that equation (36) cannot be applied to an isotherm with a knee at low p/p^0.

The concept of a universal multilayer thickness isotherm led Lippens and de Boer[284] to their *t*-method for isotherm analysis. de Boer and his

[276] A. C. Zettlemoyer, *J. Colloid Interface Sci.*, 1968, **28**, 343.
[277] E. McCafferty and A. C. Zettlemoyer, *J. Colloid Interface Sci.*, 1970, **34**, 452.
[278] E. McCafferty and A. C. Zettlemoyer, *Discuss. Faraday Soc.*, 1971, No. 52, 239.
[279] C. Pierce, *J. Phys. Chem.*, 1969, **73**, 813.
[280] M. K. Lloyd and R. F. Conley, *Clays, Clay Miner.*, 1970, **18**, 37.
[281] D. Dollimore and G. R. Heal, *J. Colloid Interface Sci.*, 1970, **33**, 508
[282] G. D. Halsey, in 'The Solid–Gas Interface', ed. E. A. Flood, Marcel Dekker, New York, 1967, vol. I, p. 503; G. D. Halsey, *J. Chem. Phys.*, 1968, **36**, 1688.
[283] C. M. Greenlief and G. D. Halsey, *J. Phys. Chem.*, 1970, **74**, 677.
[284] B. C. Lippens and J. H. de Boer, *J. Catalysis*, 1965, **4**, 319.

co-workers have claimed[285] that the 'amount of nitrogen adsorbed per unit of surface area of non-porous adsorbents is a unique function of p/p^0 for a large number of inorganic oxides and hydroxides as well as for graphitized carbon blacks'. In fact, the differences in specific and non-specific interactions discussed earlier rule out the possibility of obtaining a common multilayer thickness (t-curve) for nitrogen. This was acknowledged by de Boer[239] in one of his last papers and the hope was expressed that t-tables would be established for the adsorption of nitrogen on various groups of materials.

The t-method is a simple and effective way of comparing the characteristics of one isotherm with those of another, taken as a standard. The volume (or mass) of nitrogen adsorbed is plotted against t, the statistical thickness of the adsorbed layer of nitrogen, obtained from the 'standard' isotherm on a non-porous reference solid. A deviation in shape from the standard isotherm is detected as a departure of the t-plot from linearity.[284]

Because of its simplicity and apparent ease of application, the t-method has attracted a considerable amount of attention.[259,286-297] The controversy over the use of the t-plot has been centred on the interpretation of the data obtained with microporous solids, *i.e.* Type I isotherms. According to the approach of de Boer,[239,284] Brunauer,[236,298] and Mikhail,[287] the characteristic t-plot obtained from a Type I isotherm is interpreted in terms of monolayer coverage of a comparatively large internal surface area followed by multilayer formation on a small external surface. This interpretation has been challenged by Dubinin,[4,237,297] Marsh and Rand,[295] Nicolaon and Teichner,[296] Pierce,[256] Gregg and Sing,[45,48,262] and others,[259,288,291] who take the view that the initial uptake (*i.e.* the low p/p^0 region of a Type I isotherm) is due mainly to micropore filling rather than to surface coverage. In support of this hypothesis, it may be noted that (*a*) the overlap of the adsorption potential across the micropore (wall-to-wall) produces an enhanced energy of adsorption[4,27,47,48,237,299] and (*b*) the surface area obtained from the sub-monolayer region of the t-plot appears to be too high.[45,48,297] These

[285] J. H. de Boer, B. C. Lippens, B. G. Linsen, J. C. P. Broekhoff, A. van den Heuvel, and Th. J. Osinga, *J. Colloid Interface Sci.*, 1966, **21**, 405; see also J. C. P. Broekhoff and J. H. de Boer, ref. 48, p. 97.
[286] J. Skalny, E. E. Bodor, and S. Brunauer, *J. Colloid Interface Sci.*, 1971, **37**, 476.
[287] R. Sh. Mikhail and F. A. Shebl, *J. Colloid Interface Sci.*, 1970, **32**, 505.
[288] A. J. Tyler, F. H. Hambleton, and J. A. Hockey, *J. Catalysis*, 1969, **13**, 43.
[289] B. Lespinasse, *Bull. Soc. chim. France*, 1970, 3317.
[290] E. Koberstein and M. Voll, *Z. phys. Chem. (Frankfurt)*, 1970, **71**, 275.
[291] W. R. Smith and G. A. Kasten, *Rubber Chem. and Technology*, 1970, **43**, 960.
[292] A. Voet and P. Aboytes, *Carbon*, 1971, **9**, 135
[293] V. Wagner and S. Pizzini, *Ann. Chim. (Italy)*, 1970, **60**, 198.
[294] R. B. Fahim and A. I. Abu-Shady, *J. Catalysis*, 1970, **17**, 10.
[295] H. Marsh and B. Rand, *J. Colloid Interface Sci.*, 1970, **33**, 478.
[296] G. A. Nicolaon, *J. Chim. phys.*, 1969, **66**, 1783.
[297] M. M. Dubinin, ref. 48, p. 75.
[298] S. Brunauer, ref. 48, p. 79.
[299] B. P. Bering, V. A. Gordeeva, M. M. Dubinin, L. I. Efimova, and V. V. Serpinskii, *Izvest. Akad. Nauk S.S.S.R., Ser. khim.*, 1971, 22 (*Bull. Acad. Sci. U.S.S.R., Chem. Sci.*, 1971, 17).

arguments can only be finally settled when more data are obtained with a range of adsorptives on well-characterized microporous solids.

Unfortunately, the t-method suffers from the disadvantage that it is not independent of the BET method, since t is itself calculated from the BET monolayer capacity,[236,239] and if the C value is low ($C < 100$), the absolute values of t are uncertain.[33,45] The selection of the appropriate t-curve has presented a problem in the application of the method to adsorptives other than nitrogen. Brunauer[236,286,300] and Mikhail[287] have suggested that the correct t-curve should give the same value of C as the isotherm under analysis and that the surface area obtained from the slope of the t-plot should match the BET area. This approach overlooks the effect of pore narrowing in changing the shape of the isotherm in the BET range.[48] If the t-method is to provide an understanding of the true nature of the adsorption process, it is essential that standard isotherms (t-curves) are obtained on truly non-porous reference solids of known structure.

It is evident that the scope of the t-method is necessarily restricted to those systems with a reasonably well-defined monolayer capacity. To avoid this difficulty and to extend the application of the method to Type III isotherms, t was replaced[48] by x/x_s, termed α_s, where x_s is the amount adsorbed at a selected relative pressure (standard state). The reduced isotherm on the non-porous reference solid is therefore arrived at empirically and not *via* the BET monolayer capacity. In principle, α_s could be placed equal to unity at any convenient point on the standard isotherm; in practice, it has been found convenient[48] to place $\alpha_s = 1$ at $p/p^0 = 0.4$. With nitrogen isotherms at 77 K, monolayer coverage and micropore filling occur at $p/p^0 < 0.4$, whereas any hysteresis loop (associated with capillary condensation) is located at $p/p^0 > 0.4$.

The α_s-method has been employed for the analysis of argon, nitrogen, carbon tetrachloride, and neopentane isotherms on various forms of alumina,[48,257,301] chromia,[48,266] silica,[48,257] titania,[302] and the oxides of iron[303] and nickel oxide;[304] in each of these studies, the standard α_s data have been determined on certain carefully selected non-porous materials. Two aspects have emerged: (i) the method may be used in a similar manner to the t-method for the detection and assessment of micropore filling and capillary condensation and the determination (in appropriate cases) of the internal and external surface area; (ii) the α_s-method can reveal certain aspects of the monolayer and multilayer structure on non-porous solids.[257,266]

Carruthers *et al.*[257] reported that argon α_s-plots on low-area α-Al_2O_3 samples were linear in the multilayer range (against a non-porous γ-Al_2O_3

[300] J. Hagymassy, S. Brunauer, and R. Sh. Mikhail, *J. Colloid Interface Sci.*, 1969, **29**, 485.
[301] D. Aldcraft, G. C. Bye, and G. O. Chigbo, *Trans. Brit. Ceram. Soc.*, 1971, **70**, 19.
[302] G. D. Parfitt, D. Urwin, and T. J. Wiseman, *J. Colloid Interface Sci.*, 1971, **36**, 217.
[303] G. C. Bye and C. R. Howard, *J. Appl. Chem.*, 1971, **21**, 324.
[304] G. A. Nicolaon and S. J. Teichner, *J. Chim. phys.*, 1969, **66**, 1816.

standard) although the monolayer section deviated from linearity. The fact that the multilayer branch could be extrapolated to the origin[48, 302] confirmed the absence of microporosity and indicated localization of the monolayer. The nitrogen α_s-plots on α-Al_2O_3 did not exhibit the same range of linearity and it was suggested[257] that the formation of the localized nitrogen monolayer on the high-energy surface (with the additional $\phi \dot{r}_Q$ contribution) influenced the structure of the multilayer. (Indeed, one might expect the quadrupole interactions to extend into the multilayer.)

It is clear that the successful application of the α_s-method depends on the availability of the standard data on well-characterized reference solids. In exploratory investigations on new or composite adsorbents (*e.g.* multicomponent catalysts or pigments) a standard isotherm cannot be readily obtained. In this case, the 'comparison plot' approach of Brown and Hall[305] may be adopted, in which the amount adsorbed on one solid is plotted against the amount adsorbed at the same pressure on another, which is taken as an arbitrary reference adsorbent. The nature of the comparison plot is of course related to the respective shapes of the two isotherms and in particular cases may reflect the difference in the way the multilayers develop on the two substrates. Brown and Hall[305] construct comparison plots for various isotherms on graphitized and ungraphitized carbon black. In spite of the stepwise character of the isotherms on the graphitized material, the comparison plots are linear at high coverage and may, moreover, be back-extrapolated in several cases to the origin. These results appear to support the conclusions of Pierce[256, 279] and Zettlemoyer[276] concerning the localized structure of the monolayer on graphitized carbon and the uniformity of the multilayer at distances sufficiently removed from the surface.

Finally, we must return to the controversial subject of micropore filling and the assessment of microporosity by sorption methods. It will be recalled that the Dubinin–Radushkevich–Kaganer (DRK) equation has the same mathematical form as the earlier Dubinin–Radushkevich (DR) equation, although the underlying mechanism is in the former case based on monolayer coverage and in the latter case on micropore filling.[4, 306-309] Because of its general mathematical nature and the lack of an independent method of obtaining the micropore characteristics, the DR equation must be regarded as an empirical relation. As with so many other isotherm equations, the fact that it gives a reasonably good fit[309, 310] over a certain range of an isotherm

[305] C. E. Brown and P. G. Hall, *Trans. Faraday Soc.*, 1971, **67**, 3558.
[306] B. P. Bering, E. G. Zhukovskaya, B. Kh. Rakhmukov, and V. V. Serpinskii, ref. 81, p. 382.
[307] M. M. Dubinin and V. A. Astakhov, *Izvest. Akad. Nauk S.S.S.R., Ser. khim.*, 1971, 5 (*Bull. Acad. Sci. U.S.S.R., Chem. Sci.*, 1971, 3).
[308] M. M. Dubinin and V. A. Astakhov, *Izvest. Akad. Nauk S.S.S.R., Ser. khim.*, 1971, 11 (*Bull. Acad. Sci. U.S.S.R., Chem. Sci.*, 1971, 8).
[309] V. A. Astakhov and M. M. Dubinin, *Izvest. Akad. Nauk S.S.S.R., Ser. khim.*, 1971, 17 (*Bull. Acad. Sci. U.S.S.R., Chem. Sci.*, 1971, 13).
[310] V. Toda, M. Hatami, S. Toyoda, Y. Yoshida, and H. Honda, *Fuel*, 1971, **50**, 187.

cannot in itself provide sufficient evidence for the DR mechanism.[311, 312] At the present stage in the development of the subject, a more promising approach is likely to be one in which adsorption and presorption experiments are made on a range of well-defined microporous adsorbents using various probe molecules of different shape and polarizability.

[311] E. M. Freeman, T. Siemieniewska, H. Marsh, and B. Rand, *Carbon*, 1970, **8**, 7.
[312] H. Marsh and B. Rand, *J. Colloid Interface Sci.*, 1970, **33**, 101.

2
Adsorption at the Solid/Liquid Interface: Non-aqueous Systems

BY D. H. EVERETT

1 Introduction

Despite the great importance of adsorption from solution by solids in a wide range of natural systems and industrial processes, fundamental understanding of this phenomenon has developed relatively slowly. Kipling's[1] comprehensive survey published in 1965 was the first substantial monograph on the subject. It revealed that, although a very large number of systems of widely differing types had been studied experimentally, the underlying theory was only just beginning to emerge and had not by then exerted any major unifying influence on progress in this field. There are many reasons for this. Among the more important is the fact that most studies had, from the earliest days, been concerned mainly with adsorption from relatively dilute solutions. It was therefore not unreasonable to interpret the experimental data along lines analogous to those employed in the discussion of adsorption from the gas phase.[2] Thus it was common practice to fit experimental results to isotherm equations of the same form as the Freundlich and Langmuir equations of gas adsorption. Although recognized many years ago by Williams,[3] by Ostwald and de Izagiurre,[4] and by others,[5] the importance of regarding the phenomenon in terms of the competition of the solute and solvent molecules for adsorption sites on the solid surface was not given great prominence until more extensive work on liquid mixtures, covering the whole concentration range, drew attention to the inadequacy of the simple analogy with gas adsorption. It was realized that the experimental measurements lead to the so-called composite isotherm which, in effect, gives the surface *excess* of one component (suitably defined: see below) per unit mass of solid as a function of solution composition. But even then there has remained a strong tendency to attempt to decompose the composite isotherm into 'individual' or 'partial' isotherms which are supposed to give the actual concentrations of each

[1] J. J. Kipling, 'Adsorption from Solutions of Non-Electrolytes', Academic Press, London and New York, 1965.
[2] H. Freundlich, 'Colloid and Capillary Chemistry'. Methuen, London, 1926.
[3] A. M. Williams, *Medd. K. svenska Vetensk. acad. Nobelinst.*, 1913, 2, 23.
[4] W. Ostwald and R. de Izaguirre, *Kolloid-Z.*, 1922, 30, 279.
[5] E. Haymann and E. Boye, *Kolloid-Z.*, 1933, 63, 154.

component in the surface layer as a function of the concentration in the bulk solution. However, this calculation can be made only on the basis of certain assumptions about the nature of the adsorption process and it is clearly preferable for theoretical purposes to direct attention wherever possible to the interpretation of the experimentally observable surface excess isotherm, rather than to derived quantities of uncertain theoretical significance.

Substantial progress has been made in recent years and several general reviews have dealt with various aspects of the problem.[6-10]

Before considering in detail the work published during the period of the present Report, it is convenient to summarize briefly the position as it was at the end of 1969, and to quote in a consistent notation the basic equations which had been derived in earlier work. This is particularly necessary since it is important to be quite clear as to the status of these equations and the assumptions on which they are based before outlining recent theoretical work.

The experimental quantity which is most commonly measured is the change, from x_2^0 to x_2^l, in the mole fraction of a chosen component, 2, in the liquid phase which results from the equilibration of a mass, m, of solid with a sample of liquid containing a total amount n^0 of substance. This change, denoted by Δx_2^l, is proportional to the mass of solid and inversely proportional to n^0. The quantity $n^0 \Delta x_2^l / m$ is thus the primary experimental function characterizing adsorption equilibrium.

From a theoretical viewpoint, it is convenient to define[11] the *surface excess amount* of component 2, $n_2^{\sigma(n)}/m$ per unit mass of adsorbent, or $n_2^{\sigma(n)}/A_s = \Gamma_2^{(n)}$ per unit area of surface (the *surface excess concentration*), as the excess of the amount of 2 present in the actual system, over and above that present in a reference system containing the same total amount of substance n^0 and in which the composition of the liquid phase is uniform, and equal to that in the bulk of the real system, up to the surface of the solid. It is then readily shown that

$$\frac{n_2^{\sigma(n)}}{m} = \frac{n^0 \Delta x_2^l}{m},$$

(1)

[6] R. Aveyard, in 'Sorption and Transport in Soils', Soc. Chem. Ind. (London), 1970, Monograph no. 37, p. 3.

[7] M. J. Jaycock, in 'Dispersion of Powders in Liquids', ed. G. D. Parfitt, Elsevier, Amsterdam, 1969, p. 1.

[8] R. S. Hansen, *Croat. Chem. Acta*, 1970, **42**, 385.

[9] A. C. Zettlemoyer and F. J. Micali, *Croat. Chem. Acta*, 1970, **42**, 247; reprinted in *Croat. Chem. Acta*, Special Publication no. 1, 'Proceedings of the International Summer School on Chemistry of Solid/Liquid Interfaces', ed. B. Tezak and V. Pravdic, Zagreb, 1971.

[10] G. Schay, in 'Colloid and Surface Science', ed. E. Matijevic, Wiley-Interscience, New York, 1969, vol. 2, p. 155.

[11] See, IUPAC Information Bulletin, No. 3, January 1970, now revised and published as 'Manual of Symbols and Terminology for Physicochemical Quantities and Units, Appendix II, Definitions, Terminology and Symbols in Colloid and Surface Chemistry', Part I, *Pure Appl. Chem.*, 1973, **31**, 579.

or

$$\Gamma_2^{(n)} = \frac{n^0 \Delta x_2^l}{ma_s} = \frac{n^0 \Delta x_2}{A_s}, \qquad (2)$$

where a_s is the specific surface area of the solid, and A_s is its total area. The experimental quantity $n^0 \Delta x_2^l / m$ is thus conveniently called the *specific surface excess* of component 2, and the composite isotherm, in which $n^0 \Delta x_2^l / m$ is plotted as a function of x_2^l, is better called the *specific surface excess isotherm*.

An alternative function defining the adsorption is the (Gibbs) *relative adsorption* of component 2 with respect to component 1, $\Gamma_2^{(1)}$, which for a binary solution is related to $\Gamma_2^{(n)}$ by

$$\Gamma_2^{(1)} = \Gamma_2^{(n)} / (1 - x_2^l); \qquad (3)$$

in dilute solution of 2 in 1 as solvent, $\Gamma_2^{(1)} \approx \Gamma_2^{(n)}$.

The problems associated with the thermodynamic and statistical mechanical treatment of adsorption at the solid/liquid interface are closely analogous to those of the liquid/vapour interface. The main distinction lies in the experimental factors which, on the one hand, make it exceedingly difficult to measure by chemical analytical techniques the adsorption at the liquid/vapour interface (although the surface tension is readily measurable), whereas for a solid/liquid interface the adsorption is relatively easy to determine, although the solid/liquid interfacial tension is not amenable to direct experimental measurement. This sharp practical distinction seems to have obscured the fundamental similarities between the two phenomena: thus it is remarkable that although a thermodynamic discussion of adsorption at the liquid/vapour interface was given by Butler in 1928,[12] the analogous description of the solid/liquid interface was not clearly stated until much more recently.[13-16]

The basic concept adopted in one group of theories, and implicit in many others, is that of an 'adsorbed phase' (σ) close to, and under the influence of, the solid surface, and in equilibrium with the bulk solution. The adsorbed phase is usually supposed to be of uniform composition and autonomous in the sense that its properties are controlled entirely by its interactions with the adjacent solid, but are unaffected by molecular interactions with the bulk liquid. If the adsorbed phase consists of an amount n^σ of material in which the mole fraction of 2 is x_2^σ, then the following series of equivalent equations can be written down

$$\frac{n^0 \Delta x_2^l}{m} = \frac{n_2^{\sigma(n)}}{m} = a_s \Gamma_2^{(n)} = a_s (1 - x_2^l) \Gamma_2^{(1)}$$

$$= (x_2^\sigma - x_2^l) \frac{n^\sigma}{m} = (n_2^\sigma - n^\sigma x_2^l)/m = (n_2^\sigma x_1^l - n_1^\sigma x_2^l)/m. \qquad (4)$$

[12] J. A. V. Butler, *Proc. Roy. Soc.*, 1928, **A135**, 348.
[13] A. Schuchowitzky, *Acta Physicochim. U.R.S.S.*, 1938, **8**, 531.
[14] G. Schay, *Acta Chim. Acad. Sci. Hung.*, 1956, **10**, 281.
[15] M. Siskova and E. Erdös, *Coll. Czech. Chem. Comm.*, 1960, **25**, 1729, 3086.
[16] D. H. Everett, *Trans. Faraday Soc.*, 1964, **60**, 1803.

Many combinations of these equations appear in the literature. If n^σ is known, then x_2^σ can be calculated from $n_2^{\sigma(n)}$ (*i.e.* from Δx_2^l). However, n^σ can be derived only after some assumptions have been made about the nature of the adsorbed phase. If the adsorbent is a porous solid, then n^σ may be set equal to the amount of substance of composition x_2^σ which fills the pore space: it must be remembered that if the molecules of 1 and 2 are of different size, then n^σ will be a function of x_2^σ. Alternatively, for non-porous solids, it may be assumed that the adsorbed phase consists of t layers of molecules. The condition that the surface is completely covered is then that

$$n_1^\sigma a_1 + n_2^\sigma a_2 = A_s, \tag{5}$$

where a_1 and a_2 are the partial molar areas of 1 and 2 in the surface phase; they may be equated approximately to a_1^0/t and a_2^0/t, where a_1^0 and a_2^0 are the cross-sectional areas of the molecules. Equation (5) may be written in the alternative forms

$$x_1^\sigma a_1 + x_2^\sigma a_2 = A_s/n^\sigma \tag{6}$$

or

$$\Gamma_1 a_1 + \Gamma_2 a_2 = 1, \tag{7}$$

where

$$\Gamma_1 = n_1^\sigma/A_s \quad \text{and} \quad \Gamma_2 = n_2^\sigma/A_s \quad [\textit{N.B. } \Gamma_1 \neq \Gamma_1^{(n)}].$$

It follows from equations (4) and (5) that

$$\frac{n^\sigma}{A_s} = \frac{1 - (a_2 - a_1)\Gamma_2^{(n)}}{a_2 + (a_2 - a_1)x_2^l}, \tag{8}$$

and

$$x_2^\sigma = \frac{x_2^l + a_1\Gamma_2^{(n)}}{1 - (a_2 - a_1)\Gamma_2^{(n)}} = \frac{x_2^l + \left(\dfrac{a_1}{a_s}\right)\left(\dfrac{n^0\Delta x_2^l}{m}\right)}{1 - \left(\dfrac{a_2 - a_1}{a_s}\right)\left(\dfrac{n^0\Delta x_2^l}{m}\right)}. \tag{9}$$

Equation (9) enables the so-called individual isotherms of x_2^σ against x_2^l to be constructed from a knowledge of Δx_2^l and the surface area A_s, when values of a_1 and a_2 have been selected. The usual assumption is that the adsorbed phase consists of a single layer of molecules, so that a_1 and a_2 are taken as a_1^0 and a_2^0. These cross-sectional areas may be estimated from molecular models, from liquid densities, or from data on vapour-phase adsorption. For non-spherical molecules, assumptions must also be made about the orientation of the molecules relative to the surface.

Alternatively, the relative adsorption of component 2 with respect to component 1 may be related to x_2^σ by the following equation, which may also be derived from equations (4) and (5):

$$\Gamma_2^{(1)} = \frac{1}{a_s(1-x_2^l)} \frac{(x_2^\sigma - x_2^l)}{[a_1 + (a_2 - a_1)x_2^l]}. \tag{10}$$

In the particular case in which the two components are of the same size $(a_1 = a_2 = a)$,

$$n^\sigma = A_s/a, \tag{11}$$

and is equal to the number of 'adsorption sites' in the adsorbed phase, and

$$x_2^\sigma = x_2^l + a\Gamma_2^{(n)}$$

$$= x_2^l + \left(\frac{a}{a_s}\right)\left(\frac{n^0 \Delta x_2^l}{m}\right). \tag{12}$$

Various theories have been proposed to provide an interpretation of observed isotherms. Several of them depend on, or can be shown to be equivalent to, consideration of the phase exchange equilibrium between an adsorbed phase and the bulk liquid, which for a binary system can be written[16]

$$\left(\frac{a^\ominus}{a_1}\right)(1)^\sigma + \left(\frac{a^\ominus}{a_2}\right)(2)^l \rightleftharpoons \left(\frac{a^\ominus}{a_1}\right)(1)^l + \left(\frac{a^\ominus}{a_2}\right)(2)^\sigma. \tag{13}$$

The stoichiometric coefficients are introduced to satisfy the condition that the solid surface remains covered; they are expressed in the form $(a^\ominus/a_1)$, where $a^\ominus$ is a standard area per molecule (which may be chosen as the unit of area per molecule), rather than as $(1/a_1)$, so that they are dimensionless. The equilibrium state is then governed by an equilibrium constant which may be expressed in the alternative forms:

$$K = \left(\frac{\gamma_1^l x_1^l}{\gamma_1^\sigma x_1^\sigma}\right)^{a^\ominus/a_1} \left(\frac{\gamma_2^\sigma x_2^\sigma}{\gamma_2^l x_2^l}\right)^{a^\ominus/a_2}, \tag{14}$$

or

$$K' = \left(\frac{\gamma_1^l x_1^l}{\gamma_1^\sigma x_1^\sigma}\right)\left(\frac{\gamma_2^\sigma x_2^\sigma}{\gamma_2^l x_2^l}\right)^{1/r}, \tag{14'}$$

where the γ's are suitably defined activity coefficients and $r = a_2/a_1$ is the ratio of the areas occupied by the two molecules.

One way of deriving equation (14) is to define the chemical potentials of the molecules in the adsorbed phase by[17]

$$\mu_i^\sigma = \mu_i^{0,\sigma} + RT \ln x_i^\sigma \gamma_i^\sigma + (\sigma_i^0 - \sigma)a_i, \tag{15}$$

where σ is the interfacial tension at the solid/solution interface and σ_i^0 that at the solid/pure-liquid i interface; $\mu_i^{0,\sigma}$ is a standard potential (referred to pure

[17] D. H. Everett, *Trans. Faraday Soc.*, 1965, **61**, 2478.

i as standard state) and is equal to $\mu_i^{0,l}$, the chemical potential of pure liquid i.*

The chemical potentials in the liquid phase are

$$\mu_i = \mu_i^{0,l} + RT \ln x^l \gamma^l. \tag{16}$$

The condition of equilibrium of equation (13) is that

$$\left(\frac{a^\ominus}{a_1}\right)\mu_1^\sigma + \left(\frac{a^\ominus}{a_2}\right)\mu_2^l = \left(\frac{a^\ominus}{a_1}\right)\mu_1^l + \left(\frac{a^\ominus}{a_2}\right)\mu_2^\sigma. \tag{17}$$

Insertion of the chemical potentials [equations (15) and (16)] then leads to equation (14) with

$$K = \exp\left[(\sigma_2^0 - \sigma_1^0)a^\ominus / RT\right]. \tag{18}$$

In the case in which $a_1 = a_2 = a$, equation (14) may be rearranged to give

$$\frac{x_2^\sigma}{x_2^l} = \frac{K\gamma_1^\sigma \gamma_2^l}{K\gamma_1^\sigma \gamma_2^l x_2^l + \gamma_1^l \gamma_2^\sigma x_1^l} \tag{19}$$

and

$$a_s \Gamma_2^{(n)} = \left(\frac{n^\sigma}{m}\right)\left[\frac{\left(K\frac{\gamma_1^\sigma \gamma_2^l}{\gamma_1^l \gamma_2^\sigma} - 1\right)x_1^l x_2^l}{1 + \left(K\frac{\gamma_1^\sigma \gamma_2^l}{\gamma_1^l \gamma_2^\sigma} - 1\right)x_2^l}\right]. \tag{20}$$

A number of authors have dealt with the cases in which one or both of the co-existing phases is ideal. If both are ideal then equation (19) reduces to

$$x_2^\sigma = \frac{K x_2^l}{1 + (K-1)x_2^l} \tag{21}$$

and equation (20) becomes

$$a_s \Gamma_2^{(n)} = \left(\frac{n^\sigma}{m}\right)(K-1)\left[\frac{x_1^l x_2^l}{1 + (K-1)x_2^l}\right] = \frac{n^0 \Delta x_2^l}{m}, \tag{22}$$

while the relative adsorption is given by

$$a_s \Gamma_2^{(1)} = \left(\frac{n^\sigma}{m}\right)(K-1)\left[\frac{x_2^l}{1 + (K-1)x_2^l}\right]. \tag{23}$$

---

* An alternative definition of μ_i^σ, due to Butler[12] and used by Schay,[10,14] among others, is

$$\mu_i^{\sigma(B)} = \mu_i^0 + RT \ln x_i^\sigma \gamma_i^\sigma. \tag{15'}$$

Provided that this is associated with the correct equilibrium condition

$$\frac{\mu_i^{\sigma(B)} - \mu_i^l}{a_i} = \sigma,$$

this leads to the same results as those derived here.

Various methods of linearizing equation (22) have been suggested so that K and (n^σ/m) can be derived from experimental isotherms: the two most convenient equations are[16]

$$\frac{x_1^l x_2^l}{(n^0 \Delta x_2^l/m)} = \frac{m}{n^\sigma} \left[x_2^l + \frac{1}{K-1} \right], \tag{24}$$

and[15]

$$\frac{(n^0 \Delta x_2^l/m)}{x_1^l} = \frac{n^\sigma}{m}(K-1) - K\frac{(n^0 \Delta x_2^l/m)}{x_2^l}. \tag{25}$$

In favourable cases good straight lines are obtained when experimental data are plotted in this way. If a monolayer model is now assumed and a is set equal to a^0, the cross-sectional area of a molecule, then

$$a_s = \left(\frac{n^\sigma}{m}\right) a^0 . N_A, \tag{26}$$

where N_A is Avogadro's constant. In many instances the value of a_s derived in this way is in very satisfactory agreement with the area derived from vapour adsorption isotherms using the BET method (see below).

Schay and Nagy[18-20] suggested an alternative method of deriving the specific surface area of a solid from the adsorption isotherm. For an ideal system of molecules of equal size, the limiting slope (as $x_2^l \to 1$) of the surface excess isotherm is equal to $-(n^\sigma/m)(K-1)/K$ (Figure 1a). When $K \gg 1$ this slope is thus a good approximation to (n^σ/m), from which a_s may be calculated.

Furthermore, it is observed experimentally that, even in the case of non-ideal behaviour, many surface excess isotherms exhibit linear behaviour over part of the isotherm (Figure 1b) and Schay and Nagy suggested that the slope of this portion of the isotherm is also equal to n^σ/m. However, since at such intermediate solution concentrations the surface layer will contain two kinds of molecule, of different molecular surface areas, the calculation of a_s from the experimental data requires more careful consideration. If, over a range of x_2^l,

$$\frac{n^0 \Delta x_2^l}{m} = \alpha - \beta x_2^l, \tag{27}$$

then, from equation (4),

$$(n_2^\sigma - n^\sigma x_2^l)/m = \alpha - \beta x_2^l$$

or

$$\left(\frac{n_2^\sigma}{m} - \alpha\right) - \left(\frac{n^\sigma}{m} - \beta\right) x_2^l = 0. \tag{28}$$

[18] L. G. Nagy and G. Schay, *Magyar Kém. Folyóirat*, 1960, **66**, 31.
[19] G. Schay, L. G. Nagy, and T. Szekrenyesy, *Periodica Polytechnica*, 1960, **4**, 95.
[20] G. Schay and L. G. Nagy, *J. Chim. phys.*, 1961, 149.

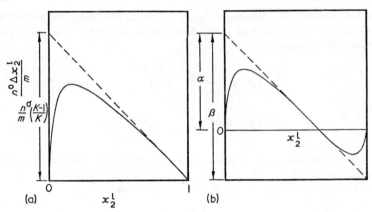

Figure 1 *Two typical forms of surface excess isotherm ($n^0 \Delta x_2^l / m$ versus x_2^l)
(a) convex upwards at all concentrations (Schay type 2), with approach
to linearity as $x_2^l \to 1$; dashed line, limiting slope $= (n^\sigma / m)(K - 1)/K$
for ideal system; (b) S-shaped isotherm (Schay type 4), showing an
azeotropic point and linear region in neighbourhood of point of
inflexion; dashed line, slope of linear region intersecting ordinates at
$x_2^l = 0$ and 1 at α and $(\alpha - \beta)$, respectively*

Clearly one *possible* solution of this equation is

$$\alpha = n_2^\sigma / m \quad \text{and} \quad \beta = n^\sigma / m \quad \text{or} \quad (\beta - \alpha) = n_1^\sigma / m. \qquad (29)$$

Thus if a monolayer is assumed, the surface area is given by

$$a_s = \frac{A_s}{m} = (\beta - \alpha)a_1^0 + \alpha a_2^0. \qquad (30)$$

This is the solution chosen by Schay and Nagy and employed by them.

The status of these methods of surface-area determination has been
discussed by Cornford, Kipling, and Wright,[21] by Larionov, Tonkonog, and
Chmutov,[22] and by Schay;[23] the position in 1969 may be summarized as
follows.

For simplicity, we consider the case in which $a_1 = a_2 = a$,* when equation
(27) can be written in the form [using equation (4) and remembering that
$n^\sigma / m = a_s / a$]

$$x_2^\sigma = \left(\frac{a}{a_s}\right)\alpha + \left[1 - \left(\frac{a}{a_s}\right)\beta\right] x_2^l. \qquad (31)$$

Thus in general x_2^σ is a linear function of x_2^l in the region in which the surface
excess isotherm is itself linear (Figure 2a). However, if, as asserted by Schay

---

* For the more general case, see Larionov *et al.*[22]

[21] P. V. Cornford, J. J. Kipling, and E. H. M. Wright, *Trans. Faraday Soc.*, 1962, **58**, 74.
[22] O. G. Larionov, L. G. Tonkonog, and K. V. Chmutov, *Zhur. fiz. Khim.*, 1965, **39**, 2226
(*Russ. J. Phys. Chem.*, 1965, **39**, 1186).
[23] G. Schay, in ref. 10, p. 173.

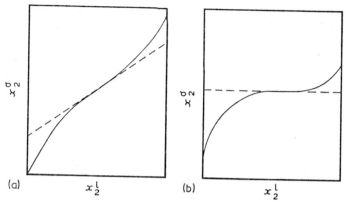

Figure 2 *Isotherms of x_2^σ versus x_2^l ('individual isotherm') (a) showing point of inflexion in neighbourhood of which x_2^σ is a linear function of x_2^l, equation (31); (b) showing point of inflexion with horizontal tangent in neighbourhood of which x_2^σ is independent of x_2^l*

and Nagy, $\beta = n^\sigma/m = a_s/a$, then x_2^σ is constant in this region and equal to α/β. It was pointed out that an adsorbed phase of constant composition can remain in equilibrium with a bulk phase of variable composition over a finite range only if the activity coefficients of the components vary in a special fashion. This criticism was accepted, but it was then suggested[23] that the apparently linear section observed experimentally really masked a point of inflexion in the isotherm, so that in reality x_2^σ was constant only over an infinitesimal range of x_2^l. Nevertheless, the assumption that $\beta = n^\sigma/m$, even over only a small range, means that the isotherm of x_2^σ against x_2^l, which is in general given in this range by equation (31) (Figure 2a), must have a horizontal tangent at a point of inflexion (Figure 2b). The thermodynamic acceptability of this form of curve will be discussed further in Section 3.

If a system of equal-size molecules is non-ideal, then the activity coefficients in the surface phase, and the equilibrium constant, may be calculated from the following expressions:[17]

$$\ln \gamma_2^\sigma = x_1^\sigma \ln \frac{x_1^\sigma x_2^l \gamma_2^l}{x_2^\sigma x_1^l \gamma_1^l} - \int_0^{x_1^\sigma} \ln \frac{x_1^\sigma}{x_2^\sigma} \cdot \frac{x_2^l \gamma_2^l}{x_1^l \gamma_1^l} \cdot dx_1^\sigma \qquad (32)$$

and

$$\ln K = \int_0^1 \ln \frac{x_1^\sigma x_2^l \gamma_2^l}{x_2^\sigma x_1^l \gamma_1^l} \, dx_1^\sigma. \qquad (33)$$

When the components of a binary solution are of different sizes, then equation (14) cannot be solved explicitly and no general form of equations

(19), (20), (21), or (22) can be derived. Nevertheless, an alternative method of analysis can be used, since at equilibrium not only is equation (17) satisfied but, in addition,

and
$$\mu_1^l = \mu_1^\sigma$$
$$\mu_2^l = \mu_2^\sigma$$
\left.\begin{array}{c}\end{array}\right\} \quad (34)

separately.

Thus, from equations (15) and (16),

$$\ln x_i^l \gamma_i^l = \ln x_i^\sigma \gamma_i^\sigma + (\sigma_i^0 - \sigma)a_i, \qquad i = 1, 2. \qquad (35)$$

Since σ is not experimentally measurable, it has to be eliminated from the equations (35) either by using equation (17), as done above, or by making use of the Gibbs adsorption isotherm for the adsorbed phase at constant T,

$$-d\sigma = \Gamma_1^{(n)} d\mu_1 + \Gamma_2^{(n)} d\mu_2, \qquad (36)$$

remembering that $\Gamma_1^{(n)} = -\Gamma_2^{(n)}$, together with the Gibbs–Duhem equation for the bulk liquid, at constant T and p,

$$0 = x_1^l d\mu_1 + x_2^l d\mu_2. \qquad (37)$$

This gives

$$d\sigma = \Gamma_2^{(n)} (d\mu_1 - d\mu_2) = [\Gamma_2^{(n)}/(1 - x_2^l)] d\mu_2$$
$$= \frac{\Gamma_2^{(n)}}{(1 - x_2^l)} \cdot RT \, d \ln x_2^l \gamma_2^l, \qquad (38)$$

so that[24]

$$(\sigma_2^0 - \sigma) a_2 = -a_2 RT \int_{x_2=1}^{x_2} \frac{\Gamma_2^{(n)}}{(1 - x_2^l)} \, d \ln x_2^l \gamma_2^l. \qquad (39)$$

Hence, from equation (35),

$$\ln \gamma_2^\sigma = \ln [x_2^l \gamma_2^l / x_2^\sigma] + a_2 RT \int_0^{x_2} \frac{\Gamma_2^{(n)}}{(1 - x_2^l)} \, d \ln x_2^l \gamma_2^l, \qquad (40)$$

where x_2^σ is calculated from equation (9).

Integration of equation (39) across the whole concentration range and comparison with equation (18) leads to

$$\ln K = \frac{a^\ominus}{a_1} \ln K' = (\sigma_2^0 - \sigma_1^0) a^\ominus / RT = a^\ominus \int_0^1 \frac{\Gamma_2^{(n)}}{(1 - x_2^l)} \, d \ln x_2^l \gamma_2^l. \qquad (41)$$

We recall in passing that $\Gamma_2^{(n)}/(1 - x_2^l) = \Gamma_2^{(1)}$.

[24] G. Schay, L. G. Nagy, and T. Szekrenyesy, *Periodica Polytechnica*, 1962, **6**, 91.

Equations (40) and (41) do not involve any assumptions as to the relative sizes of the two components, although in any calculations based on equation (40) suitable values of a_1 or a_2 must be inserted. Nor are these equations dependent on the assumption of monolayer adsorption: they do, however, embody, through equation (35), the concept of an adsorbed phase. Full experimental studies of the applicability of these equations have yet to be carried out.

For many practical purposes, a quantity of interest in characterizing adsorption from solution is the *separation factor*, analogous to that of importance in fractional distillation. This is defined by

$$S = \frac{x_2^\sigma x_1^l}{x_1^\sigma x_2^l} \qquad (42)$$

which, as shown by Schay,[10] is given in general by the equation

$$S = \frac{\gamma_1^\sigma \gamma_2^l}{\gamma_2^\sigma \gamma_1^l} \exp\left[\frac{\sigma}{RT}(a_2 - a_1) + \frac{1}{RT}(a_1\sigma_1^0 - a_2\sigma_2^0)\right]. \qquad (43)$$

In the particular case of an ideal system of molecules of equal size, $S = K$. In general, however, its value is determined by four factors: the ratio of the bulk activity coefficients of components 1 and 2, the ratio of the surface activity coefficients, the difference between the molar surface free energies ($a\sigma^0$) of the two pure components, and the variation of σ with composition. Of these only the third factor is independent of composition.

The validity of the concept of an autonomous adsorbed phase, and in

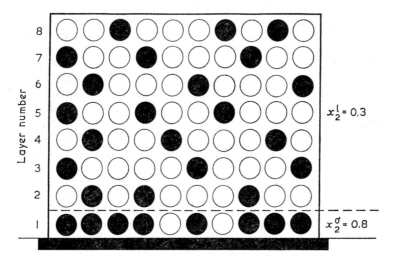

Figure 3 *Schematic representation of monolayer model of adsorption from solution*

particular that of the monolayer model, has been discussed several times and it has been generally accepted that models of this kind must be rather crude approximations. The physically unrealistic nature of this model may be appreciated from consideration of Figure 3, which represents the adsorbed phase as a monolayer. Thus layer 1 has a composition x_2^σ whereas layers 2, 3, . . . ∞ have a composition x_2^l and are supposed to have properties identical with those of the bulk liquid. However, a layer such as 5, well away from the surface, is flanked above and below by layers of the same composition, whereas layer 2 has above it a layer of composition x_2^l and below it one of x_2^σ. If the energy of interaction between unlike molecules, ε_{12}, differs from the arithmetic mean of the energies of interaction between like molecules, $\frac{1}{2}(\varepsilon_{11} + \varepsilon_{22})$, then the energy of layer 2 must be different from that of layer 5 and it cannot therefore have the same thermodynamic properties as the liquid. The composition of layer 2 must therefore be modified to bring it into equilibrium both with the surface layer 1 and the layers above it. This in turn will bring about a secondary influence on the composition of layer 3. In principle, therefore, any physically realistic model must allow for the lack of autonomy of the surface region, which implies a gradual change in composition from $x_2^{\sigma(1)}$ in the first layer to x_2^l in the bulk.

This shortcoming of the monolayer model, when applied to the liquid/ vapour interface of a regular solution {whose properties are characterized by the quantity $w \propto [\varepsilon_{12} - \frac{1}{2}(\varepsilon_{11} + \varepsilon_{22})]$} was first pointed out by Defay and Prigogine,[25] who showed that the equations obtained from a statistical mechanical treatment of the model[26] were incompatible with the Gibbs adsorption isotherm. They showed that this thermodynamic inconsistency (reflecting a physically impossible assumption in the model) could to a large extent be eliminated by adopting a two-layer model, and it was presumed that by taking a sufficient number of layers into account, the inconsistency could be completely removed. This was shown to be so by Ono,[27] who solved the system of difference equations relating the compositions of successive layers at equilibrium by a numerical technique, and calculated the profile of $x\binom{i}{2}$, the mole fraction in the ith layer, as a function of i for some typical values of the parameter w. A fuller study using an alternative computational method was carried out by Lane.[28] This work showed that the overall shape of the surface excess isotherm was not appreciably altered by assuming multilayer adsorption, but that the parameters required to fit experimental data depended on the model chosen.

The monolayer model becomes particularly unrealistic when applied to mixtures of molecules of widely different sizes. If one of the molecules is a chain molecule made up of segments of about the size of the other component, then the only simple monolayer model is the so-called parallel-layer model

[25] R. Defay and I. Prigogine, *Trans. Faraday Soc.*, 1950, **46**, 199.
[26] E. A. Guggenheim, *Trans. Faraday Soc.*, 1945, **41**, 150.
[27] S. Ono, *Mem. Fac. Eng. Kyushu Univ.*, 1947, **10**, 195; S. Ono and S. Kondo in 'Hanpbuch der Physik', ed. S. Flügge, Springer, Berlin 1960, vol. 10, p. 134.
[28] J. E. Lane, *Austral. J. Chem.*, 1968, **21**, 827.

in which all segments of the chain molecule lie in the layer adjacent to the surface.[29] A multilayer model must take account of the possible orientations of the adsorbed molecules, any one of which may have segments in several different layers. The case of adsorption of dimers from solution in a monomer, originally studied by Mackor and van der Waals,[30] was investigated in greater detail by Ash, Everett, and Findenegg,[31] who took explicit account of all possible orientations of the dimer on a hexagonal close-packed lattice extending out from the surface. Recent work on polymer adsorption is reviewed in Chapter 3.

2 Thermodynamic and Statistical Mechanical Studies: Monolayer Model

During the period under review, several papers have appeared dealing with the thermodynamic and statistical mechanical aspects of adsorption from solution. Sircar and Myers[32] have discussed the case of the ideal adsorbed phase in equilibrium with a non-ideal bulk phase. That this might be a reasonable approximation had been previously indicated by the work of Kiselev[33-35] and of Nagy and Schay,[36] who had shown that in many cases experimental data could be well represented by equations obtained from equation (19), by setting $\gamma_1^\sigma = \gamma_2^\sigma = 1$, and that when equation (40) was employed the values of the surface activity coefficients calculated for a given system were often much closer to unity than the corresponding bulk activity coefficients. Sircar and Myers used a cell model for both the adsorbed and bulk phases, and limited consideration to a monolayer model with molecules of equal size. The statistical mechanical discussion, in terms of grand partition functions, is in effect a more sophisticated form of that used by Everett.[16] They obtain, in the present notation,

$$\exp\left(-\sigma a/RT\right) = x_1^l \gamma_1^l \exp\left(-\sigma_1^0 a/RT\right) + x_2^l \gamma_2^l \exp\left(-\sigma_2^0 a/RT\right), \quad (44)$$

which is a generalization of the equation derived by Guggenheim[26] for the liquid/vapour interface. Combination of equation (44) with the Gibbs adsorption isotherm [equation (36)] and the Gibbs–Duhem equation [equation (37)] then gives

$$\frac{A_s \Gamma_1^{(n)}}{n^\sigma} = \frac{x_1^l \gamma_1^l - x_1^l (x_1^l \gamma_1^l + K x_2^l \gamma_2^l)}{x_1^l \gamma_1^l + K \gamma_2^l x_2^l}, \quad (45)$$

[29] R. Defay, I. Prigogine, A. Bellemans, and D. H. Everett, 'Surface Tension and Adsorption', Longmans, London, 1966, Chap. XIII.
[30] E. L. Mackor and J. H. van der Waals, *J. Colloid. Sci.*, 1952, **7**, 535.
[31] S. G. Ash, D. H. Everett, and G. H. Findenegg, *Trans. Faraday Soc.*, 1968, **64**, 2645.
[32] S. Sircar and A. L. Myers, *J. Phys. Chem.*, 1970, **74**, 2828.
[33] A. V. Kiselev and L. F. Pavlova, *Izvest. Akad. Nauk S.S.S.R., otdel. khim. Nauk*, 1965, 18 (*Bull. Acad. Sci. U.S.S.R., Chem. Ser.*, 1965, 15).
[34] A. V. Kiselev and I. V. Shikalova, *Doklady Akad. Nauk S.S.S.R.*, 1966, **171**, 1361 (*Doklady Phys. Chem.*, 1966, **171**, 808).
[35] A. V. Kiselev and V. V. Khopina, *Trans. Faraday Soc.*, 1969, **65**, 1936.
[36] L. G. Nagy and G. Schay, *Acta Chim. Acad. Sci. Hung.*, 1963, **39**, 365 (*cf.* ref. 24).

which also follows immediately from equation (20) when $\gamma_1^{\sigma} = \gamma_2^{\sigma} = 1$. They also derive equation (41). Attention is drawn to a test of the self-consistency of measurements on the adsorption by the same solid surface from the three possible pairs of binary mixtures drawn from three single components. This follows immediately from equation (18): if components 1, 2, and 3 are taken in pairs then, if the data are self-consistent,

$$\log K_{12} + \log K_{23} + \log K_{31} = 0, \tag{46}$$

or

$$K_{12} . K_{23} . K_{31} = 1. \tag{46'}$$

A second paper by Sircar and Myers[37] covers much the same ground, but with particular emphasis on the importance of establishing the thermodynamic consistency of experimental data.

Larionov and Myers[38] restate the thermodynamics with particular reference to the separation factor: equation (43) is rederived and it is emphasized that, in general, S is a function of composition. It is also shown that, for molecules of different size, the surface excess is given in terms of the separation factor by

$$n_1^{\sigma(\text{n})} = A_{\text{s}} \Gamma_1^{(\text{n})} = \frac{A_{\text{s}} x_1^l x_2^l (S-1)/a_1}{S x_1^l + r x_2^l}, \tag{47}$$

where r is the ratio of the areas of the two molecules: a slightly rearranged form of this equation was also given by Schay.[39] Methods are also proposed for the calculation of the excess Gibbs energy of the adsorbed phase:

$$\frac{\Delta g^{\text{E},\sigma}}{RT} = x_1^{\sigma} \ln \gamma_1^{\sigma} + x_2^{\sigma} \ln \gamma_2^{\sigma}, \tag{48}$$

and for obtaining individual activity coefficients.

The thermodynamics of adsorption from binary mixtures has also been considered by Robert[40] and by Robert and Kessaissia.[41] They arrive at an equation which is essentially equation (14). They then obtain, for the case of equal size molecules, equations equivalent to (21), (22) and (23) except that their parameter α which appears in place of K includes an activity coefficient quotient:

$$\alpha = K \left(\frac{\gamma_2^l}{\gamma_2^{\sigma}} \cdot \frac{\gamma_1^{\sigma}}{\gamma_1^l} \right). \tag{49}$$

In the second paper, molecules of different sizes are considered: the bulk phase is supposed to be a regular binary solution and the surface phase to

[37] S. Sircar and A. L. Myers, *Amer. Inst. Chem. Engineers J.*, 1971, **17**, 186.
[38] O. G. Larionov and A. L. Myers, *Chem. Eng. Sci.*, 1971, **26**, 1025.
[39] G. Schay, in 'Proceedings of the International Symposium on Surface Area Determination, 1969', Butterworths, London, 1970, pp. 282, 288.
[40] L. Robert, *Compt. rend.*, 1971, **272**, C, 1957.
[41] L. Robert and Z. Kessaissia, *Compt. rend.*, 1971, **273**, C, 1681.

be ideal, an assumption which had been investigated for molecules of equal size by Nagy and Schay.[36] By putting $\gamma_1^\sigma = \gamma_2^\sigma = 1$, $\gamma_1^l = \alpha(1 - x_1^l)^2$, and $\gamma_2^l = \alpha(1 - x_2^l)^2$ in equation (14′)* and solving (presumably by a suitable computer method), x_2^σ was found as a function of x_2^l for various values of the parameters α and r at a fixed value of K ($= e^2$); the results are displayed graphically in terms of $\Gamma_2^{(1)}/a_2$ as a function of x_2^l. Although the calculations have not been carried beyond $x_2^l \sim 0.8$, so that the limiting behaviour of $\Gamma_2^{(1)}/a_2$ as $x_2^l \to 1$ cannot be assessed exactly, it appears that for $r < 1$ and α in the range 0—0.52, the limiting value of $\Gamma_2^{(1)}/a_2$ approaches unity; for $r \sim 1$ the limiting value is about 0.8, while for $r > 1$ this value falls rapidly. At low concentrations ($x_2^l < 0.15$), $\Gamma_2^{(1)}/a_2$ varies with r in the opposite sense, decreasing as r decreases. Recalling that the limiting value of $\Gamma_2^{(1)}/a_2$ is also equal to the limiting slope as $x_2^l \to 1$ of the plot of $\Gamma_2^{(1)}$ against x^l, these calculations would suggest that Schay and Nagy's method[18-20] of finding surface areas from this limiting slope may be justified for non-ideal solutions provided that the preferentially adsorbed molecule is much smaller than the other component. Some rather incomplete data for the systems benzene + n-pentane, n-hexane, n-heptane, and n-octane, on silica gel, were obtained by a chromatographic technique and appear to support these general trends. However, these calculations cover a relatively small range of size ratios, from 0.3 to 1.6, and take no account of the deviations from ideality of the bulk phase arising from size differences. It will be interesting to compare these results with those previously obtained by Everett[17] and by Ash, Everett, and Findenegg,[42] who studied a model in which both bulk and surface phases are non-ideal, and where the Flory–Huggins equation was used to take account of the influence of the size ratio in the activity coefficients: it was found that for a given value of K, the adsorption at a given mole fraction passed through a minimum when $r = 1$. The experimental data of Parfitt and Willis[43] seemed to bear out this prediction, which is contrary to the conclusions of Robert and Kessaissia, who found a steady variation of adsorption with size ratio. Further work on this problem is indicated.

Problems continue to arise from the use of different definitions of chemical potentials by various authors (see footnote on p. 54). Unless the definition is unambiguous and related correctly to the equilibrium conditions, confusion and apparent contradictions are all too easily produced. The latest example arises from the discussion by Nassonov[44, 45] of the so-called Gibbs and Thomson adsorption equations. The former is usually written as[46]

$$d\sigma = - S^\sigma \, dT - \sum_{i=1}^{i=o} \Gamma_i \, d\mu_i^\sigma, \qquad (50)$$

---

* α is here the parameter of regular solution theory and is not to be confused with α in equation (49).

[42] S. G. Ash, D. H. Everett, and G. H. Findenegg, *Trans. Faraday Soc.*, 1968, **64**, 2639.
[43] G. D. Parfitt and E. Willis, *J. Phys. Chem.*, 1964, **68**, 1780.
[44] P. M. Nassonov, *Zhur. fiz. Khim.*, 1971, **45**, 2813 (*Russ. J. Phys. Chem.*, 1971, **45**, 1593).
[45] P. M. Nassonov, *Zhur. fiz. Khim.*, 1971, **45**, 2818 (*Russ. J. Phys. Chem.*, 1971, **45**, 1595).
[46] Ref. 29, Chap. VII.

where σ is the interfacial tension, S^σ the excess interfacial entropy (defined by $S^\sigma = S - S' - S''$, where S is the total entropy of the system, and S' and S'' the entropies attributed to homogeneous bulk phases separated by a chosen dividing surface), Γ_i is the surface excess concentration of i defined relative to the same dividing surface, and μ_i^σ, the chemical potential of i in the surface, is defined by

$$\mu_i^\sigma = (\partial F / \partial n_i^\sigma)_{T, V, A_s, n_j^\sigma}. \tag{51}$$

Here F is the Helmholtz energy $(U - TS)$ of the whole system, and A_s is the surface area of the interface. The condition of equilibrium with respect to diffusion of i to or from the interface may be shown to be

$$\mu_i' = \mu_i^\sigma = \mu_i'', \tag{52}$$

where μ_i' and μ_i'' are the chemical potentials in the bulk phases. Equation (52) can be combined with (50) and the Gibbs–Duhem equation for the two bulk phases to obtain[46]

$$d\sigma = -S^{\sigma(1)} \, dT - \sum_{i=2}^{i=c} \Gamma_i^{(1)} \, d\mu_i, \tag{53}$$

where $S^{\sigma(1)}$ and $\Gamma_i^{(1)}$ are the *relative* excess entropy and *relative* excess surface concentration with respect to component 1: *i.e.* they are defined relative to the dividing surface which makes $\Gamma_1 = 0$; μ_i is now the common value of the chemical potential throughout the system.

Nassonov challenges the correctness of equation (52) on the grounds that Gibbs defined an impossible equilibrium process. His statement seems to imply that the definition of equation (51) is inappropriate. He then presents a modified argument in which a chemical potential is introduced without an explicit definition, but in terms which imply that it is defined by

$$\mu_i^a = (\partial F / \partial n_i^\sigma)_{T, V, \sigma, n_j^\sigma}, \tag{54}$$

where the superscript a is used to differentiate this chemical potential from that defined by equation (51). This is the same definition as used by Butler[12] (see footnote on p. 54). Equilibrium between a liquid and an interface is then shown to exist when

$$\mu_i^l = \mu_i^a - W_i, \tag{55}$$

where W_i is a non-zero quantity. Nassonov then claims that equations (50) and (53) are incorrect and that the right-hand side of each is identically zero. However, this conclusion results from the replacement of μ_i^σ and μ_i by μ_i^a in these equations. Indeed, if equation (50) is written in terms of μ_i^a, then his contention is correct: thus it is readily shown[16] that

$$\mu_i^l = \mu_i^\sigma = \mu_i^a - \sigma a_i, \tag{56}$$

where $a_i = (\partial A_s / \partial n_i^\sigma)_{T, V, \sigma, n_j^\sigma}$ is the partial molar area of i in the surface. Thus the quantity W_i introduced by Nassonov is simply σa_i.

Then, at equilibrium,

$$\frac{\mu_i^a - \mu_i^\sigma}{a_i} = \sigma \tag{57}$$

(agreeing with Butler) and hence

$$\frac{\sum n_i^\sigma (\mu_i^\sigma - \mu_i^\sigma)}{\sum n_i^\sigma a_i} = \sigma, \tag{58}$$

or

$$\sigma = \sum \Gamma_i (\mu_i^a - \mu_i^\sigma) = \sum \Gamma_i W_i, \tag{59}$$

since $\sum n_i^\sigma a_i$ is equal to the total area of surface.

Now equation (50) can be rewritten

$$\begin{aligned} d\sigma &= -S^\sigma \, dT - \sum \Gamma_i [d\mu_i^a - dW_i] \\ &= -S^\sigma \, dT - \sum \Gamma_i \, d\mu_i^a + \sum \Gamma_i \, dW_i \\ &= -S^\sigma \, dT - \sum \Gamma_i \, d\mu_i^a + \sum \Gamma_i \, d(a_i\sigma). \end{aligned} \tag{60}$$

However, in the last summation a_i is constant and $\sum \Gamma_i a_i = 1$, so that the last term on the right is just $d\sigma$, which cancels with the left-hand side. Nassonov's claim that the right-hand side of equation (50) is identically zero is thus only true if μ_i^σ is inserted in place of μ_i^σ. When due account is taken of the two possible definitions of surface chemical potentials, no inconsistency arises. Similarly, Nassonov's claim that equation (50) is incompatible with Thomson's adsorption equation:[47]

$$RT \ln \frac{c}{c^a} - A \frac{d\sigma}{dn_i^\sigma} = 0 \tag{61}$$

is again only true if μ_i^a is inserted into equation (50). On the other hand, if

$$\mu_i^\sigma = \mu_i^\ominus + RT \ln c^a - \sigma a_i \tag{62}$$

is used (where $\mu_i^\ominus$ is a standard chemical potential for both liquid and interface, c^a is the concentration in the interface, and c the bulk concentration), then the equations are compatible [when proper account is taken of the fact that equation (61) is strictly speaking appropriate only to a "surface phase" formulation].

One hopes that the distinctions emphasized above will become more widely recognized and that it will be realized that alternative formulations of the problem are possible. Each, when used self-consistently, is correct and there is no fundamental conflict between them: the choice between them is largely a matter of taste.

---

[47] J. J. Thomson, 'Applications of Dynamics to Physics and Chemistry', Macmillan, London, 1888 (reprinted, Dawson, London 1968), p. 190.

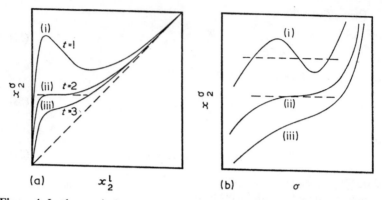

Figure 4 *Isotherms (schematic) of* (a) x_2^σ *versus* x_2^l, *showing the effect of changing the thickness* (t) *assumed for the adsorbed phase on the isotherm calculated from equation* (63); *curve* (i) $t=1$; (ii) $t=2$; (iii) $t=3$; (b) σ *as a function of* x_2^σ *for the three cases of Figure* 4a

3 The Validity of the Monolayer Model

As indicated in the Introduction, the validity of the monolayer model of adsorption has been open to question for a long time. There are general theoretical grounds for doubting its adequacy on several counts: the assumption that a monolayer is autonomous is physically unrealistic, except for ideal systems, and equations derived on the basis of this model can be shown to be inconsistent with general thermodynamic equations. A third and somewhat more direct test is to use experimental data in conjunction with the model to calculate the composition of the monolayer and to consider the acceptability of the results so obtained. In effect, equation (9) is used to calculate x_2^σ from experimental adsorption measurements, and the thermodynamic acceptability of the curves of x_2^σ against x_2^l is examined. This test seems first to have been suggested by Rusanov[48] and applied by him to the liquid/vapour interface of the n-hexane + acetone system. The simplest criterion is that $(\partial x_2^\sigma / \partial x_2^l)_{T,p}$ must be positive for systems in which $\Gamma_2^{(n)}$ is positive. The basis of this test is readily deduced from a consideration of the Gibbs adsorption isotherm [equation (36)], which leads to the conclusion that if $\Gamma_2^{(n)}$ is positive, $(\partial \sigma / \partial x_2^\sigma)_{T,p}$ has the same sign as, and is inversely proportional to, $(\partial x_2^\sigma / \partial x_2^l)_{T,p}$. This means (see Figure 4a) that if there is a region of the x_2^σ against x_2^l curve in which $(\partial x_2^\sigma / \partial x_2^l) < 0$, there will be a corresponding region in the curve of σ against x_2^σ in which $(\partial \sigma / \partial x_2^\sigma)$ is negative (Figure 4b), and σ will be multi-valued for a surface phase of constant composition. This state of affairs is thermodynamically unacceptable and must mean that the method of calculation of x_2^σ from Δx_2^l *via* equation (9) is physically meaningless. We note that

[48] A. I. Rusanov, 'Phase Equilibrium and Surface Phenomena', Chimia, Leningrad, 1967 (in Russian), Chap. VI.

equation (9) can be written in the form

$$x_2^\sigma = \frac{tx_2^l + a_2^0 \Gamma_2^{(n)}}{t - (a_2^0 - a_1^0)\,\Gamma_2^{(n)}} \qquad (63)$$

from which it is seen [*cf.* equation (4)] that the calculation of x_2^σ from Δx_2^l involves the use of the four parameters t, a_1^0, a_2^0, and A_s. Of these, a_1^0 and a_2^0 are usually calculated from independent data. If the monolayer model is employed, then $t=1$, while A_s may be derived either from gas adsorption measurements or from the solution adsorption data themselves, when it becomes a dependent parameter. For the liquid/vapour interface the question of A_s does not arise since $\Gamma_2^{(n)}$ may be obtained from surface-tension data, remembering that $\Gamma_2^{(n)} = (1 - x_2^l)\,\Gamma_2^{(1)}$ [equation (3)]. If for a given system the choice of $t=1$ gives an unacceptable form for x_2^σ as a function of x_2^l, it may be concluded either that the values of a_1^0 and a_2^0 are unacceptable, or that the monolayer model is invalid, or even that the concept of a surface phase is meaningless. Since, when such discrepancies arise, it is often found that variation of a_1^0 and a_2^0 over the range of physically reasonable values fails to eliminate the thermodynamic inconsistency, it is generally concluded that the choice of $t=1$ is incorrect. By setting $t=2, 3 \ldots$ successively the form of the curve of x_2^σ against x_2^l can be changed until the region of negative slope is eliminated (Figure 4a). On these grounds it is therefore possible to establish the *minimum* thickness which leads to thermodynamically consistent results. Thus for the n-hexane + acetone liquid/vapour interface $t=1$ gives an unacceptable curve, whereas $t=2$ or 3 both eliminate the inflexion.

The use of equation (9) in its monolayer form [*i.e.* equation (63) with $t=1$] to calculate x_2^σ is also criticized by Bering and Serpinskii,[49] who conclude that the only correct way of calculating the isotherm $x_2^\sigma(x_2^l)$ using equation (9) is by using values of a_1^0, a_2^0, and A_s derived independently from gas-phase adsorption measurements. They criticize the use of the liquid adsorption data themselves to obtain A_s, and in particular Schay's graphical method. A somewhat different argument from that given earlier is used to show that if the slope of a linear segment in the surface excess isotherm is used to obtain A_s, then x_2^σ must remain constant in the linear region. Thus, from equation (4),

$$-\frac{\mathrm{d}\Gamma_2^{(n)}}{\mathrm{d}x_2} = \frac{\mathrm{d}}{\mathrm{d}x_2^l}(n_1^\sigma x_2^l - n_2^\sigma x_1^l) = n_1^\sigma + n_2^\sigma - x_1^l\frac{\mathrm{d}n_2^\sigma}{\mathrm{d}x_2^l} + x_2^l\frac{\mathrm{d}n_1^\sigma}{\mathrm{d}x_2^l}$$

$$= n^\sigma + \Delta. \qquad (64)$$

Thus only when $\Delta = 0$, *i.e.* both $\mathrm{d}n_2^\sigma/\mathrm{d}x_2^l = 0$ and $\mathrm{d}n_1^\sigma/\mathrm{d}x_2^l = 0$, is the slope of the isotherm equal to the total amount of adsorbed material at the surface. Thus, as pointed out above, if the slope is equated to n^σ, and a_s is calculated from this value, then the x_2^σ against x_2^l isotherm will automatically have at

⁴⁹ B. P. Bering and V. V. Serpinskii, *Izvest. Akad. Nauk S.S.S.R., Ser. khim.*, 1970, 1232 (*Bull. Acad. Sci. U.S.S.R., Chem. Ser.*, 1970, 1169).

least a point of inflexion with a horizontal tangent (Figures 2 and 4); in effect, the surface phase is bound to exhibit a behaviour which lies on the boundary between thermodynamic acceptability and inconsistency. This limiting behaviour is perhaps not expected to be shared by all systems showing a linear portion (or point of inflexion) on the excess isotherm, so that in general one may expect the surface areas calculated by this method to represent a lower limit of the true values as suggested by Dubinin,[50] but refuted by Schay[39] (see below).

Two papers by Tóth[51,52] discuss this problem further. He derives equation (63), but by a somewhat more lengthy argument: in his equation, x_2^σ is the mean mole fraction of component 2 in the n layers of adsorbed phase. He concludes that when the monolayer model is applied to the ethanol + water liquid/vapour interface, the curve of σ against x_2^σ exhibits a point of inflexion with a vertical tangent (*i.e.* $\partial\sigma/\partial x_2^\sigma \to \infty$; $\partial x_2^\sigma/\partial x_2^l \to 0$), which he regards as thermodynamically impossible. As pointed out above, however, this represents the limiting case between what is and what is not thermodynamically permissible. The author does not, however, quote the values adopted for the molecular areas in his calculations, and it is to be noted that Kipling,[53] in a similar analysis of the same experimental data, found that although $\partial x_2^\sigma/\partial x_2^l$ fell to a value close to zero, a small change in the basis of the calculation removed the horizontal inflexion. By taking $n = 2$, Tóth shows that a thermodynamically acceptable situation is achieved.

He then considers data for the systems (benzene + ethanol)/charcoal and (benzene + acetic acid)/charcoal, each of which exhibits a linear section in the excess isotherm. These are then said to be associated with regions in which $\partial x_2^\sigma/\partial x_2^l$ is zero and the monolayer model is therefore rejected. This conclusion, however, illustrates the danger of becoming involved in a circular argument. Tóth does not say how the surface areas of the charcoals were derived: if they were obtained from the slope of the linear portion of the excess isotherm, then the horizontal tangent in x_2^σ against x_2^l follows automatically; a small increase in the value of A_s would then ensure that the calculated x_2^σ curve was thermodynamically acceptable.

Thus, though on general grounds one is inclined to reject the monolayer model, care must certainly be taken in assessing the strength of the rejection 'on thermodynamic grounds' until the reliability of the parameters used in the calculation can be assessed.

However, Tóth prefers to regard the systems he considers as containing a bimolecular adsorbed layer, the mole fractions in the two layers being different. Since equation (9) governing the stoichiometry of the system is of the same form both for monolayer and for multilayer adsorption, he describes these systems as exhibiting 'equivalent multilayer' adsorption. He stresses

[50] M. M. Dubinin, in 'Proceedings of the International Symposium on Surface Area Determination, 1969', Butterworths, London, 1970, p. 288.
[51] J. Tóth, *Acta Chim. Acad. Sci. Hung.*, 1970, **63**, 67.
[52] J. Tóth, *Acta Chim. Acad. Sci. Hung.*, 1970, **63**, 179.
[53] J. J. Kipling, *J. Colloid Sci.*, 1963, **18**, 502.

that the analysis cannot lead to any conclusions about the relative concentrations in the two layers, but gives a graphical method of deriving self-consistent pairs of mole fractions $x_2^{\sigma(1)}$ and $x_2^{\sigma(2)}$ in the two layers. Two possible types of system are then postulated: those where $x_2^{\sigma(1)} > x_2^{\sigma(2)} > x_2^l$, called $(+, +)$ adsorption, in which both adsorbed layers are enriched in component 2; and those where $x_2^{\sigma(1)} > x_2^l > x_2^{\sigma(2)}$ which exhibit $(+, -)$ adsorption, the first layer being enriched in component 2, and the second in component 1. These concepts are developed in the second paper, in which it is argued that systems in which the molecules are of similar polarity are likely to show $(+, +)$ adsorption, whereas if the components have widely different polarities then $(+, -)$ adsorption may occur. One possible consequence of $(+, -)$ adsorption is that the excess isotherm may pass through zero, *i.e.* exhibit an azeotropic point. Below this concentration component 2 may be preferentially adsorbed, whereas at higher concentrations component 1 is adsorbed. Tóth argues that on a homogeneous surface the difference between the energies of adsorption of, say, benzene and ethanol, is unlikely to change sign as the concentration of the bulk solution changes. Thus the adsorption of benzene in the first layer, if initially positive, will remain so over the whole concentration range. The situation in the second layer (under relatively weak control from the solid surface) will then be determined by the nature of the first adsorbed layer; it is suggested that on a slightly polar charcoal surface benzene is subjected to positive adsorption at all concentrations, and that the more polar hydrophilic ethanol is expelled into the second layer where it accumulates. The overall excess isotherm results from the summation of $x_2^{\sigma(1)}$, which for benzene [component (2)] is always positive, with $x_2^{\sigma(2)}$, which is negative, in the second layer; these two contributions can therefore cancel exactly at one particular concentration leading to an S-shaped excess isotherm with an azeotropic point (Figure 1b). Furthermore, the total surface tension σ of the interface is regarded as the sum of contributions from the two layers $\sigma^{(1)}$ and $\sigma^{(2)}$; each of these when plotted against $x_2^{\sigma(1)}$ or $x_2^{\sigma(2)}$ is thermodynamically acceptable. Tóth presents a series of tables which indicate that S-shaped isotherms [which he associates with $(+, -)$ adsorption] are most commonly found with systems in which the components differ widely in polarity. Pairs of strongly polar or of non-polar molecules tend to exhibit isotherms which do not change sign.

It seems not unlikely that this interpretation of isotherms showing a change of sign may have some validity in certain cases, although the mechanism which Tóth envisages is open to question: 'the excess mass in the boundary phase is due to "first-order" surface forces (monomolecular adsorption) while the "second-order" forces distribute this excess among several layers'.

One must recall, moreover, the other suggested explanations of sign inversion in $\Gamma_2^{(n)}$. Siskova and Erdös[54] postulated that in such cases the surface phase was heterogeneous. In one region one component was more strongly adsorbed and in the other the second component was preferred.

⁵⁴ M. Siskova and E. Erdös, *Coll. Czech. Chem. Comm.*, 1960, **25**, 2599.

These authors clearly thought of these regions as being associated with different parts of the solid surface. In effect, Tóth regards this as an unlikely situation and identifies the two regions with the first and second adsorption layers. Isotherms of this kind can also be interpreted in terms of the non-ideality of the system. Thus various treatments[17,20] of adsorption from regular solution show that for systems having a relatively high positive interchange energy, w, a change of sign of the excess isotherm is to be expected. The mechanism by which this occurs, though not quite that envisaged by Tóth, does nevertheless reflect the way in which the 'demixing' tendency associated with positive deviations from Raoult's law can be transmitted and amplified by concentration changes at the surface. It may be commented that the system (benzene + ethanol)/charcoal to which Tóth devotes his main attention has apparently been adequately analysed by Nagy and Schay[20,36] in terms of the non-ideality of the bulk phase.

The present position is thus that although there are compelling reasons for rejecting the monolayer model as a satisfactory general description of adsorption from solution, it is impossible on the basis of experimental measurements of excess isotherms to reach any more detailed picture.

4 Statistical Mechanical Studies: Multilayer Model

The fundamental difficulties encountered by the monolayer model when applied to adsorption from non-ideal solutions, and in particular to mixtures of molecules of different size, have led to a series of studies in which attempts have been made to calculate the concentration profile as a function of distance from the adsorbing surface.

The theory is usually developed[27] on the basis of a lattice model in which it is supposed that the system can be regarded as a succession of layer planes, each of N adsorption sites, stacked one above the other, the first plane being in contact with the solid adsorbing surface. The adsorption forces are usually supposed to be short range, affecting directly only the molecules in the first layer. Concentration changes in successive layers then arise from the lack of symmetry of the intermolecular forces experienced by a given layer: the ith layer experiences forces from the $(i-1)$th layer different from those exerted on it by the $(i+1)$th layer. The molecules are assumed to be of equal size and each is supposed to have z nearest neighbours, of which lz lie in the same lattice plane and mz in each of the adjacent planes: $l + 2m = 1$; most calculations have been based on a close-packed lattice with $z = 12$, $l = \frac{1}{2}$, and $m = \frac{1}{4}$. The interaction energies between like molecules are denoted by ε_{11} and ε_{22}, while that between unlike molecules is characterized by the interchange energy $w = z[\varepsilon_{12} - \frac{1}{2}(\varepsilon_{11} + \varepsilon_{22})]$. The mole fraction of component 2 in the ith layer is written as $x_2^{(i)}$.

The free energy of the system can now be calculated by one of several standard statistical mechanical techniques. For example, the energy and entropy may be evaluated separately,[16,27,55] or the problem can be handled using the partition function[27,28] or grand partition function techniques.[56-58] In the former method, the energy of each layer is summed. That of the first layer is equal to the sum of the interactions of $Nx_1^{(1)}$ molecules of type 1 and $Nx_2^{(1)}$ molecules of type 2 with the solid surface, U^{ads}, plus the interaction, $U^{1,1}$, between molecules in the first layer with other molecules in the same layer, plus one half the interaction energy, $U^{1,2}$, between molecules in the first layer with those in the second. Similar arguments are applied to successive layers to give the total potential energy in the form

$$\Phi = U^{ads} + U^{1,1} + \tfrac{1}{2}U^{1,2} + \sum_{i=2}^{i=l} [U^{i,i} + \tfrac{1}{2}(U^{i,i-1} + U^{i,i+1})], \qquad (65)$$

where $U^{i,j}$ is the energy of interaction between molecules in plane i with those in plane j. By assuming random arrangements in all planes, $U^{i,i}$ can be written as a function of $x_2^{(i)}$, whereas $U^{i,i-1}$ and $U^{i,i+1}$ are functions of (x_2^i, x_2^{i-1}) and (x_2^i, x_2^{i+1}), respectively.[55] The configurational entropy of the system is calculated from the number of ways of arranging the equilibrium set of molecules of types 1 and 2 on the lattice:

$$g = \prod_{i=1}^{i=l} \frac{N!}{(Nx_1^{(i)})!(Nx_2^{(i)})!}. \qquad (66)$$

The configurational free energy, F_c, is then given by

$$F_c = \Phi - kT \ln g. \qquad (67)$$

The condition for equilibrium in the system is that the $x_2^{(i)}$ values shall be distributed in such a way that interchange of a pair of molecules between any pair of planes (Figure 5a) shall leave F_c unaltered:

$$\frac{\partial F_c}{\partial n_2^{(i)}} = \frac{\partial F_c}{\partial n_2^{(j)}}, \qquad \text{(all } i,j) \qquad (68)$$

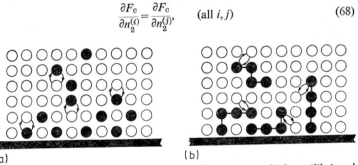

(a)   (b)

Figure 5 *Illustrating exchange processes with respect to which equilibrium has to be established (a) for molecules of equal size; (b) for tetramer molecules*

[55] S. Ono, *Mem. Fac. Eng. Kyushu Univ.*, 1950, **12**, 1.
[56] T. Murakami, S. Ono, M. Tamura, and M. Kurata, *J. Phys. Soc., Japan*, 1951, **6**, 309.
[57] G. Delmas and D. Patterson, *J. Phys. Chem.*, 1960, **64**, 1827.
[58] G. A. H. Elton, *J. Chem. Soc.*, 1954, 3813.

subject to

$$\sum_k \delta n_2^k = 0. \tag{69}$$

Solution of these equations, or maximization of the grand partition function, leads to the following difference equations:[55,59]

$$x_2^{(2)} - x_2^{(1)} = \frac{kT}{2mzw} \left[\ln \frac{x_1^l x_2^{(1)}}{x_2^l x_1^{(1)}} \right] + \frac{1}{m} [x_2^l - x_2^{(1)}]$$

$$- \tfrac{1}{2}[1 - 2x_2^{(1)}] - \frac{1}{zw} [(\chi_1 - \tfrac{1}{2}zm\varepsilon_{11}) - (\chi_2 - \tfrac{1}{2}zm\varepsilon_{22})] \tag{70}$$

and

$$2x_2^{(i)} - [x_2^{(i+1)} + x_2^{(i-1)}] = \frac{kT}{2mzw} \left[\ln \frac{x_2^l x_1^{(i)}}{x_1^l x_2^{(i)}} \right] - \frac{1}{m} (x_2^{(i)} - x_2^l), \tag{71}$$

where x_2^l is the limiting value of $x_2^{(i)}$ for large i. χ_i is the energy of interaction of a single molecule of i with the surface, while $\tfrac{1}{2}zm\varepsilon_{ii}$ is the loss of molecular interaction energy when the interaction between a molecule of i in the pure liquid with its neighbours in an adjacent plane is replaced by interaction with solid when that plane is replaced by solid. Thus $(\chi_1 - \tfrac{1}{2}zm\varepsilon_{11})$ is the energy of adsorption of component 1 from its pure liquid.

These equations may be solved[27,28,55] by starting with trial values of $x_2^{(1)}$ and given parameters of x_2^l, m, z, w, χ_1, and χ_2, and calculating $x_2^{(i)}$ for increasing i. When the correct value of $x_2^{(1)}$ is chosen, $x_2^{(i)} \to x_2^l$ as i increases and the total adsorption is then given by

$$\Gamma_2^{(n)} = N \sum_{i=1}^{i=t} (x_2^{(i)} - x_2^l), \tag{72}$$

where t is chosen such that $(x_2^{(t)} - x_2^l)$ is less than some chosen small number (*e.g.* 10^{-6}).

This iterative method of solution is tedious, and recently a new method of solving these equations has been proposed by Altenberger and Stecki.[59] They consider first the case of an ideal mixture ($w = 0$) for which the solution of equations (70) and (71) is simply

$$x_2^{(1),\text{id}} = \frac{x_2^l}{x_2^l + K'(1 - x_2^l)} \tag{73}$$

$$x_2^{(i),\text{id}} = x_2^l \text{ for all } i > 1, \tag{74}$$

where

$$K' = \exp \{[(\chi_1 - \tfrac{1}{2}zm\varepsilon_{11}) - (\chi_2 - \tfrac{1}{2}zm\varepsilon_{22})]/kT\}$$

$$= \exp Q. \tag{75}$$

[59] A. R. Altenberger and J. Stecki, *Chem. Phys. Letters*, 1970, **5**, 29.

This solution is, of course, in agreement with equation (21) (with $K' = 1/K$).

Altenberger and Stecki then examine the possibility of expressing the solution for $w \neq 0$ as a Maclaurin series starting from the ideal case:

$$x_2^{(i)}(w) = x_2^{(i),\text{id}} + \sum_{n=1}^{\infty} \frac{B^n}{n!} x_2^{(i)(n)}(0), \qquad (76)$$

where $B = zw/kT$ and $x_2^{(i)(n)}(0)$ is the nth derivative of $x_2^{(i)}$, obtained by differentiating equations (70) and (71), taken at the point $w = 0$. They find that the series (76) converges quite rapidly and provided that $|B| < 1$ only a few terms are needed. Figure 6 shows $x_2^{(i)}$ as a function of i for a bulk mole fraction of 0.5 for $K = e^{-1}$ and $B = \pm 1$. An interesting feature of these results (originally noted by Lane[28]) is that for $B = -1$ there is an alternation of preferential adsorption in successive layers; the physical origin of this[28] is

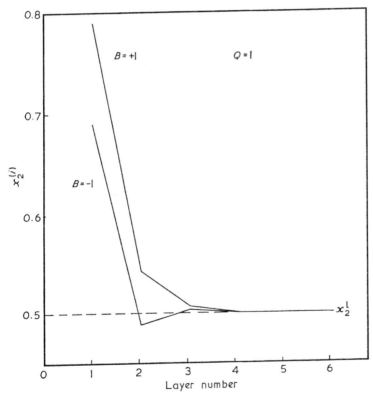

Figure 6 *Variation of mole fraction of component 2 in successive layers ($x_2^{(i)}$) as a function of layer number (i) according to the regular solution model with $\log K = 1$ ($Q = 1$) and B ($= zw/kT$) taking values of $+1$ and -1, showing oscillation of $x_2^{(i)}$ about x_2^l for negative values of B*

D

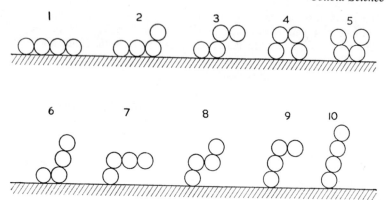

Figure 7 *Possible configurations of flexible homogeneous tetramers (aaaa).*
The set shown is $R_{1,s}$ where s goes from 1 to 10; the set $R_{2,s}$ has
the nearest segment in the second layer and all other segments in
that layer or further from the surface. For 'end-group active' tetramers
(baaa), s goes from 1 to 27, since configurations 2, 3, 6, 7, 8, 9, and
10 can then be arranged in two ways
(Redrawn with permission from *Trans. Faraday Soc.*, 1970, **66**, 708)

that the alternation makes possible an increase in the number of $(1-2)$
contacts which are energetically favourable. However, although this provides
theoretical support for the existence of the $(+, -)$ type of behaviour suggested by Tóth, the systems to which he ascribes this behaviour are expected
to have $B > 0$ (*i.e.* to show positive deviations from Raoult's law).

In the case of mixtures of molecules of different size, the statistical problem
is more complex. Ash, Everett, and Findenegg[31] considered the case of
monomer + dimer mixtures, and have recently[60] extended the calculation
to monomer + trimer and monomer + tetramer mixtures. In the case of
monomer + tetramer mixtures the ten configurational species shown in
Figure 7 for symmetrical tetramers (*aaaa*) were considered (if the tetramers
are 'end-group active', *baaa*, then seventeen configurations have to be considered). Each configurational species is denoted by a symbol $R_{n,s}$. The first
subscript indicates that one or more of the segments is located in layer n,
while the rest of the molecule is further away from the surface; the second
subscript denotes the species numbered in Figure 7. The number of molecules
of type (n, s) is $N_{n,s}$ and their concentration, expressed in volume fractions,
is $\phi_{n,s}$. The configurational free energy can now be calculated as before in
terms of the configurational entropy $k \ln g$, and the total potential energy Φ:

$$F_c = -kT \ln g (A_s, \phi_{1,1}, \ldots \phi_{n,s}, \ldots \phi_{t,c})$$

$$+ \Phi(A_s, \phi_{1,1}, \ldots \phi_{n,s}, \ldots \phi_{t,c}). \quad (77)$$

[60] S. G. Ash, D. H. Everett, and G. H. Findenegg, *Trans. Faraday Soc.*, 1970, **66**, 708.

The equilibrium condition is now that any change of configuration arising from the movement of a segment from one layer to another (Figure 5b) must take place without changing F_c; this may be written

$$\delta F_c = \sum_{R_{n,s}} \left(\frac{\partial F_c}{\partial N_{n,s}}\right)_{A_s, N_{1,1} \cdots N_{t,c}} \delta N_{n,s} = 0, \tag{78}$$

subject to

$$\sum_{R_{n,s}} \delta N_{n,s} = 0. \tag{79}$$

These equations are solved by the method of Lagrange multipliers: multiplying equation (79) by the Lagrange multiplier μ and subtraction from equation (78) leads to *tc* simultaneous equations of the form

$$\left(\frac{\partial F_c}{\partial N_{n,s}}\right)_{A_s, N_{1,1} \cdots N_{t,c}} = \mu, \qquad \text{all } R_{n,s}. \tag{80}$$

The multiplier μ is thus essentially a 'chemical potential per segment' which must be constant throughout the system and be equal to that in the bulk liquid.

The resulting equations can be solved by computer techniques, and the surface excess concentrations of tetramer calculated. The results of this work provide interesting examples of the interplay of energy and entropy terms in determining the preferred configurations of adsorbed molecules and confirm the tendencies observed previously with the dimer + monomer system. For example, Figure 8 shows, for the case of $\log K = 1$, $w = 0$, the concentrations of homogeneous tetramers with one, two, three, and four segments adsorbed in the first layer. At low concentrations (less than a volume fraction in solution of 0.1) the concentrations of the different configurations follow the order of number of adsorbed segments. However, since a tetramer wholly in the first layer occupies four lattice sites, its presence at higher concentrations tends to reduce the number of configurations available in that layer; the entropy can be increased by reducing the number of adsorbed segments and this offsets the increase in potential energy accompanying the desorption of a segment. Thus, at higher concentrations configurations with 1 or 2 adsorbed segments become relatively more stable than those with 3 or 4 segments in layer 1. This effect becomes even more marked if the tetramer has only one end group active in adsorption: reduction in the number of segments in the first layer is achieved without an increase in energy, so that 'vertical' orientations of the molecules are more readily achieved. However, within the range of parameters studied, no evidence was found for any first-order phase changes in the adsorbed layers. Unfortunately, extension of calculations of this kind to larger size ratios would involve excessive computer times.

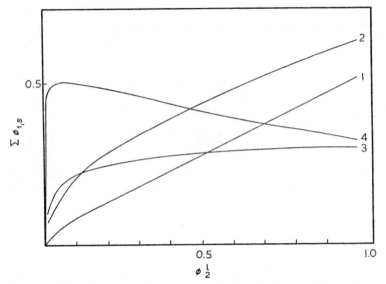

Figure 8 *Volume fractions of tetramer molecules having one* $(R_{1,7} + R_{1,8} + R_{1,9} + R_{1,10})$, *two* $(R_{1,3} + R_{1,4} + R_{1,5} + R_{1,6})$, *three* $(R_{1,2})$, *and four* $(R_{1,1})$ *segments in contact with the surface as function of volume fraction of tetramer in liquid*
(Redrawn with permission from *Trans. Faraday Soc.*, 1970, **66**, 708)

5 Determination of Surface Areas from Adsorption from Solution

The use of measurements of adsorption from solution for the determination of the surface areas of solids has a long history, of which a survey up to 1965 is given by Kipling.[1] The subject was also discussed critically at the IUPAC/SCI Symposium on the Determination of Surface Areas.[61] Schay's introductory paper[62] reviews the various systems which have been used for this purpose, and the choice of methods of analysis.

Much early work was concerned with the adsorption of relatively large solute molecules, which are often incompletely soluble in the commonly used solvents. If, in the range of solubility of the solute, the measured adsorption reached a plateau value, then it was assumed that under those conditions the surface was completely covered with a monolayer of adsorbed solute. It was to systems of this kind that isotherm equations more relevant to gas adsorption than solution adsorption were applied. The adsorption was often measured not by $n^0 \Delta x_2^l / m$, but by $V^0 \Delta c_2^l / m$ where V^0 is the volume of initial solution and Δc_2^l the change in concentration when the solid is equilibrated with the solution; in sufficiently dilute solution these two measures of adsorp-

[61] 'Proceedings of the International Symposium on Surface Area Determination, 1969', Butterworths, London, 1970.
[62] G. Schay, ref. 61, p. 272.

tion are equal but at higher concentrations $V^0 \Delta c_2^l / m$ differs from $n^0 \Delta x_2^l / m$ by a factor of $[1 + (r-1)x_2^l]^{-1}$, where r is the ratio of molar volumes of the components of the mixture:

$$a_s \Gamma_2^{(n)} = \frac{n^0 \Delta x_2^l}{m} = \left[1 + (r-1) \, x_2^l\right] \frac{V^0 \Delta c_2^l}{m}. \tag{81}$$

Thus the older measurements should have been corrected in this way before being used to obtain surface areas. Bearing this in mind, we see that the plateau observed experimentally was probably more often than not the maximum of the surface excess isotherm; because of the limited solubility of many of the solutes employed, it was not possible to test this possibility experimentally. To convert such plateau values to the actual amount of solute in the (assumed monolayer) surface phase one should consider

$$\frac{a_s \Gamma_2^{(n)}}{(1 - x_2^l)} = \frac{n_2^\sigma}{m} - \frac{n_1^\sigma}{m} \cdot \frac{x_2^l}{x_1^l} = a_s \Gamma_2^{(1)}, \tag{82}$$

the relative adsorption of 2 with respect to 1.

If the plateau occurs at a sufficiently low concentration, then $a_s \Gamma_2^{(n)} / (1 - x_2^l)$ [or even $a_s \Gamma_2^{(n)}$] is a reasonable measure of n_2^σ / m. The observation that $\Gamma_2^{(n)}$ has reached a plateau value does not therefore mean that n_2^σ has reached its maximum value; indeed, from equation (82), even though $\Gamma_2^{(n)}$ remains constant, n_2^σ will continue to rise as x_2^l increases. If n_2^σ reaches a maximum value, then it is generally assumed that n_1^σ is zero. This is often hard to justify unless the solute is very strongly adsorbed.

Despite the uncertainties associated with the interpretation of the adsorption of large molecules (often of limited solubility) from dilute solution, these methods are of considerable practical importance. Because of the convenience of optical methods of analysis, dyestuff adsorption has been extensively studied and applied to surface area determinations. The method has been critically reviewed by Giles, D'Silva, and Trevedi[63] and by Padday.[64] The main criteria which must be satisfied if reliable data are to be obtained are summarized by Padday as (i) establishment of thermodynamic equilibrium, (ii) accessibility of the whole surface of the powder, (iii) presence of the dye in one form only and not in several aggregation states, (iv) formation of a physically adsorbed monolayer, and not multilayers, and (v) constant site area occupied by a dye ion or molecule at saturation adsorption. Inconsistent results can also result from other causes, among which Giles *et al.* list specific dye–substrate bonding; interference by chemical reaction of the dye with the solid or with adventitious ions introduced into solution by the solid; molecular sieve action or porosity in the solid; and ageing effects on the solid surface. The choice of suitable dyes is therefore a matter of some difficulty and Giles gives details of a number of dyes which are recommended for surface area determination: when used to examine a wide variety of oxide,

[63] C. H. Giles, A. P. D'Silva, and A. S. Trivedi, ref. 61, p. 317.
[64] J. F. Padday, ref. 61, p. 331.

carbon, and halide surfaces, ranging in area from 2 to 100 m² g⁻¹, they are said to give results (based on molecular cross-sections obtained by calibration against BET N_2 measurements) lying within 10% of those obtained by N_2-adsorption. Padday shows that, particularly for silver halides, the site area may depend quite markedly on the composition of the substrate and concludes that, to be reliable, surface area measurements by the dye-adsorption technique must be carefully calibrated against some more reliable method.

More recently, attention has moved towards the determination of surface areas using adsorption from mixtures of pairs of completely miscible liquids comprising molecules of comparable size. Despite the shortcomings of the monolayer model of adsorption from solution discussed in Sections 3 and 4 there are nevertheless circumstances in which measurements of this kind can give apparently reliable information about the surface areas of solids. In particular, if the solutions do not depart seriously from ideality, and the components are of about the same size, then the monolayer model is likely to be a reasonably good approximation, and a suitable analysis of experimental data may be expected to give useful results.

The analysis can be carried out in several alternative, equivalent ways, based on equations (23), (24), and (25).

As indicated earlier, the limiting slope of the surface excess isotherm as $x_2^l \rightarrow 1$, for ideal systems of molecules of equal size, is equal to $-(n^\sigma/m)(K-1)/K$ which, when K is large, is a good approximation to $-(n^\sigma/m)$, where n^σ is the amount of component 2 required to cover the surface. One way of finding this limiting slope is to plot $(n^0 \Delta x_2^l/m)/(1-x_2^l)$ against x_2^l and extrapolate the curve to $x_2^l = 1$; reference to equation (3) shows this this means that $\Gamma_2^{(1)}$ is extrapolated to $x_2^l = 1$. Thus,

$$\text{Limiting slope of } \left(\frac{n^0 \Delta x_2^l}{m}\right) \text{ against } x_2^l = \underset{x_2^l \rightarrow 1}{\text{Lt}} \ \Gamma_2^{(1)} = \left(\frac{K-1}{K}\right) n_2^{\sigma(0)}, \quad (83)$$

where to be quite explicit n^σ is written as $n_2^{\sigma(0)} = A_s/a_2$.

Alternatively, the data may be plotted according to equations (24) or (25). According to equation (24) the linear graph has*

$$\text{Slope} = \frac{1}{n_2^{\sigma(0)}}, \tag{84}$$

and

$$\underset{(x_2^l=1)}{\text{Intercept}} = \frac{1}{n_2^{\sigma(0)}} \left(\frac{K}{K-1}\right). \tag{85}$$

If K is large the intercept itself is a good approximation to $1/n_2^{\sigma(0)}$. Schay[39] gives a supposedly more general equation for molecules of different size, based on the assumption that, provided the sizes do not differ too greatly, the separation factor S is essentially constant. Comparison of equations (14')

---

* A useful relationship which does not seem to have been pointed out previously is that (Intercept at $x_2^l=1$)/(Intercept at $x_2^l=0$) $= K$.

and (42) shows that, strictly speaking, this assumption is not compatible with the conditions of thermodynamic equilibrium, although the errors involved may not be too serious. Schay's equation has the form [*cf.* equation (47)]

$$\frac{x_1^l x_2^l}{(n^0 \Delta x_2^l / m)} = \frac{1}{n_2^{\sigma(0)}(S-1)} [r + (S-r) x_2^l], \tag{86}$$

so that the graph of the left-hand side against x_2^l now has

$$\text{Slope} = \frac{1}{n_2^{\sigma(0)}} \left(\frac{S-r}{S-1}\right), \tag{87}$$

and

$$\underset{(x_2^l = 1)}{\text{Intercept}} = \frac{1}{n_2^{\sigma(0)}} \left(\frac{S}{S-1}\right). \tag{88}$$

Once again, if S is much larger than r the slope of the graph will be an acceptable approximation to $1/n_2^{\sigma(0)}$.

Thus, unless r is an appreciable fraction of S, this method of analysis is not seriously affected by inequality in molecular sizes.

Schay gives a number of examples of the analysis of experimental data in these various ways: Figure 9 shows the relationship between the methods outlined above, and Table 1 gives a selection of data which illustrate the

Table 1 *Specific surface areas determined by gas adsorption (BET, N_2 method) and by adsorption from solution (graphical method)[39]*

Adsorbent	Gas adsorption	$a_s/\text{m}^2\,\text{g}^{-1}$ Adsorption from solution[a]	
Silica gel	540	560	toluene + n-heptane
Silica gel	450	530 ⎫ 460 ⎪	
Silica gel	210	202 ⎬	benzene + n-heptane[b]
Alumina gel	240	236 ⎪	
Alumina gel	186	190 ⎭	
Charcoal	612	587 ⎫	
Charcoal	840	870 ⎬	ethanol + benzene[c]
Charcoal	1100	1200 ⎪	
Carbon black	68	72 ⎭	

[a] $a(\text{CH}_3\text{C}_6\text{H}_5) = 34$ Å^2 molecule$^{-1} = 204$ m^2 mmol^{-1}; $a(\text{C}_6\text{H}_6) = 30$ Å^2 molecule$^{-1} = 180$ m^2 mmol^{-1}; $a(\text{C}_2\text{H}_5\text{OH}) = 20$ Å^2 molecule$^{-1} = 120$ m^2 mmol^{-1}; $a(\text{N}_2) = 16.2$ Å^2 molecule^{-1}.
[b] Isotherms of the form in Figure 1a.
[c] Isotherms of the form in Figure 1b.

agreement found between surface areas determined by these techniques and those from gas-adsorption data. Eltekov[65] gives another example, for the adsorption of benzene from n-hexane by macroporous hydroxylated silica

[65] Yu. A. Eltekov, ref. 61, p. 291.

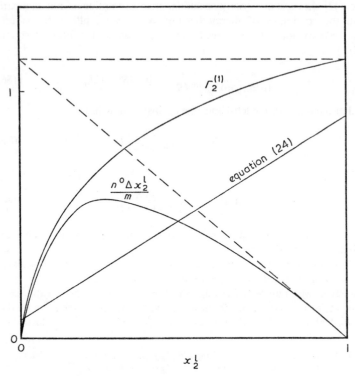

Figure 9 *Relationship between various methods proposed for determination of surface areas from solution adsorption measurements (schematic)*

gel, in which the surface areas derived from solution adsorption and gas adsorption agree to within 2%.

Much greater controversy surrounds the method proposed by Nagy and Schay for isotherms which exhibit an S-shape with a substantial linear section. Some of the problems associated with this method have been discussed in Section 4. They were raised again by Dubinin,[50] who presented arguments similar to those of Bering and Serpinskii.[45] Schay is, however, of the opinion that provided dx_2^{σ}/dx_2^l is zero only at a point of inflexion (Figure 2b), then the analysis is not thermodynamically inconsistent. However, the success of this method, for which Schay recommends the system ethanol+benzene for carbon adsorbents, must be regarded as largely empirical: some typical results are given in the lower part of Table 1.

Two further papers by Nagy and Schay,[66, 67] in which methods proposed

[66] L. G. Nagy and G. Schay, *Magyar Kém. Lapja*, 1970, **25**, 439 (*Chem. Abs.*, 1970, **73**, 123 858).

[67] L. G. Nagy and G. Schay, *Magyar Kém. Folyóirat*, 1971, **77**, 113 (*Chem. Abs.*, 1971, **75**, 10 608).

for surface area determination using adsorption from solution are critically evaluated, are abstracted and appear to cover much of the ground dealt with above.

The problem of the determination of the surface areas of porous substances must in many instances be discussed in relation to the practical purposes to which such measurements are to be applied. Thus when one is concerned with the study of reactions in the liquid phase catalysed by solids, the surface area determined by gas adsorption may not be a very relevant quantity since it includes the area of the surface of micropores from which the reactants may be excluded. In these circumstances surface area measurements using adsorption from solution of organic molecules may be of more value; phenol has been often used for this purpose. Schwuger[68] has considered the use of adsorption of phenol from aqueous solution for the determination of the surface areas of various porous hydrophobic, carbonaceous adsorbents. He employs a form of the BET equation in which the relative pressure p/p^0 is replaced by the relative concentration c/c^0, where c^0 is the saturation concentration of phenol in water. This equation is shown to fit the experimental data satisfactorily and (using an area for phenol of $2.64 \text{ m}^2 \text{ g}^{-1} = 41.0 \text{ Å}^2$ molecule^{-1}) leads to surface areas which agree well with those from the BET N_2 method for carbon blacks, but are in all cases substantially lower in the case of active carbons. It is also shown that for certain surfaces (for which the BET constant is large) an adequate measure of the surface area for practical purposes can be obtained from a single adsorption measurement.

6 Some Alternative Theoretical Treatments

Polanyi Potential Theory.—Although Polanyi envisaged that his potential theory of adsorption of gases by solids[69] could be applied to the case of adsorption from solution,[70] this approach was not developed in detail, nor subjected to any experimental tests, until comparatively recently. The relationship between the potential theory and the more conventional approaches outlined in the Introduction may be developed in the following manner, which is similar to that employed by Hansen and Fackler.[71] Consider an element of the adsorption space of volume dv, at a point (or on an equipotential contour) ϕ, in which the mole fractions of the components are x_1^ϕ and x_2^ϕ. Equilibrium between this element of volume and its neighbours, and eventually with the bulk liquid, requires that equilibrium shall be achieved in the following exchange process:

$$\frac{v^\ominus}{v_1^\phi}(1)^\phi + \frac{v^\ominus}{v_2^l}(2)^l \rightleftharpoons \frac{v^\ominus}{v_1^l}(1)^l + \frac{v^\ominus}{v_2^\phi}(2)^\phi, \tag{89}$$

where $v^\ominus$ is a standard volume per mole and v_i^ϕ and v_i^l are the partial molar

[68] M. J. Schwuger, *Kolloid-Z.*, 1969, **234**, 1048.
[69] M. Polanyi, *Verh. Deut. Physik. Ges.*, 1914, **16**, 1012.
[70] M. Polanyi, *Verh. Deut. Physik. Ges.*, 1916, **18**, 55.
[71] R. S. Hansen and W. F. Fackler, *J. Phys. Chem.*, 1953, **57**, 634.

volumes of the two components in the adsorption space and in the bulk liquid. This equation is clearly analogous to equation (13), which is expressed in terms of area rather than volume occupancy. For simplicity it may be assumed that $v_i^\phi = v_i^l = v_i$, the molar volume of pure liquid i. Furthermore, the adsorption potential $\varepsilon_i(\phi)$, of i in the element of volume under consideration is defined by

$$\varepsilon_i(\phi) = \mu_i^{0,\phi}(T, p) - \mu_i^{0,l}(T, p), \tag{90}$$

where $\mu_i^{0,\phi}(T, p)$ is the chemical potential of an element of pure liquid brought up to ϕ *at constant T and p* (*cf.* Chapter 4, p. 146). It is assumed, moreover, that $\varepsilon_i(\phi)$ is equal to the adsorption potential of i, at the same ϕ, as determined by vapour-phase adsorption.

Equilibrium in the above exchange reaction requires that

$$\frac{\mu_1^\phi - \mu_1^l}{v_1} = \frac{\mu_2^\phi - \mu_2^l}{v_2}, \tag{91}$$

where again the μ values are all taken at the same T, p. This condition may be compared with that given in the footnote on p. 54.

Now

$$\mu_i^l = \mu_i^{0,l} + RT \ln x_i^l \gamma_i^l$$

and

$$\mu_i^\phi = \mu_i^{0,\phi} + RT \ln x_i^\phi \gamma_i^\phi, \tag{92}$$

so that equation (91) leads to

$$\frac{x_1^\phi \gamma_1^\phi}{x_1^l \gamma_1^l} \left(\frac{x_2^l \gamma_2^l}{x_2^\phi \gamma_2^\phi} \right)^{v_2/v_1} = \exp\left[-\left\{ (\mu_1^{0,\phi} - \mu_1^{0,l}) - \frac{v_1}{v_2}(\mu_2^{0,\phi} - \mu_2^{0,l}) \right\} / RT \right], \tag{93}$$

which is analogous to equation (14) with equation (18).

If it is assumed that the activity coefficient terms cancel, then this may be written

$$\frac{x_1^\phi}{x_1^l} \left(\frac{x_2^l}{x_2^\phi} \right)^{v_2/v_1} = \exp\left[-\left\{ \varepsilon_1(\phi) - \left(\frac{v_1}{v_2} \right) \varepsilon_2(\phi) \right\} / RT \right]. \tag{94}$$

If $\varepsilon_1(\phi)$ and $\varepsilon_2(\phi)$ are known from gas-adsorption data, then x_2^ϕ can be found as a function of x_2^l. The excess amount of component 2 in the element of volume dv $(dn_2^{\sigma(n)})$ is given by

$$dn_2^{\sigma(n)} = (x_2^\phi - x_2^l) \frac{dv}{v^\phi}, \tag{95}$$

where v^ϕ is the mean molar volume at ϕ. Hence, the total excess amount of component 2 in the whole adsorption space in equilibrium with a mole fraction x_2^l in the liquid is given by [*cf.* equation (4)]

$$\frac{n_2^{\sigma(n)}}{m} = \frac{n^0 \Delta x_2^l}{m} = \int_0^\infty \frac{x_2^\phi - x_2^l}{v^\phi} \, dv; \tag{96}$$

similarly,

$$\frac{V\Delta c_2^l}{m} = \int_0^\infty \left(\frac{x_2^\phi}{v^\phi} - \frac{x_2^l}{v^l}\right) dv, \tag{97}$$

where the integration is taken over the adsorption space of unit weight of adsorbent. The mean molar volume v^ϕ can be calculated from the equation

$$v^\phi = v_1 x_2^\phi + v_2 x_2^\phi, \tag{98}$$

where the partial molar volumes v_1 and v_2 can be approximated by the molar volumes of pure liquid.

According to this treatment, which breaks away completely from the monolayer concept and from that of a discrete adsorbed phase, the surface excess isotherm should be calculable from the vapour-adsorption isotherms of the individual components. Unfortunately, few attempts have been made to test the applicability of equations (96) or (97). Hansen and Fackler[71] attempted to calculate the adsorption of propan-1-ol and butan-1-ol from their aqueous solutions by Spheron-6, but the agreement between the observed and calculated isotherms was not entirely satisfactory. This is perhaps not surprising, since the adsorption isotherms of water and the lower alcohols by carbons do not conform to the Polanyi theory for vapour adsorption. However, it was found that the addition of an extra term, evaluated empirically but having some qualitative rational basis, to the adsorption potentials of the alcohols from the solution could remove at least part of the discrepancy. Further more extensive and detailed study of the applicability of the Polanyi theory in this form is long overdue.

An alternative application of the potential theory of adsorption[72] returns to Polanyi's original concept that when adsorption occurs from a liquid mixture showing incomplete miscibility, the adsorption potential can be related to the solubility of the solute, just as in the gas case the adsorption potential is related to the saturation vapour pressure. This implies that the adsorbed phase consists essentially of the pure insoluble component. If component 2 is liquid then

$$\mu_2^\phi = \mu_2^{0,l} = \mu_2^{\ominus,\phi} + RT \ln a_2(\text{sat}), \tag{99}$$

while

$$\mu_2^l = \mu_2^{\ominus,l} + RT \ln a_2^l. \tag{100}$$

Similar equations hold if component 2 separates out in the adsorption space as a solid phase.

Hence, from equations (91) and (90)

$$\frac{\mu_1^{\ominus,\phi} - \mu_1^{\ominus,l}}{v_1} - \frac{\mu_2^{\ominus,\phi} - \mu_2^{\ominus,l}}{v_2} = \frac{\varepsilon_1(\phi)}{v_1} - \frac{\varepsilon_2(\phi)}{v_2}$$

$$= -RT\left[\frac{1}{v_1}\ln\frac{a_1^\phi}{a_1^l} - \frac{1}{v_2}\ln\frac{a_2(\text{sat})}{a_2^l}\right]. \tag{101}$$

[72] M. Manes and L. J. E. Hofer, *J. Phys. Chem.*, 1969, **73**, 584.

We introduce $\varepsilon_{12}(\phi)$ defined by

$$\varepsilon_{12}(\phi) = \varepsilon_2(\phi) - \frac{v_2}{v_1}\,\varepsilon_1(\phi); \tag{102}$$

this is equal to the potential energy change when one mole of component 2 displaces (v_2/v_1) moles of component 1 from the region ϕ of adsorption space.

If the solution is sufficiently dilute, then $x_1^\phi \approx x_1^l \approx 1$, and the activity coefficients cancel so that

$$\left[\frac{\varepsilon_1(\phi)}{v_1} - \frac{\varepsilon_2(\phi)}{v_2}\right] = \frac{RT}{v_2}\ln\frac{c_2(\text{sat})}{c_2^l} = \frac{\varepsilon_{12}(\phi)}{v_2}. \tag{103}$$

Thus, for any degree of filling of the adsorption space ϕ, $\varepsilon_{12}(\phi)$ can be calculated from the equilibrium concentration, c_2^l, and a knowledge of the solubility of the solute, $c_2(\text{sat})$. Since the theory implies complete expulsion of solvent from the surface, ϕ is calculated from the amount of 2 adsorbed (which in sufficiently dilute solution is adequately given by $V\Delta c_2^l/m$) by multiplying by the molar volume of pure 2 in the liquid or solid state as is appropriate.

A fundamental assumption of the potential theory is that $\varepsilon(\phi)$ is of the same form for all substances, so that curves of $\varepsilon(\phi)$ can be brought into coincidence with that of a reference substance by multiplying all values of $\varepsilon(\phi)$ for a given substance by a common factor (Dubinin's affinity coefficient[73]). For closely similar groups of substances (*e.g.* hydrocarbons) the affinity coefficient is inversely proportional to the molar volume, so that curves of $\varepsilon(\phi)/v$ superimpose. More generally, however, such curves have to be adjusted by multiplying by a further factor which is approximately equal to the ratio of the polarizability per unit volume (p_i) of the liquid to that of the reference substance. The polarizability is usually calculated from

$$p_i = \left(\frac{n_i^2 - 1}{n_i^2 + 2}\right), \tag{104}$$

where n_i is the refractive index of the liquid. Manes and Hofer denote $\varepsilon(\phi)/v$ by α, and α_i/α_s, where s is the standard substance, by γ_i. Using this notation we then have

$$\gamma_{21} = \gamma_2 - \gamma_1 = \frac{p_2 - p_1}{p_s}. \tag{105}$$

Experimental verification of the theory in this form requires first that $\varepsilon_{21}(\phi)/v$ derived from measurements of adsorption from solution can be superimposed on the vapour-phase adsorption potential derived from gas-adsorption measurements on the reference substance: the multiplying factor is γ_{21}. If gas-adsorption data were available also for components 1 and 2, then the first equality of equation (105) could be tested. However, if such data

[73] M. M. Dubinin, *Chem. Rev.*, 1960, **60**, 235.

are not available, and in particular if component 2 is an involatile solid, alternative tests must be devised.

Manes and Hofer[72] studied the adsorption by activated charcoal of the dyes Sudan III and Butter Yellow from a range of organic solvents. The charcoal had been characterized by adsorption of n-heptane from the vapour and it was found that superimposition of $\varepsilon_{12}(\phi)/v$ on $\varepsilon_s(\phi)/v$ was possible only at low concentrations, although the $\varepsilon_{12}(\phi)/v$ curves themselves were approximately conformal. Although no other vapour-adsorption isotherms were measured, it was possible to test the first equality of equation (105) by considering the correlation between the variation in γ_{21} when component 1 is varied as determined with the two alternative choices of component 2. Manes and Hofer's data are plotted in Figure 10, and show reasonable

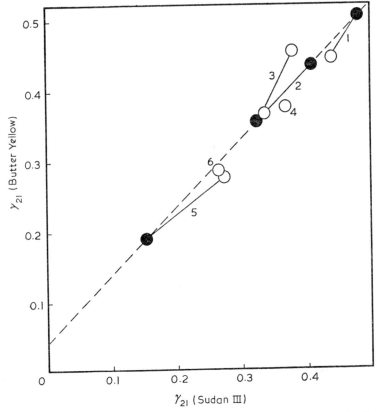

Figure 10 γ_{21} (2 = *Butter Yellow*) *as function of* γ_{21} (2 = *Sudan III*) *where component* 1 *is:* 1, *acetone*; 2, *n-heptane*; 3, *cyclohexane*; 4, *carbon tetrachloride*; 5, *benzene*; 6, *carbon disulphide. Open circles, experimental; filled circles, calculated from polarizabilities, using equation* (105)

correlation, bearing in mind the variety of chemical types represented by the solvents. The line of unit slope intersects the axis at 0.042, which should be equal to γ(Butter Yellow)$-\gamma$(Sudan III). A further test is to compare the experimental values of γ_{21} with those calculated from the right-hand side of equation (105), using in the case of the dyes the values of p calculated from atomic refractivities. The calculated values are also included in the figure and, although the experimental and theoretical values of γ_{21} for a given solvent are in only moderate agreement (as indicated by the length of the heavy lines joining the observed and calculated values), the general correlation is remarkably good.

Systems of a different kind were studied by Wohleber and Manes[74] using mixtures of sparingly soluble organic compounds in water. The active carbon was the same as that used previously. The curves of ε_{12}/v against ϕ had the same form as that for heptane vapour adsorption, and in particular exhibited the same saturation volume, thus confirming, apparently, that water is completely expelled from the adsorption space. The factors γ_{21} required to bring the observed curves into coincidence with the gas-phase heptane curve are given in Table 2. The method of analysis then used was to

Table 2 *Values of γ_{2,H_2O}, γ_2, and γ_{H_2O} for the adsorption of the substances listed from aqueous solution by activated carbon, taking gas-phase adsorption of heptane as reference system*[74]

	γ_{2,H_2O}	$\gamma_2{}^a$	γ_{H_2O}
1,2-Dichloroethane	0.80	1.12	0.32
Diethyl ether	0.62	0.91	0.29
Ethyl acetate	0.69	0.96	0.27
Methylene chloride	0.83	1.08	0.25
Propionitrile	0.70	0.95	0.25

^{a} Calculated from refractive index.

calculate γ_2 for each liquid from refractive index measurements, and hence for each system to calculate γ_{H_2O}. The data in Table 2 show that the values of γ_{H_2O} obtained for various systems are in reasonable agreement with a mean value of 0.28. However, this value is only about one-third of that calculated from the refractive index of water, and reflects the anomalous nature of intermolecular forces in systems involving water (*cf.* Chapter 5).

It is concluded that the potential theory in the form expressed by equation (105) applies to adsorption of strongly bound adsorbates from dilute solution, and once the value of γ_1 has been established, either theoretically or empirically, the theory can be used to predict the adsorption behaviour of a range of solutes.

Significant Structures Theory.—The significant structures theory of liquids[75] sets out to distinguish between various types of molecule characterized by

[74] D. A. Wohleber and M. Manes, *J. Phys. Chem.*, 1971, **75**, 61.
[75] H. Eyring, T. Ree, and N. Hirai, *Proc. Nat. Acad. Sci. U.S.A.*, 1958, **44**, 683.

their local environment: thus in the simplest case molecules can be described as 'gas-like' and 'solid-like'. This theory has been applied to gas/solid adsorption by McAlpin and Pierotti[76] and recently to adsorption from solution by Baret.[77] The development is based on calculating the rates of adsorption and desorption in terms of the fractions of gas-like and solid-like molecules in the liquid and adsorbed phases, and the energy barriers hindering the desorption of these two kinds of molecule from the surface. The condition of equilibrium is then deduced and shown to lead to an isotherm equation of the form

$$c = \frac{B n_0 n + (E - B) n^2}{A(n_0 - n)}, \tag{106}$$

where c is the solution concentration, n the amount adsorbed, n_0 the saturation capacity, and A the surface area; B and E are coefficients controlling the rates of desorption of gas-like and solid-like molecules, respectively. The change of form of this equation is discussed in terms of the variation of the parameter $\beta = (B - E)/B$. When $\beta = 1$, *i.e.* $E = 0$ and only 'gas-like' molecules take part in the equilibration process, the isotherm reduces to the linear (Henry's law) isotherm

$$c = \frac{B}{A} n. \tag{107}$$

(In the paper this form is attributed, wrongly in the view of the Reporter, to the case of $\beta = -\infty$.) Alternatively, when $\beta = 0$, and the probabilities of desorption of gas-like and solid-like molecules are equal, a Langmuir-type equation results:

$$c = \frac{B n_0}{A} \cdot \frac{n}{n_0 - n}. \tag{108}$$

These conclusions are consistent with a virial-type equation for the spreading pressure, which the author derives:

$$\frac{\pi a}{kT} = 1 + \tfrac{1}{2}(1 - \beta)\,\theta + \tfrac{1}{3}(1 - \beta^2)\,\theta^2 + \tfrac{1}{4}(1 - \beta^3)\,\theta^3 + \ldots, \tag{109}$$

(where π is the spreading pressure and $\theta = n/n_0$). However, there is some ambiguity in the author's discussion, since equation (109) is said to give the ideal equation of state when $\beta \to -\infty$, while the case of $\beta = 1$ is said to occur when the 'adsorption is saturated', thus implying that in some way β changes with the extent of adsorption. It seems to the Reporter that the case $\beta \to -\infty$ must correspond to $B \to 0$, $E \neq 0$, which means that gas-like molecules have zero probability of desorption, whereas solid-like molecules can desorb: this case seems physically improbable. These, and one or two other apparent

[76] J. J. McAlpin and R. A. Pierotti, *J. Chem. Phys.*, 1964, **41**, 68.
[77] J. F. Baret, *Kolloid-Z.*, 1971, **246**, 636.

inconsistencies in this paper, need to be cleared up before the usefulness of this approach can be assessed.

7 Studies of Specific Systems

We first review papers which provide additional experimental material to illustrate and develop the theoretical ideas outlined in Sections 2—4.

Sircar and Myers[32] have determined isotherms for adsorption on silica gel from mixtures of benzene + cyclohexane, benzene + n-heptane, cyclohexane + n-heptane, and benzene + 1,2-dichloroethane at 30 °C. The first three systems provide data which may be used to test equation (46); the first two deviate only slightly from ideality, and analysis in terms of equation (22) leads to values of K of 0.107 and 0.119, respectively, for the preferential adsorption of the first-named component. From equation (46) the value predicted for K for the third system is 1.11: the observed value is, within experimental error, equal to 1.0; no preferential adsorption from the cyclohexane + n-heptane mixture could be detected. The same experimental data are analysed by Sircar and Myers[37] in a second paper. Here K is obtained by the integration of equation (41) and values of $(\sigma_i^0 - \sigma_j^0)$ are calculated. According to this analysis, $\sigma^0(\text{benzene}) - \sigma^0(\text{cyclohexane}) = -24.5 \text{ mJ m}^{-2}$, while $\sigma^0(\text{n-heptane}) - \sigma^0(\text{benzene}) = +24.5 \text{ mJ m}^{-2}$, which again, since the sum of these is zero, predicts $K=1$ for the cyclohexane + n-heptane system. Larionov and Myers[38] present data for the three components benzene (1), iso-octane (2), and carbon tetrachloride (3) adsorbed by aerosil. Analysis was carried out in terms of equation (41) and the values of $(\sigma_i^0 - \sigma_j^0)/RT$ so obtained are given in Table 3. These figures satisfy exactly the criterion of thermodynamic con-

Table 3 *Surface tension differences* $(\sigma_i^0 - \sigma_j^0)$ *between pure liquid i/aerosil and pure liquid j/aerosil calculated from adsorption data*[37]

System[a]	$(\sigma_i^0 - \sigma_j^0)/RT$
1+2	1.325
1+3	0.754
3+2	0.571

[a] 1 = benzene; 2 = iso-octane; 3 = carbon tetrachloride.

sistency. An important feature of this work is the observation that whereas the isotherms for adsorption from benzene + carbon tetrachloride and carbon tetrachloride + iso-octane indicated that the adsorbed phase is ideal, the data for benzene + iso-octane could not be described by equation (47) 'irrespective of the non-ideality of the adsorbed phase'. This statement is, however, difficult to understand since the activity coefficients in the surface phase can, in principle, always be calculated from the experimental data through equation (40); these coefficients must then reproduce the measured values. It is of considerable interest to note, however, that even though the data cannot all be represented by a simple equilibrium constant equation, the thermodynamic consistency test, through equation (46), remains valid. The anomalous behaviour of the (benzene + iso-octane)/aerosil system is also indicated by the fact that the monolayer capacity for benzene, calculated from the limiting slope of the surface excess isotherm as x(benzene)$\rightarrow 1$ of 0.8 mmol g^{-1} does not agree with the value of 0.6 mmol g^{-1} derived from benzene-vapour adsorption studies. This discrepancy 'stresses the elusive character of the size of the monolayer' (which is) . . . 'an approximation to the actual multi-layer structure of the surface'. Even so it is concluded that the adsorbed-phase model is a reasonable and useful one.

Measurements of the adsorption by silica gel from benzene + ethanol, cyclohexane + ethanol, and benzene + cyclohexane have also been made by Sircar and Myers.[37] No numerical results are given, but it is reported that these data are not thermodynamically consistent, possibly because of chemi-sorption of ethanol by the silica gel. Further work is promised on these systems.

Sircar and Myers[32] have also studied adsorption by activated carbon from benzene + 1,2-dichlorobenzene, benzene + ethanol, and cyclohexane + ethanol mixtures. In the two latter cases, S-shaped isotherms similar to those previously reported by Schay and his co-workers were obtained. Although the fit was not perfect, these data could be represented reasonably well by employing the bulk activity coefficients and assuming the surface phase to be ideal [equation (20) with $\gamma_1^\sigma = \gamma_2^\sigma = 1$]. This confirms Schay's interpretation and indicates that Tóth's ideas of $(+, -)$ adsorption in a double-layer model are not necessary to explain the behaviour of systems of this kind.

Similar measurements, on adsorption at 30 °C by Cab-O-Sil from mixtures of ethanol + cyclohexane, benzene + cyclohexane, and ethanol + benzene are reported by Matayo and Wightman:[78] the importance of eliminating traces of water from the ethanol if meaningful results are to be obtained is stressed. All isotherms, which are of moderate precision, show preferential adsorption of the first-named component over the whole concentration range. The excess isotherms have been resolved into individual isotherms on the basis of the monolayer model, using the surface area of the solid determined by N$_2$ adsorption and cross-sectional areas of the molecules taken as benzene 32, cyclohexane 46, and ethanol 18 Å^2 molecule^{-1}. No test of thermodynamic consistency was applied; nor was any attempt made to fit the data quantitatively to any theoretical equations. The temperature coefficient of the excess isotherm for cyclohexane + ethanol was found to be zero in the range 25—35 °C.

An extensive study of adsorption by chromatographic silica gel (of unstated origin) from mixtures of cyclohexane + aromatic compounds and cyclo-hexane + cyclic ethers at 30 °C has been carried out by Kagiya, Sumida, and Tachi[79] with the objective of establishing a relationship between the extent of preferential adsorption and the chemical nature of the second component. They also set out to compare liquid-phase adsorption with that from the gas phase and to evaluate the orientation angles of the components adsorbed on

[78] D. R. Matayo and J. P. Wightman, *J. Colloid Interface Sci.*, 1971, **35**, 354.
[79] T. Kagiya, Y. Sumida, and T. Tachi, *Bull. Chem. Soc. Japan*, 1971, **44**, 1219.

the surface. They define the selective adsorptive capacity (a_2) of the preferentially adsorbed component by

$$a_2 = \frac{v_2}{(1-x_2^l)}\left(\frac{n^0\Delta x_2^l}{m}\right),\tag{110}$$

where v_2 is the molar volume of component 2. Comparison with equations (2) and (3) shows that a_2 is in fact simply $v_2 a_s \Gamma_2^{\prime(1)}$, where a_s is the specific surface area of the solid. The statement that equation (110) holds only when the component 2 is so strongly adsorbed that the surface is almost entirely covered by component 2 even at low concentrations is incorrect. They then show that x_2^l/a_2 is a linear function of x_2^l and assume that this means that the adsorption from the mixtures used was of the Langmuir type. They then show also that the data fit equation (24), apparently overlooking the fact that, apart from the factor v_2, the linear relation between x_2^l/a_2 and x_2^l is equivalent to equation (24). The parameters for equation (24) found in this work are given in Table 4.

Table 4 *Parameters[79] of equation (24) together with effective molecular areas and calculated orientation angles (between aromatic or cyclic ether ring and surface), for components preferentially adsorbed from cyclohexane by silica gel[a]*

Component	K	$n^\sigma/m \times 10^4$ mol g^{-1}	a_2 Å^2 molecule^{-1}	θ deg
Chlorobenzene	14.1	15.4	41.3	3
Benzene	16.0	18.5	34.4	0
Toluene	18.6	14.8	43.0	5
p-Xylene	19.4	15.0	42.2	11
Anisole	49.2	19.0	33.5	17
Dioxolan	37.8	40.3	15.8	77
1,4-Dioxan	63.2	27.7	23.0	90
Tetrahydrofuran	80.0	23.5	27.1	86

[a] For this silica gel, $a_s = 383$ m^2 g^{-1}.

Since presumably the adsorption of the substances under consideration depends on nucleophilic interaction of the solute with the silanol groups on the silica surface, correlations were sought between log K [which is proportional to the standard free energy of differential adsorption, equation (18)] and the ionization potential (I_p) of the aromatic hydrocarbon, and with the Hammett constant (σ_p). The Hammett constants provided a good linear correlation except for anisole, but although the ionization potentials of benzene, toluene, and p-xylene were roughly related to log K, neither chlorobenzene nor anisole showed good correlation. The failure of anisole to fit in with the correlation was taken to mean that in this case adsorption was controlled more by the oxygen atom than by the benzene ring.

Furthermore, if adsorption takes place through silanol groups, a correlation with the shift of the OH frequency in the i.r. spectrum, caused by interaction with the solute, may be expected. This was tested in two ways, first by comparing $\log K$ with the shift of the OH frequency $(\Delta\nu_{OH}^S)$ of surface hydroxy-groups of silica gel caused by adsorption of the compounds concerned from the gas phase and secondly by considering the frequency shift $(\Delta\nu_{OH}^L)$ of the OH frequency of trimethylsilanol when dissolved in the liquid concerned. In both cases excellent correlation is found, as shown in Figure 11.

Finally, these authors attempted to deduce the orientation of the adsorbed molecules at the surface by calculating the area occupied by the molecule

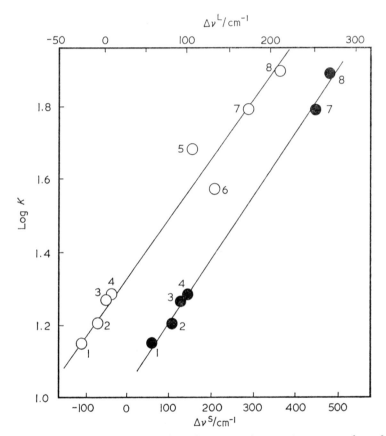

Figure 11 *Dependence of* $\log K$ *for adsorption of aromatic compounds and cyclic ethers from cyclohexane solution by silica gel on (filled circles and lower scale) shift of OH infrared band of surface silanol groups* $(\Delta\nu^S)$ *caused by adsorption of the corresponding vapour by silica gel, and (open circles and upper scale) shift of the OH infrared band of trimethylsilanol* $(\Delta\nu^L)$ *when it is dissolved in the corresponding liquid: 1, chlorobenzene; 2, benzene; 3, toluene; 4, p-xylene; 5, anisole; 6, dioxolan; 7, 1,4-dioxan; 8, tetrahydrofuran*

(Redrawn with permission from *Bull. Chem. Soc. Japan*, 1971, **44**, 1219)

from the values of n^σ/m in Table 4 and the BET N_2 surface area of the silica gel (383 m² g⁻¹): the calculated values are included in the Table. By approximating the shape of the various molecules by a cylinder, it was concluded that all the aromatic molecules except anisole lie nearly flat on the surface, whereas the cyclic ethers adsorb with the ring approximately normal to the surface. The intermediate orientation for anisole may again reflect the participation both of the O atom and the aromatic ring in the adsorption bond: that both interactions may occur is confirmed by the presence of two shifted OH bands in the infrared spectrum when trimethylsilanol is dissolved in anisole.

Kiselev and his co-workers have continued their work on the chemical factors which control adsorption phenomena by a study of the adsorption of hydrocarbons from solutions by aerosils and carbon blacks.[80] The adsorption of two hydrocarbons, 6-methylhept-1-ene and phenanthrene, from solution in n-heptane was studied. The four adsorbents were a hydroxylated (8.5 μmol OH m⁻²) and a dehydroxylated (0.7 μmol OH m⁻²) silica, and an oxidized (by $HNO_3 + H_2SO_4$ mixture) and a graphitized channel black. Measurements were performed in relatively dilute solution (to $x = 0.2$ for 6-methylhept-1-ene and to $x = 0.025$ for phenanthrene) on the rising segments of the surface excess isotherms. For the alkene the orders of adsorption were: hydroxylated $SiO_2 >$ dehydroxylated $SiO_2 >$ oxidized $C >$ graphitized $C = 0$, thus showing clearly the importance of interactions between the π-electrons of the double bond of the adsorbate molecule and surface OH and oxide groups. Phenanthrene exhibited somewhat similar behaviour, except that the order was now oxidized $C >$ hydroxylated $SiO_2 \approx$ graphitized $C >$ dehydroxylated SiO_2; this is taken to indicate that the aromatic rings of phenanthrene interact more strongly with the oxidized C and graphitic surfaces than does the single double-bond of the alkene. Evidence for the participation of aromatic rings in the adsorption is derived from measurements of the adsorption by hydroxylated silica of benzene, naphthalene, and phenanthrene from n-heptane: the fraction of surface covered at a given mole fraction in solution increases with the number of aromatic rings in the adsorbed molecule. No attempt is made in this paper, however, to analyse the data in terms of isotherm equations, or to establish any quantitative basis for comparing the various systems studied.

Somewhat similar conclusions were reached by Budkevich, Slinyakova, and Neimark,[81] who compared the adsorption of acetone, dioxan, and benzene from n-hexane solutions and of benzene, nitrobenzene, toluene, phenol, and chlorobenzene from carbon tetrachloride on silica gel and on silica xerogels containing different surface groups. Evidence for the interaction of the aromatic ring with Si—O—Si units is claimed, and this is attributed to the donor–acceptor properties of this surface grouping.

[80] A. V. Kiselev and I. V. Shikalova, *Kolloid. Zhur.*, 1970, **32**, 702 [*Colloid. J.* (*U.S.S.R.*), 1970, **32**, 588].
[81] G. B. Budkevich, I. B. Slinyakova, and I. E. Neimark, *Kolloid. Zhur.*, 1970, **32**, 17 [*Colloid. J.* (*U.S.S.R.*), 1970, **32**, 12].

An extensive study of the adsorption of nitromethane and nitrobenzene by chromatographic silica gel ($a_s = 412$ m² g⁻¹) at 20 °C and alumina ($a_s = 65.2$ m² g⁻¹) at 30 °C from solutions in benzene, cyclohexane, and dioxan is reported by Suri and Ramakrishna.[82] This work was intended to provide a test of the theoretical analyses of Schay and of Everett, outlined in the Introduction.

The data were analysed to obtain n^{σ}/m and K using the linear plot according to equation (24), and n^{σ}/m by Schay's graphical method using linear sections of the isotherms (Figure 1). Of the ten systems studied, six showed a reversal of sign of the adsorption on traversing the concentration range: for these the analysis in terms of equation (24) was limited to the branches of the isotherm at low concentrations of each component. Despite the diversity of types of behaviour exhibited by these systems, it is perhaps surprising that the values of n^{σ}/m derived from the data by the alternative routes agreed moderately well (roughly $\pm 20\%$). Certainly one does not expect equation (24) to be valid for isotherms showing a change in sign. However, in contrast to many of the systems studied by Schay, the surface areas derived from n^{σ}/m show poor agreement with those from BET N₂ isotherms, the discrepancies being by factors between 0.38 and 11.1 for the alumina sample and 0.53 to 1.66 for the silica. Most of the values are greater than unity and are interpreted by Suri and Ramakrishna as indicative of multilayer adsorption. Those less than unity occur when dioxan is one of the components. Calculations are also presented of x_1^{σ}. These were made both using equation (12), where $a_s/a = n^{\sigma}/m$ was derived from the linear plots of equation (23), and from equation (9) using values of n^{σ}/m from Schay's graphical method. Since the molecules concerned do not differ greatly in size, the surface mole fractions calculated in these two ways are in reasonable agreement.

Because of the labour involved in analysing the data according to equations (40) or (41), and the lack of reliable data on the bulk activity coefficients, no values of K calculated by this procedure are reported.

Measurements were also made of the surface tensions of these mixtures and a comparison was made of the adsorption at the free liquid surface with that at solid surfaces. Figure 12a shows the results for the benzene + nitrobenzene system (the cyclohexane + nitrobenzene system behaves somewhat similarly). The preferential adsorption of benzene at the free surface becomes progressively less marked at the alumina and silica surfaces, and in each case the surface excess isotherm shows a reversal of sign. By contrast, in the dioxan + nitrobenzene (Figure 12b) and benzene + nitromethane systems, the sign of the adsorption at the free surface is opposite to that at the oxide surface, and both oxides give almost identical isotherms when plotted as surface excess concentrations ($\Gamma_2^{(n)}$). The dioxan + nitromethane system (Figure 12c) behaves in a quite anomalous fashion, the two oxide surfaces showing opposite preferences; no explanation for this is offered.

Attention has been directed earlier to the importance of checking that values

[82] S. K. Suri and V. Ramakrishna, *Acta Chim. Acad. Sci. Hung.*, 1970, **63**, 301.

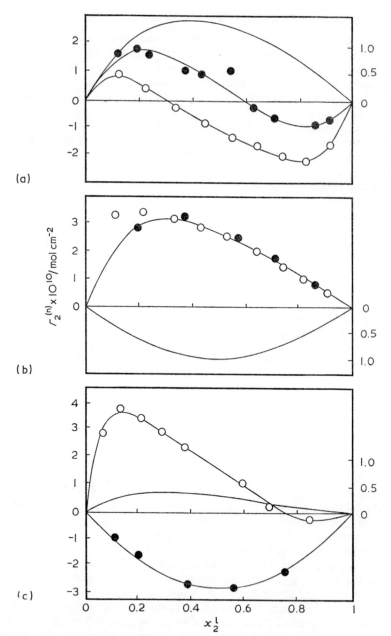

Figure 12 *Comparison of adsorption at the liquid/vapour interface (full line, no experimental points, right-hand scale), at the liquid/silica gel interface (open circles), and at the liquid/alumina interface (filled circles) for the systems: (a) nitrobenzene (1) + benzene (2); (b) nitrobenzene (1) + dioxan (2); (c) nitromethane (1) + dioxan (2)*

(Redrawn with permission from *Acta Chim. Acad. Sci. Hung.*, 1970, **63**, 301)

of x_i^σ calculated from the surface excess isotherms are thermodynamically acceptable, and it was pointed out that an inappropriate choice of a_1 and a_2 could lead to unacceptable values. This point is brought out forcibly in a paper by Suri,[83] who re-analyses the data on [benzene (1) + nitromethane (2)]/silica gel obtained in the preceding paper. He finds that when a_1 and a_2 for benzene and nitromethane are equated to molecular areas calculated from liquid-density measurements, the calculated values of x_1^σ obtained from equation (9) (with components 1 and 2 interchanged, of course) are *negative*. This arises from the fact that for this system $(n^0 \Delta x_1^l / m)$ is negative and the second term in the numerator of equation (9) exceeds the first. Choosing a different value for the molecular area of benzene (corresponding to a flat molecular orientation) does not improve the situation [this is indeed obvious from equation (9)], whereas halving the molecular area of nitromethane gives acceptable positive values of x_1^σ. This, it is suggested, could mean that the nitromethane is adsorbed as dimers from benzene and that these dimers are oriented normal to the surface and occupy, per molecule of nitromethane, one half the molecular area. Strictly speaking, a somewhat more elaborate calculation is needed to take proper account of the effect of dimerization.

The adsorption on to a series of chromium oxide (Cr_2O_3) samples from binary mixtures of cyclohexane and benzene at 30 °C has been studied by Narang and Ramakrishna.[84] In all cases S-shaped isotherms were found. These were analysed to obtain values of (n^σ / m) using the graphical method, and through equation (24) using results at each end of the concentration range. Surprisingly, as Suri and Ramakrishna had found previously, equation (24) gives linear graphs over limited ranges at each end of the concentration range and the values of (n^σ / m) derived from the two limiting regions for a given isotherm are mostly in very reasonable agreement ($\pm 25\%$), and the same is true of the results using the graphical intercept method. The data show a moderate degree of scatter, and the ratio of the surface areas derived from the adsorption from solution measurements and those from gas-adsorption measurements (BET N_2) vary from 0.62 to 1.54 for the first method of analysis and from 0.78 to 1.87 for the second. Whether this is sufficient evidence to conclude, as the authors do, that adsorption in this system extends from 1.5 to 2 monolayers may be doubted.

Other reported measurements of adsorption from completely miscible pairs of liquids include the work of Radulescu, Raseev, and Ilea[85] on the adsorption of paraffins from solution in benzene, n-heptane, and iso-octane by activated carbon. The data are said to have been correlated by calculating the free surface energy of the solid in contact with pure solvent, these values giving the trend of adsorption from this solvent. A formula is quoted for the

[83] S. K. Suri, *J. Colloid Interface Sci.*, 1970, **34**, 100.
[84] K. C. Narang and V. Ramakrishna, *Indian J. Chem.*, 1970, **8**, 923.
[85] G. A. Radulescu, S. Raseev, and M. Ilea, *Petrol. Gaze (Bucharest)*, 1970, **21**, 666 (*Chem. Abs.*, 1971, **74**, 116300).

evaluation of this free energy from measurements of the heat of immersion of the solid in the solvent, and the surface tension of the solvent, but in the abstract not all the terms in the equation are defined.

An extensive study of adsorption from ethanol + benzene, ethanol + water, acetone + water, and acetone + chloroform by two types of active carbon, a channel black, and a zeolite, has been made by Rozin and Komarov.[86] One of the main objects of the work was to investigate pore size effects in determining the selectivity of the adsorption from various systems; only an abstract of this work is readily available.

The adsorption of n-aliphatic acids by silica gel from solutions in non-polar solvents has been examined by Zeliznyi, Litkovets, and Grushchak,[87] who have evaluated, by methods not explained in the abstract, the area per molecule, the thickness of the adsorbed layer, and the limiting adsorption volume for the acids $CH_3(CH_2)_nCO_2H$ with $n = 0, 1, 2, 5$, or 8 from n-heptane, carbon tetrachloride, and benzene. Each acid exhibited its greatest area in n-heptane and smallest area in carbon tetrachloride; the spread of areas was greatest for the acids of lowest and highest molecular weight, and passed through a minimum at intermediate molecular weights. These effects are discussed in terms of the competition of the acids and solvents of varying polarity for adsorption sites.

Reliable data on the adsorption from mixtures of molecules of different size, which might be employed to study in more detail the theoretical concepts outlined in sections 1, 2, and 4, are few. The recent study by Parfitt and Thompson[88] on the systems (n-heptane + n-hexadecane)/Graphon and (n-heptane + n-hexadecane)/rutile goes some way to making good this deficiency. Special attention was paid to the factors which lead to inaccuracies in experimental measurements of adsorption from solution and a technique was devised to overcome most of them. Because of the rather small adsorptions in these systems, a high solid:liquid ratio was used in the adsorption bulbs and long times (several days) were needed to achieve equilibrium. Data were obtained at 25, 35, and 45 °C for adsorption by Graphon and at 25 and 45 °C for rutile; the maximum of the surface excess concentration isotherms reached $1.1 - 1.3 \times 10^{-6}$ mol m^{-2} for Graphon and $0.3 - 0.6 \times 10^{-6}$ mol m^{-2} for rutile; the mean deviation of the experimental points from a smooth curve amounted to roughly $\pm 3 \times 10^{-8}$ or $\pm 2.5 \%$ of the maximum adsorption for Graphon, $\pm 7 \%$ for rutile. The data were analysed according to equation (41) to obtain the equilibrium constants (K') given in Table 5; these were then used to obtain the enthalpy difference ΔH^0 which, according to equation (18), should be equal to the differences between the enthalpies of formation of the n-heptane/solid and n-hexadecane/solid interfaces of area a_1, and hence to the difference in the enthalpies of immersion of solid in the

[86] A. T. Rozin and V. S. Komarov, 'Org. Katal.', ed. N. S. Kozlov, Nauka Teknika, Minsk, U.S.S.R., 1970, p. 43 (*Chem. Abs.*, 1971, **74**, 116263).
[87] A. M. Zeliznyi, E. A. Litkovets, and V. T. Grushchak, *Ukrain. khim. Zhur.*, 1971, **37**, 337 (*Chem. Abs.*, 1971, **75**, 67854).
[88] G. D. Parfitt and P. C. Thompson, *Trans. Faraday Soc.*, 1971, **67**, 3372.

Table 5 *Equilibrium constants for adsorption of hexadecane from n-heptane by Graphon and by rutile*[88]

$t/^{\circ}C$	K'	
Graphon		
25	9.7 ± 0.4	
35	7.5 ± 0.9	$\Delta H^0 = -28$ kJ mol^{-1}
45	4.9 ± 0.3	
Rutile		
25	2.0 ± 0.1	$\Delta H^0 = -6.3$ kJ mol^{-1}
45	1.7 ± 0.3	

two pure liquids. The values obtained* for Graphon are -6.6 kcal mol^{-1} (28 kJ mol^{-1}) and 1.5 kcal mol^{-1} (6.3 kJ mol^{-1}); when converted to unit area of surface (taking a_1 for n-heptane as 50 Å^2) these give -93 and -21 mJ m^{-2} (or erg cm^{-2}), respectively. If we take these values to refer to the mean temperature of 35 $^{\circ}$C, then the difference between them may be compared with that between the enthalpies of immersion at this temperature. For Graphon at 25 $^{\circ}$C, $\Delta_w H(C_{16}) - \Delta_w H(C_7) = -103$ mJ m^{-2};[89] this quantity is, however, expected to decrease in magnitude with increasing temperature and may be estimated to have a value of about -80 mJ m^{-2} at 35 $^{\circ}$C, in satisfactory agreement with ΔH^0 obtained by Parfitt and Thompson. For rutile, no data are available for n-C$_{16}$H$_{34}$, but Parfitt and Thompson quote figures which suggest that $\Delta_w H$ for the immersion of rutile is much less dependent on chain length than for Graphon, so that the lower value of ΔH^0 seems reasonable. Since the differences between the enthalpies of immersion of Graphon in heptane and hexadecane show a strong dependence on temperature, whereas the present data if anything indicate an opposite dependence of ΔH^0 on temperature, the authors conclude that the heat of adsorption calculations cannot be compared directly with the heat of immersion data and they speculate as to the influence of the structure of the interfacial layer on such a comparison. However, in view of the relatively large uncertainty in the values of ΔH^0, this discussion seems to be somewhat premature, and further conclusions must await the availability of more extensive data.

The use of calorimetric methods for the study of adsorption from solution is exemplified by recent papers by Groszek on the adsorption of n-dotria-contane and n-butanol from n-heptane and iso-octane by a variety of graphites,[90] and by Allen and Patel[91] on the adsorption of n-alkanols by several inorganic powders. In both cases a flow-microcalorimeter was employed to determine enthalpies of adsorption from dilute solutions ($x_2 < 0.02$), and these results were combined with adsorption isotherm measurements.

---

* The data are insufficient to derive, as the authors do, any information regarding the temperature dependence of ΔH^0.

[89] J. H. Clint, J. S. Clunie, J. F. Goodman, and J. R. Tate, *Nature*, 1969, **223**, 51; D. H. Everett and G. H. Findenegg, *ibid.*, p. 52; *J. Chem. Thermodynamics*, 1969, **1**, 573.
[90] A. J. Groszek, *Proc. Roy. Soc.*, 1970, **A314**, 473.
[91] T. Allen and R. M. Patel, *J. Appl. Chem.*, 1970, **20**, 165.

Groszek found that, on graphite, adsorption of n-$C_{32}H_{66}$ from n-heptane followed the Langmuir-type equation to which equation (21) reduces in very dilute solutions. The monolayer capacity of the graphites determined in this way, when compared with the surface areas of the solids as obtained by N_2 adsorption, led to a molecular area of 1.78 nm^2 for the $C_{32}H_{66}$ molecule. It was then shown that by only a relatively small distortion of bond angles (leading to a shortening of the molecule), an n-paraffin molecule in the extended zig-zag form fits exactly on to the basal plane of graphite with one hydrogen of each CH_2 group exactly over the centre of a hexagon ring. The area occupied by $C_{32}H_{66}$ on this basis is 1.77 nm^2, in close agreement with the observed value. That there should be preferential adsorption of the longer-chain hydrocarbon is explained by supposing that (as indicated by molecular models) the adsorption of $C_{32}H_{66}$ (involving the formation of 32 H/graphite ring contacts) is accompanied by the desorption of about $3\frac{1}{2}$ molecules of n-heptane and to the breaking of 24.5 H/graphite ring contacts. If the solvent is iso-octane, which can only make 3 contacts per molecule, and which occupies about $\frac{3}{8}$ the area of $C_{32}H_{66}$, the differential energy of adsorption of $C_{32}H_{66}$ from iso-octane should be some three times that from n-heptane.

Flow-calorimetric data on the integral enthalpy of adsorption as a function of solution concentration also followed a Langmuir-type isotherm, the enthalpy being directly proportional to the amount adsorbed over much of the concentration range. However, at very low concentrations the enthalpies were a function of concentration, and in the low-coverage range that for iso-octane was about four times the value for n-heptane. Adsorption of the hydrocarbon was assumed to be on the basal plane of graphite and the enthalpy data for graphites of various kinds enabled estimates to be made of the surface area of basal plane exposed; for ground graphites these were substantially lower than the BET areas. However, these graphites are expected to possess a substantial proportion of 'polar' sites which adsorb n-butanol preferentially. Measurements of the enthalpy of adsorption of n-butanol from n-heptane may therefore be used to estimate the area attributable to 'polar sites'. Graphon, oleophilic graphite, and synthetic graphite before grinding exhibited enthalpies of adsorption of n-butanol of less than 0.3 J g^{-1}, whereas graphites which had been ground *in vacuo* had enthalpies rising to 15 J g^{-1}, depending on the duration of the grinding process. These enthalpies were directly proportional to the amounts of n-butanol adsorbed, thus indicating that the chemical nature of the 'polar sites' is independent of the duration of grinding. It was concluded that ground graphites possess distinct polar and non-polar sites with characteristic adsorptive properties capable of adsorbing specific solutes independently from dilute solutions. The non-polar sites are identified with the basal plane of graphite, but no firm suggestions are made as to the nature of the polar sites.

The work of Allen and Patel[91] follows broadly the same general pattern as that of Groszek. Enthalpy of adsorption measurements were made using a flow microcalorimeter, after a series of experiments had checked the performance and calibration procedures for the instrument. Following Groszek,[92] the flow microcalorimeter was also used in the determination of the adsorption isotherms. Four powders, α-Fe_2O_3, $BaSO_4$, TiO_2 (anatase), and NaF were studied in relation to their adsorbent properties towards a series of n-alkanols containing from 1 to 12 carbon atoms, from dilute solution in n-heptane.

It was found that the enthalpy of adsorption was always directly proportional to the amount adsorbed, and that both quantities were related to the solution concentration by a Langmuir-type equation. For all four powders the enthalpy of adsorption per unit mass of powder, when plotted as a function of the chain length of the alcohol (n = number of carbon atoms), exhibited a maximum for $n = 6$. This appears to be the consequence of the superposition of two factors: the amount of solute adsorbed (expressed in mol g^{-1}) falls off with increase in n while the enthalpy of adsorption (in energy mol^{-1}) increases with increase in n. Both curves have a maximum gradient for n in the region of 6—8. Table 6 gives the areas occupied by one molecule and the integral energies of adsorption, for adsorption on α-Fe_2O_3; detailed data are not reported for the other powders.

Table 6 *Adsorption[91] of alcohols by α-Fe_2O_3 (BET, N_2 area = 3.45 m^2 g^{-1}) from n-heptane[a]*

	n^σ/m mol $g^{-1} \times 10^{-5}$	a Å^2 molecule^{-1}	$\Delta_a h$ kJ mol^{-1}	$\Delta_a h$ mJ g^{-1}
C_2H_5OH	2.72	21.1	15.3	416
n-C_4H_9OH	2.56	22.4	19.8	506
$C_5H_{11}OH$[b]	2.23	25.7	24.6	549
n-$C_6H_{13}OH$	2.08	27.6	28.5	593
n-$C_7H_{15}OH$	1.63	35.2	31.3	510
n-$C_8H_{17}OH$	1.20	47.8	35.8	429
n-$C_{10}H_{21}OH$	1.08	53.1	36.9	398
$C_{16}H_{23}OH$[b]	0.96	59.6	37.4	359
n-$C_{18}H_{37}OH$	0.95	60.5	37.4	355

[a] Data are for monolayer coverage at unstated temperature (presumed to be room temperature).
[b] Isomer not stated.

Vorontsov, Kuznetsova, and Eltekov[93, 94] have studied the adsorption of the 2,3-dinitrophenylhydrazones of acetone, methyl ethyl ketone, and methyl propyl ketone from solution in ethanol and carbon tetrachloride by graphitized carbon black. Experiments were carried out at 0, 20, and 40 °C: the solubility of the solids did not exceed mole fractions greater than 2.5×10^{-3}. At the highest temperature the isotherms were of the Langmuir-type, approaching a saturation value at a mole fraction about 0.4 of the saturation value; at lower temperatures the isotherms exhibited a point of inflexion with a

[92] A. J. Groszek, S.C.I. Monograph No. 28, 1968, p. 174.
[93] V. V. Vorontsov, L. P. Kuznetsova, and Yu. A. Eltekov, *Kolloid. Zhur.*, 1970, **32**, 32 [*Colloid. J. (U.S.S.R.)*, 1970, **32**, 25].
[94] V. V. Vorontsov, L. P. Kuznetsova, and Yu. A. Eltekov, *Kolloid. Zhur.*, 1970, **32**, 354 [*Colloid. J. (U.S.S.R.)*, 1970, **32**, 293].

nearly horizontal tangent before rising sharply as the solubility limit was reached. The plateau adsorption, or that corresponding to the inflexion, was identified with monolayer coverage and, taking the surface area of the carbon from gas adsorption, the molecular areas of the hydrazones were calculated and compared with values estimated in various ways using molecular models. Only at low temperatures does a dense monolayer form; with the MEK hydrazone there is evidence for the formation of two adsorbed layers from ethanol. At higher temperatures the plateau or inflexion point indicates that the adsorbed layer is a mixture, the amount of solvent present being greater in the case of carbon tetrachloride: this is consistent with the fact that carbon tetrachloride is preferentially adsorbed by graphitized carbon from mixtures with ethanol. A correlation is suggested between solubility and the adsorption: the adsorption of a given substance is greater from the solvent in which it is least soluble; on the other hand, for this series of chemically similar solutes in the same solvent, the most soluble substance is the most strongly adsorbed.

Approximate values of the isosteric enthalpies of adsorption were derived from the temperature coefficient of the adsorption at constant solution mole fraction: the values at about 35% monolayer coverage for the systems studied are given in Table 7. Thus the most soluble of the three substances studied, MEK hydrazone, exhibits both the highest adsorption and highest enthalpy of adsorption.

Table 7 *Isosteric enthalpies of adsorption of hydrazones from solution by graphitized carbon black*

Hydrazone	$\Delta_a h/\text{kJ mol}^{-1}$ Solvent	
	C_2H_5OH	CCl_4
Acetone	21.8 ± 2.9	17.6 ± 1.7
Methyl ethyl ketone	36.8 ± 2.9	27.6 ± 2.5
Methyl n-propyl ketone	25.1 ± 2.5	18.8 ± 1.7

Comparison of the adsorption of MEK 2,4-dinitrophenylhydrazone on graphitized carbon black and on ungraphitized oxidized material[93] showed that graphitization increases the adsorption from ethanol: it is suggested that this sharp increase in adsorption arises from the decrease in the part played by specific interaction of solvent molecules with the carbon black, and to an increase in the role of non-specific dispersion force interactions of hydrazone molecules with the graphitized surface.

The dependence of the adsorption of sparingly soluble substances on the nature of the system is controlled by the interplay between solute–surface, solvent–surface, and solute–solvent interactions. According to Martynov and Maevskaya[95] and Kurlyandskaya[96] the adsorption in the linear region

[95] Y. M. Martynov and B. M. Maevskaya, *Zhur. fiz. Khim.*, 1970, **44**, 1534 (*Russ. J. Phys. Chem.*, 1970, **44**, 855).
[96] I. I. Kurlyandskaya, *Zhur. fiz. Khim.*, 1971, **45**, 1169 (*Russ. J. Phys. Chem.*, 1971, **45**, 655).

of the isotherm at low concentrations can be related to the properties of the components of the liquid phase by an equation of the form

$$RT \ln \frac{\Gamma_2}{\gamma_2} + \lambda_2 - r\lambda_1 = b_1(T_{b,2}^{\frac{1}{2}} - T_{b,1}^{\frac{1}{2}}) + b_2, \tag{111}$$

where b_1 and b_2 are empirical constants. Γ_2 is the adsorption coefficient (*i.e.* limiting slope of the adsorption isotherm) and γ_2 the activity coefficient in infinitely dilute solution; λ_1 and λ_2 are the latent heats of evaporation of the components (if component 2 is a solid, then λ_2 is the heat of sublimation) and $T_{b,1}$ and $T_{b,2}$ their normal boiling points. The parameters b_1 and b_2 are characteristic of the solid and the solvent concerned, so that knowledge of the heat of evaporation of the solute, and its normal boiling point, should enable the initial slope of the adsorption isotherm to be calculated. This equation was derived earlier by Kurlyandskaya and Martynov,[97] and is based on the ideas of regular solution theory.

Martynov and Maevskaya have now analysed earlier reported data for the adsorption of the halides of Ca, Al, Sb, and Fe by silica gel from solution in $SiCl_4$, $TiCl_4$, and $SnCl_4$ in the concentration range up to about 10^{-3} mass percent and have obtained values of b_1 and b_2 for these solvents. Kurlyandskaya has extended the measurements to the adsorption of $AsCl_3$, $SbCl_3$, and $POCl_3$ by activated carbon from solution in $SiCl_4$ at temperatures in the range -23 to $+20$ °C. On the basis of the parameters b_1 and b_2 derived for this solvent, the adsorption of a series of chlorides is predicted. The existing experimental evidence suggests that equation (111) gives a satisfactory account of the adsorption of chlorides from solution in $SiCl_4$ by silica gel and by activated carbon, and of alcohols and aliphatic acids from aqueous solution by activated carbon.

Ershova and Martynov[98] have studied the adsorption of the complexes PCl_5AlCl_3 and PCl_5BCl_3 from silicon tetrachloride solutions by silica gel in the temperature range -10 to $+22$ °C and have derived enthalpies of sorption from their measurements. The enthalpy of sorption data are summarized in Table 8.

Table 8 *Isosteric enthalpies of adsorption of chlorides and complexes by silica gel*[96-98]

Solvent	$\Delta_a h$/kJ mol^{-1}						
	BCl_3	$AsCl_3$	$SbCl_3$	$POCl_3$	PCl_5AlCl_3	PCl_5BCl_3	$TiCl_4$
$SiCl_4$	12.1	17.6	23.0	18.8	67.0	43.9	6.3
CCl_4	—	—	—	—	77.4	40.2	—

8 Adsorption from Multicomponent Systems

In view of the difficulties still to be overcome in the understanding of adsorption by solids from binary solutions, it is not surprising that relatively little

[97] I. I. Kurlyandskaya and Y. M. Martynov, *Zhur. fiz. Khim.*, 1966, **40**, 872 (*Russ. J. Phys. Chem.*, 1966, **40**, 468).

[98] G. G. Ershova and Y. M. Martynov, *Zhur. priklad. Khim.*, 1971, **44**, 1882.

progress has so far been made in fundamental work on multicomponent systems. Because of the potentially very large number of measurements needed to characterize such systems, it is important that experimental work in this field should be carefully planned to achieve maximum effectiveness. The problems involved are discussed by Gryazev *et al.* in a series of papers.[99-101] The first two are concerned with the planning of experiments on the adsorption from the ternary mixture of propionic acid + stearic acid + decalin by diatomite, and from acetic acid + lauric acid + n-decane by silica gel. The data obtained enabled three-dimensional adsorption isotherms to be constructed using in each case the concentrations of the two acids as the independent parameters. In general the addition of a second acid leads to a decrease of the adsorption of the other acid. The third paper deals with the four-component system acetic acid + lauric acid + isopropylbenzene + n-decane adsorbed by silica gel, with special reference to the adsorption of isopropylbenzene, which is decreased strongly by the addition of acetic acid, although addition of isopropylbenzene is said to have relatively little effect on the adsorption of acetic acid or lauric acid. Empirical equations are derived to represent the experimental data.

The use of multicomponent systems is suggested empirically for certain purposes. For example, Poon and Sung[102] find that the adsorption of glucose from aqueous solution is enhanced in the presence of phenylphosphoric acid. Studies were also made of the quaternary system including peptone.

A fuller understanding of adsorption from multicomponent liquid mixtures is clearly desirable and may well suggest other practical applications.

[99] N. N. Gryazev, M. N. Rakhlevskaya, and L. P. Shepeleva, *Zhur. fiz. Khim.*, 1970, **44**, 491 (*Russ. J. Phys. Chem.*, 1970, **44**, 272).

[100] N. N. Gryazev, G. A. Rumyantseva, and M. N. Rakhlevskaya, *Zhur. fiz. Khim.*, 1970, **44**, 2107 (*Russ. J. Phys. Chem.*, 1970, **44**, 1198).

[101] N. N. Gryazev, M. N. Rakhlevskaya, and G. A. Rumyantseva, *Doklady Akad. Nauk S.S.S.R.*, 1971, **198**, 876 (*Doklady Phys. Chem.*, 1971, **198**, 480).

[102] C. P. C. Poon, and Maw-Hao Sung, *Water Resources Bull.*, 1969, **5**, 39.

3
Polymer Adsorption at the Solid/Liquid Interface

BY S. G. ASH

1 Introduction

Interest in polymer adsorption stems partly from the diversity of current and potential applications of the adsorption process, in industry, technology, and medicine, and partly from an academic fascination with systems which lend themselves to experimental investigation and theoretical modelling.

Typical fields of application include flocculation and dispersion of colloidal material (*e.g.* effluents, pigments, emulsions) (see Chapters 5 and 6), adhesion,[1] lubrication,[2] surface treatment (*e.g.* modifying wetting properties, imparting anti-static properties), and membrane technology.

An understanding of adsorption is important in chromatography, either for the separation of polymers, or for the separation of smaller molecules, where polymers adsorbed on a substrate may be used as the stationary phase.[3]

An increasing awareness of the role of interfaces in biology and medicine has been accompanied by a rapid growth of research into the adsorption of, *e.g.*, proteins, polysaccharides, lipids on to, *e.g.*, cell walls, capillary walls, and artificial vessels.[4]

Fundamental studies of polymer adsorption give insight into the nature of the forces acting both within the bulk solution and between polymer or solvent and the solid adsorbent.

Adsorption of polymers is distinguished from adsorption of smaller molecules in several ways. First, the number of conformations which a molecule can adopt at an interface increases rapidly with increasing size. Small molecules may be regarded as non-flexible, or as having limited flexibility, such that only their orientations relative to the surface are counted as different configurations. With polymer molecules, flexibility is very important and the conformations of polymer molecules at an interface are a key feature in polymer adsorption. These conformations largely determine the configurational entropy and enthalpy of adsorbed polymers, and these in turn are

[1] M. E. Schrader and A. Block, *J. Polymer Sci., Part C, Polymer Symposia*, 1971, **34**, 281.
[2] E. S. Forbes, A. J. Groszek, and E. L. Neustadter, *J. Colloid and Interface Sci.*, 1970, **33**, 629.
[3] A. V. Kiselev, N. V. Kovaleva, O. G. Kryukova, and V. V. Khopina, *Kolloid. Zhur.*, 1970, **32**, 527 [*Colloid J. (U.S.S.R.)*, 1970, **32**, 438].
[4] L. Vroman and A. L. Adams, *J. Polymer Sci., Part C, Polymer Symposia*, 1971, **34**, 159.

needed to calculate the extent of adsorption, experimentally the most easily accessible parameter.

Secondly, the larger the molecule the greater the number of possible adsorption contacts with the surface per molecule. This results in a large net adsorption energy for polymer molecules, even if the individual contacts are weak.

Thirdly, the large size of polymer molecules in solution results in slow diffusion to the surface and magnifies the importance of adsorbent porosity.

Finally, because of the topological nature of polymers, there is a high probability that adsorbed polymers remain in metastable entangled conformations at the surface, rather than adopt conformations of lowest energy.

2 Theory

Theories of adsorption of polymers at interfaces have been reviewed by Patat *et al.*[5] and by Stromberg.[6] Some aspects of the theory have been summarized by Silberberg.[7]

The majority of theories for polymer adsorption include the following assumptions: (i) The detailed structure of a polymer is represented by a flexible chain of statistical elements (segments). (ii) All or some of these segments can adsorb at the interface. (iii) The interface is plane and structureless. (iv) An adsorbed polymer consists of an alternating sequence of two-dimensional trains of segments in contact with the surface and three-dimensional loops of segments extending away from the surface. (v) The system is in thermodynamic equilibrium. This last assumption applies only to those theories which are solved by equilibrium statistical mechanics. In a Monte Carlo study of polymer adsorption[8] true equilibrium is not necessarily achieved, and metastable entangled polymer conformations can be studied. Many theories are concerned with the adsorption of isolated polymer molecules, while others include assumptions about the excluded volume effect, polymer–solvent interactions, and the distribution of surface adsorption sites.

The problem is to calculate the most probable configuration or configurations of the adsorbed polymer molecules, thence their chemical potential. Equating this potential to the chemical potential of the polymer molecules in the bulk solution gives the adsorption isotherm.

Much of the debate in the 1960's has centred on two problems: correct counting of the number of *a priori* equally probable states of an adsorbed polymer, and the approximations in evaluating the partition function for the system. There is now considerable agreement concerning these two points.

For an isolated adsorbed polymer molecule, theories predict (i) there is a wide distribution of train and loop sizes; (ii) low adsorption energies favour

[5] F. Patat, E. Killmann, and C. Schliebner, *Fortschr. Hochpolymer. Forsch.*, 1964, **3**, 332.
[6] R. R. Stromberg, in 'Treatise on Adhesion and Adhesives', ed. R. L. Patrick, Marcel Dekker, New York, 1967, Vol. 1.
[7] A. Silberberg, *J. Polymer Sci., Part C, Polymer Symposia*, 1970, **30**, 393.
[8] E. J. Clayfield and E. C. Lumb, *J. Colloid and Interface Sci.*, in the press.

larger loops, and shorter trains; high energies favour smaller loops and longer trains; (iii) the number of segments in contact with the surface rises sharply with increase in energy at low adsorption energies and increases much less rapidly at higher adsorption energies; (iv) loop and train size are independent of molecular weight; (v) high polymer flexibility favours more shorter trains and loops.

However, it is also predicted that the adsorption isotherm has a very steep initial gradient, and that in most practical cases measurements are made not on molecules which can be considered isolated, but on adsorbed layers of high surface coverage. It is in this region of high surface coverage, where polymer–polymer interactions are important, that much of the progress has been made in the theory of polymer adsorption during the period covered by this Report. These interactions comprise two parts: a repulsive interaction arising from excluded volume effects, and an attractive force arising from the van der Waals attraction between atoms.

Hoeve has extended his treatment of polymer adsorption to include excluded volume effects[9] and heat of mixing.[10] The change in free energy for the adsorption of N_p polymer chains is composed of three contributions:

$$\Delta F = \Delta F_1 + \Delta F_2 + \Delta F_3 \tag{1}$$

ΔF_1 arises from polymer–adsorbent (and solvent–adsorbent) interactions, and the changes in internal configurational entropy on adsorption. ΔF_2 is the change in the free energy of mixing of the first adsorbed layer of segments relative to infinite dilution, and ΔF_3 is the free energy of mixing of the loops relative to infinite dilution, and includes a contribution from excluded volume effects.

In a good solvent, when the Flory expansion factor α is large, the loop size is small and most of the segments are in the first adsorbed layer. Under these conditions the theory is inaccurate because the entropy of adsorption then depends on the finer details of the molecular configuration of the adsorbed polymer which are omitted from the model. The model is applicable only for values of α close to unity, when the contribution of excluded volume effects to ΔF_3 is small and may be neglected.

The partition function, Q, for the system containing N_s polymer molecules in a solution of volume V and N_p adsorbed molecules in an adsorption volume $A\delta$, where A is the surface area and δ is the surface layer thickness (*i.e.* thickness of adsorbed trains) is written as:

$$Q = \sum_{\text{all configs.}} \exp\left(-\Delta F/kT\right)(A\delta)^{N_p} V^{N_s}/N_p! N_s! \tag{2}$$

and ΔF is given by equation (1). Selecting the maximum term in Q, subject to the condition $N_p + N_s = \text{constant}$, gives the average characteristics of the adsorbed polymers and the adsorption isotherm in terms of the energy parameters and polymer flexibility.

[9] C. A. J. Hoeve, *J. Polymer Sci., Part C, Polymer Symposia*, 1970, **30**, 361.
[10] C. A. J. Hoeve, *J. Polymer Sci., Part C, Polymer Symposia*, 1971, **34**, 1.

E

The trends for adsorption at low coverages, where chain interaction is neglected, have been studied previously and are summarized below:

1. Adsorption is proportional to solution concentration. The logarithm of the initial gradient of the adsorption isotherm is proportional to the polymer molecular weight.

2. For a given differential free energy of adsorption per segment ($-kT \ln \sigma$ in Hoeve's notation) there is a critical flexibility, c, beyond which adsorption takes place. Likewise for a given value of c there is a critical value of σ.

3. Immediately above the critical value of σ, the fraction of adsorbed segments, p, in a molecule increases rapidly before levelling off at higher values of σ. At these higher values of σ the average loop size is small and train length large.

4. The number of segments adsorbed is proportional to molecular weight at constant c and σ.

Results for the region of surface coverage where chain interactions become important are:

1. Adsorption rises very steeply at low solution concentrations and then tends to level off, although no horizontal plateau is reached and the gradient remains positive.

2. The adsorption of segments in the surface layer reaches a limiting value, and the slowly increasing adsorption with increasing solution concentration results from an increase in the number of segments situated in loops.

3. The higher the molecular weight the greater the weight of polymer adsorbed in the 'levelled-off' region. This arises from an increased volume of polymer in the loops.

4. At the θ point the size of the loops increases beyond bounds with increasing molecular weight. The mean-square distance of segments from the surface, $\langle z^2 \rangle$, is approximately proportional to the molecular weight. For better-than-θ solvents there is a limiting value of $\langle z^2 \rangle$ as the molecular weight increases. The better the solvent or the more flexible the molecule the smaller the limiting value of $\langle z^2 \rangle$ and weight adsorbed. For higher molecular weight polymers adsorbed from better-than-θ solvents the isotherms resemble Langmuir isotherms. The limiting adsorption in this case is caused by the mutual repulsion of polymer loops giving rise to an osmotic effect.

Hesselink[11,12] has studied the density distribution of segments of an adsorbed polymer. By combining the results of Hoeve for the loop size distribution of an adsorbed homopolymer with a previously obtained expression for the density distribution of segments within a loop, Hesselink obtained for the density distribution $\rho_h(x)$:

$$\rho_h(x) = 2a\sqrt{6}(\bar{i}l)^{-1} \exp(-2ax\sqrt{6}/\bar{i}l) \tag{3}$$

where $\bar{i}l$ is the average loop length and a is a numerical constant ≈ 0.7. The exponential decrease of ρ_h with distance from the surface agrees with other theoretical treatments.

[11] F. Th. Hesselink, *J. Phys. Chem.*, 1969, **73**, 3488.
[12] F. Th. Hesselink, *J. Phys. Chem.*, 1971, **75**, 65.

Hesselink also calculated the density distribution for a random copolymer consisting of random length chains of non-adsorbed segments separated by firmly anchored adsorbing segments. The distribution shows a maximum slightly nearer the surface than that found for a single loop. The tail of the distribution at larger distances from the surface is much less steep than for single loops.

Hoffmann and Forsman[13] have modified an earlier theory[14] taking into consideration new evidence concerning the segment density distribution. These authors criticize previous theories on the grounds that the parameters of the theories are not easily related to experimentally measurable parameters. The free energy of an adsorbed polymer, ΔF, is equated to the sum of three parts:

$$\Delta F = \Delta F_m + E_i - T\Delta S_c \qquad (4)$$

where ΔF_m is the free energy of mixing of solvent and segments for an adsorbed molecule, E_i is the surface interaction energy, and ΔS_c is the configurational entropy. Two different models of this segment density distribution, ρ, are used:

1. A double-Gaussian model giving:

$$\rho = \rho_0 [\exp\{-\beta^2(x^2+y^2)/\alpha_{xy}\}][A \exp\{-z^2/B^2\} + (1-A) \exp\{-z^2/C^2\}] \quad (5)$$

where z is the distance normal to the surface, x, y are distances parallel to the surface, β is a scaling factor, α_{xy} is the expansion factor to account for solvent interactions, and A, B, and C are independent parameters which can alter the profile form. When $A = 0$ the profile is Gaussian; when B is large and C small the distribution resembles that used by Hoeve. At high surface coverages the segment density parallel to the surface was assumed to be uniform ($\beta = 0$).

2. A two-step function, also assuming a uniformly covered surface, giving:

$$\rho = \rho_1 \quad \text{for} \quad 0 \leqslant z \leqslant a$$
$$\rho = \rho_2 \quad \text{for} \quad a < z \leqslant b$$
$$\rho = 0 \quad \text{for} \quad b < z$$

where ρ_1, ρ_2, and a are independent parameters.

Expressions for ΔF_m were derived from Flory–Huggins theory, and ΔS_c was calculated from the probability density function for the radius of gyration of the random flight chain, using the above models to express the z-component of the radius of gyration in terms of independent parameters. The configurational integral was maximized with respect to the model parameters. The parameter values giving the minimum free energy were used to calculate the equilibrium segment density profile and chemical potential of adsorbed polymers.

[13] R. F. Hoffmann and W. C. Forsman, *J. Polymer Sci.*, Part A-2, *Polymer Phys.*, 1970, **8**, 1847.
[14] W. C. Forsman and R. Hughes, *J. Chem. Phys.*, 1963, **38**, 2130.

At low surface coverages the model predicts a dense layer of segments close to the surface and a diffuse layer extending out into solution, in agreement with other theories. There is no correlation between the number of segments adsorbed and molecular weight, adsorption energy, or Flory parameter, possibly because of limitations in the numerical methods used to solve the equations. Adsorption rises very steeply at low solution concentrations, in agreement with experimental observations.

At high surface coverages the density distribution normal to the surface tends to a single Gaussian form and the molecular dimensions of the adsorbed molecules approach those of polymers in solution. The increased ratio of segments-in-loops to segments-in-trains with increasing coverage is predicted by other theories. The isotherm continues to rise slowly at higher solution concentrations.

The adsorption of polymer molecules, either in slits or pores has received considerable theoretical interest during the period of this review. This work has a bearing on flocculation (see Chapter 7), adhesion and gel-permeation chromatography.

A study of the configuration of an isolated polymer molecule adsorbed from athermal solution between two parallel planes with short-range adsorption forces[15] has shown that there is no critical value of the differential energy of adsorption per segment required to bring about adsorption. This is in contrast to the result at a single interface. The segment density profile between the planes depends very much on the adsorption energy. At low energies the segment density is highest midway between the planes, whereas at high energies the density is greatest near the surfaces. No adsorption isotherms were calculated.

Pouchly[16] has extended the work of Casassa on the adsorption of polymers in a pore. Cylinders, spheres, and slits are treated. The model is developed to calculate the partition of polymer between the pores and bulk solution and considers the balance between the adsorption forces and the loss of configurational entropy when a polymer enters a pore. Casassa considered only the entropy change. Expressions for the partition coefficient are derived in terms of the pore characteristics, polymer size, and adsorption energies, but are not evaluated.

3 Experimental Techniques

A critical evaluation of models for polymer adsorption would require experimental data on the dimensions of adsorbed polymer loops and trains. Recently introduced techniques are heading in this direction, but classical measurements of adsorption isotherms still form the majority of the published experimental data. These isotherms are frequently incomplete since the sharp initial gradient of the isotherms predicted and observed makes difficult the

[15] E. A. DiMarzio and R. J. Rubin, *J. Chem. Phys.*, 1971, **55**, 4318.
[16] J. Pouchly, *J. Chem. Phys.*, 1970, **52**, 2567.

measurement at surface coverage $< ca.$ 80%. Little use has been made of modern analytical tools such as radio-tracer techniques to probe the area of low surface coverage. Unfortunately, conclusions based on these incomplete isotherms usually depend very much on explicit or implicit unproven assumptions, *e.g.* that the adsorbed molecule retains its solution conformation on adsorption.

A combination of classical techniques together with more direct observation of layer conformation is a more powerful approach. Some techniques used during the period of the review are described below.

I.R. Spectroscopy.—The displacement of an i.r. band of a particular group in a polymer when it adsorbs has been used quantitatively[17-19] to measure the fraction of i.r.-active adsorbable groups actually adsorbed[17,18] ('i.r.-bound fraction'). For instance, free carbonyl groups of methyl methacrylate residues absorb at 5.78 μm, bound carbonyl groups at 5.85 μm, free phenyl groups of styrene residues at 14.30 μm, and bound phenyl groups at 14.23 μm. Measurements are made in double-beam spectrometers with either a dispersion of adsorbent in polymer solution or a gel of adsorbent, separated from the supernatant solution by centrifugation, in the sample beam. Polymer solution in a variable-path-length cell is inserted into the reference beam and the absorption at the unbound group frequency is compensated. A discussion of the techniques and errors has been published.[18]

Adsorption on to a hydroxylated surface can be followed by observing the i.r. absorption band of the surface hydroxy-groups which act as adsorption sites.[20]

N.M.R. and E.S.R. Spectroscopy.—The use of n.m.r. and e.s.r. spectroscopy for the determination of the structure and study of Brownian motion of adsorbed macromolecules has been briefly discussed.[21]

Ellipsometry.—For adsorption on to reflecting surfaces, ellipsometry provides data on the 'thickness' of the adsorbed polymer layer. Ellipsometry alone or a combination of ellipsometry and isotherm measurement yields information on the conformation of adsorbed macromolecules.[22,23]

Hydrodynamic Barrier Thickness.—Hydrodynamic barrier thickness[24,25] has been measured for polymers adsorbed on to sols. The intrinsic viscosity of

[17] R. Botham and C. Thies, *J. Polymer Sci., Part C, Polymer Symposia,* 1970, **30**, 369.
[18] J. M. Herd, A. J. Hopkins, and G. J. Howard, *J. Polymer Sci., Part C, Polymer Symposia,* 1971, **34**, 211.
[19] R. Kapler and L. I. Nekrasov, *Zhur. fiz. Khim.,* 1971, **45**, 750 (*Russ. J. Phys. Chem.,* 1971, **45**, 419).
[20] H. Rupprecht and H. Liebl, *Kolloid-Z.,* 1970, **239**, 685.
[21] J. Turkevich, *Croat. Chem. Acta,* 1970, **42**, 383.
[22] E. Killmann and H.-G. Wiegand, *Makromol. Chem.,* 1970, **132**, 239.
[23] R. R. Stromberg, L. E. Smith, and F. L. McCrackin, *Symposia Faraday Soc.,* 1970, no. 4, p. 192.
[24] D. W. J. Osmond and D. J. Walbridge, *J. Polymer Sci., Part C, Polymer Symposia,* 1970, **30**, 381. S. Rohrsetzer, E. Wolfram, M. Nagy, and M. Kubicza, *Magyar Kém. Folyóirat,* 1970, **76**, 92.
[25]

a sol plus polymer solution is the weight fraction mean of the viscosities for the two components separately, in the absence of adsorption. With adsorption but no flocculation the intrinsic viscosity is less than the mean, whereas flocculation increases it above the mean. The intrinsic viscosity is related to the particle size and therefore to the thickness of the polymer layer on the surface of the sol particles.

The inaccuracy of adsorption isotherm measurements deters workers from studying the temperature dependence of adsorption[26] and from calculating isosteric heats of adsorption. Calorimetry has been used in a practical study,[2] but neglected in more basic investigations.

In conclusion, there is much scope for the use of new spectroscopic techniques and ellipsometry, and for improvements in the accuracy of conventional isotherm measurements. The use of several techniques bearing on the same problem greatly increases the value of the study.

4 Experimental Results

Adsorption Isotherms.—As predicted theoretically, adsorption isotherms in general show a steep initial gradient at concentrations which are too low to be easily studied experimentally (typically $c < 0.1$—1 g l^{-1}). For those systems in which the initial part of the isotherm is not too steep, experimental inaccuracy precludes useful comparison with theoretical isotherms.

The adsorption, Γ, as measured directly by the change in solution concentration, Δc, ($\Gamma = V\Delta c/A$, where V is the volume of solution and A the area of adsorbent) is an excess quantity. Thus, at high solution concentrations, if the quantity of polymer in contact with the solid remains constant, Γ must decrease with increasing concentration. This is not frequently observed experimentally; instead, as the solution concentration increases the isotherms become much less steep, in some cases levelling off to a plateau, in other cases continuing to rise slowly with increasing solution concentration. The polymer adsorption parameter most frequently measured is the extent of adsorption in the plateau or diminished gradient region, and this will be referred to as the plateau adsorption, despite the fact that the gradient may not be zero.

Although the quality of published mass adsorption isotherms is disappointing, isotherms of polymer-layer thickness as measured by ellipsometry are of considerable interest. A study of adsorption by ellipsometry of polyvinylpyrrolidone ($M_w = 38\ 000$) from aqueous solution and from methanol and of poly(ethylene oxide) ($M_w = 6130$ and $40\ 000$) from aqueous solution on to chrome plate[22] reveals that the plateau of the thickness isotherm is reached at a much higher solution concentration (~ 10 mg cm^{-3}) than the plateau of the mass isotherm (< 1 mg cm^{-3}). These concentration values are higher than those suggested by Silberberg.[7] Thus the polymer concentration in the interfacial region decreases with increasing solution concentration

²² E. Hamori, W. C. Forsman, and R. E. Hughes, *Macromolecules*, 1971, **4**, 193.

from 1 to 10 mg cm^{-3}, passes through a minimum, and then slowly increases at higher concentrations. This implies that the thin compact layer of adsorbed polymer at low concentrations re-orientates to give a thicker less concentrated layer as the solution concentration increases. The plateau thickness for poly(ethylene oxide) ($M_w = 40\,000$) lies close to the r.m.s. dimension of the polymer coil in solution, but for the other polymers the thicknesses approach the uncoiled polymer length.

This study underlines the danger of deducing the conformation of adsorbed polymers from adsorption mass isotherms alone for, throughout much of the concentration range when the layer thickness is changing, the mass of adsorbed polymer remains constant within experimental error.

In an ellipsometric study[23] of polystyrene ($M_w = 5.4 \times 10^5$) adsorbed from cyclohexane on to chrome ferrotype plate, the extent of adsorption and film thickness reach maximum values at approximately the same solution concentration. The results indicate a relatively flat orientation of the polymer molecules at low concentrations, transforming at higher concentrations into a layer about 40 nm thick.

Influence of Molecular Weight.—*Non-porous Adsorbent.* Three isotherms for the adsorption of a polyelectrolyte of different molecular weights from aqueous salt solution[27] are sufficiently precise to calculate reliable values of their initial gradients. The initial gradient increases with molecular weight, but not according to the logarithmic relation given by Hoeve.[28] Several workers have studied the influence of molecular weight on the plateau adsorption. These results are usually correlated using the relationship:

$$\Gamma_{\text{plateau}} = K[M]^\alpha_{}$$

where Γ_{plateau} is the mass adsorption in the region of small or zero gradient, M the molecular weight, K and α constants for a given system. If the polymer molecules are attached to the surface by one end-group and stretch out from the surface then the value of α would be unity. If the polymer molecules were to adsorb at the interface in a random-coil configuration, and act as impenetrable spheres or were to lie flat on the surface, then α would take the value zero. Unfortunately, deviations of α from the value unity cannot be unambiguously interpreted in terms of structure at the surface. Silberberg[29] predicts that α decreases as M increases.

For several systems (see Table) the adsorption is independent of molecular weight, and this has been interpreted in terms of a flat polymer configuration.[30,31] The decrease in adsorption with increasing molecular weight observed for the system [random poly(methyl methacrylate–styrene) + benzene]/

[27] B. W. Greene, *J. Colloid Interface Sci.*, 1971, **37**, 144.
[28] C. A. J. Hoeve, E. A. DiMarzio, and P. Peyser, *J. Chem. Phys.*, 1965, **42**, 2558.
[29] A. Silberberg, *J. Chem. Phys.*, 1968, **48**, 2835.
[30] G. J. Howard and P. McConnell, *J. Phys. Chem.*, 1970, **71**, 2974.
[31] K. Hara, I. Mizuhara, and T. Imoto, *Kolloid-Z.*, 1970, **238**, 438.

Table   *Values of α for adsorption on to non-porous adsorbents[a]*

System	Tempera-ture/°C	Coverage/ μg cm^{-2}	α	Ref.
Polyethers–benzene–silica	20	0.07→0.13	0.11	30
Polyethers–methanol–silica	20	0.02→0.04	0.31	30
Polyethers–water–silica	20	0·04	∼0	30
Polyethers–chloroform–silica	20	0.04→0.06	0.08	30
Trimethylsilylated poly(dimethyl-siloxane)–benzene–glass	20	0.02→0.07	0.70	33
Trimethylsilylated poly(dimethyl-siloxane)–n-hexane–glass	20	0.10→0.15	0.91→0.06	33
Commercial poly(dimethylsiloxane)–benzene–glass	20	0.08	0.41	33
Poly(ethylene–vinyl acetate)–benzene–glass	30	n.d.	0.33→0	34
Ethyl cellulose–benzene–glass	30	n.d.	∼0	31
Ethyl cellulose–1,2-dichloroethane–glass	30	n.d.	∼0	31
Poly(methyl methacrylate)–benzene–silica	25	n.d.	∼0	18
Poly(methyl methacrylate-styrene)–benzene–silica	25	n.d.	∼0.08	18
Poly(methyl methacrylate)–toluene–silica	25	0.03	∼0	26
Poly(methyl methacrylate)–1,2-dichloroethane–silica	25	0.015	0	26
Poly(methyl methacrylate-styrene)–benzene–Graphon	25	<0.07	negative	32

[a] n.d. = not determined.

Graphon[32] is attributed to flocculation of the graphitized carbon black and consequent pore formation. The majority of results do not cover a sufficiently wide range of molecular weights to permit assessment of Silberberg's prediction, but the systems [poly(dimethylsiloxane) + n-hexane]/glass[33] and [poly-(ethylene–vinyl acetate) + benzene]/glass[34] correlate qualitatively. The results for the adsorption of poly(dimethylsiloxane)[33] are higher than the average.

Plateau adsorption values of low molecular weight poly(ethylene glycol) from water on to silica indicate an almost constant number of molecules adsorbed,[20] whereas the molar adsorption of polyethoxynonylphenol on to silica from water or carbon tetrachloride decreased with increasing molecular weight, indicating an area of 0.15 nm^2 per additional ethoxy-group.

The r.m.s. thickness in the plateau region of polystyrene adsorbed on to chrome ferrotype from cyclohexane at about the theta point, as determined ellipsometrically,[23] is proportional to the square root of the molecular weight up to approximately 3×10^6, beyond which the thickness rises less steeply. This dependence is predicted by Hoeve (see above). For adsorption on to a mercury surface the thickness is independent of molecular weight in the

[31] A. Hopkins and G. J. Howard, *J. Polymer Sci., Part A-2, Polymer Phys.*, 1971, **9**, 841.
[33] B. V. Ashmead and M. J. Owen, *J. Polymer Sci., Part A-2, Polymer Phys.*, 1971, **9**, 331.
[34] K. Hara and T. Imoto, *Kolloid-Z.*, 1970, **237**, 297.

range $0.5—3 \times 10^6$. This independence is attributed to higher interaction energies on mercury than on the chrome ferrotype.

Porous Adsorbents. The molecular sieve effect of porous adsorbents opposes any relationship between the amount adsorbed and molecular weight that would be observed on a plane surface. Frequently the net result is that adsorption decreases with increasing molecular weight, as observed in the systems [poly(vinyl acetate) + benzene]/cellulose,[35] [poly(methyl methacrylate) + benzene]/silicic acid,[26] and [poly(methyl methacrylate–styrene) + benzene]/charcoal.[32] However, for the system [poly(ethylene oxide) + various solvents]/charcoal,[36] adsorption was independent of molecular weight, except when the solvent was methanol and adsorption increased with molecular weight. This increase is explained in terms of the dramatic decrease in solubility of poly-(ethylene oxide) in methanol as molecular weight increases, because of the decreasing relative importance of the terminal hydroxy-groups which impart the solubility in methanol.

Influence of Polymer Structure.—Copolymers of varying composition and different monomer unit sequences provide interesting material for adsorption/ studies. For the system [random poly(methyl methacrylate–styrene) + benzene] silica, Howard *et al.*[18] observed a plateau adsorption which was independent of composition from pure poly(methyl methacrylate) (PMMA) up to 90% styrene content. At this point, the adsorption rapidly drops to the level of the polystyrene (PS) homopolymer. However, the introduction of short styrene blocks into the centre of a PMMA polymer, to form an ABA block copolymer, increases the adsorption above that for PMMA homopolymer from benzene solution, whereas at higher PS content the adsorption again falls. In contrast, when the solvent is trichloroethylene, the block and random copolymers give the same adsorption as a function of composition, greater than that for PMMA homopolymer, up to about 90% PS content. For this system data are available on the i.r.-bound fraction (p). These show for both random and block copolymers that increasing the styrene content raises the value of p for the carbonyl groups of the methyl methacrylate units from $p = 0.25$ to $p = 0.50$ by the 50% composition point. However, the bound fraction of carbonyl groups plus phenyl residues remains constant at $p = 0.25$ up to high styrene content. Thus the number of bound carbonyl groups per molecule remains constant until the styrene content is over 50% at which point bound styrene residues start to replace bound carbonyl groups at the surface sites. This system demonstrates the much greater affinity of methacrylate groups for the silica surface than that of phenyl groups.

Results for the system [random and block P(MMA–S) + benzene]/porous charcoal[32] were similar to those for the [P(MMA–S) + benzene]/silica system,[18] notably the rapid fall in adsorption at high styrene content. Random co-

[35] B. Alince and A. A. Robertson, *J. Appl. Polymer Sci.*, 1970, **14**, 2581.
[36] G. J. Howard and P. McConnell, *J. Phys. Chem.*, 1970, **71**, 2981.

polymers were slightly more adsorbed than block copolymers of the same overall composition. However, when the adsorbent is graphitized carbon black (Graphon) and the solvent is benzene it is the PS rather than the PMMA which is adsorbed. The sharp fall in adsorption now appears at the methyl methacrylate-rich end of the composition range. This system shows also considerable difference in behaviour between block copolymers and random copolymers, the former being the more strongly adsorbed. A possible explanation is particle bridging by the random copolymers giving rise to pores.

The sudden reduction in adsorption of these copolymers at high content of the non-preferentially adsorbed component is in contrast to the results for the systems {poly[ethylene oxide (EO)–propylene oxide (PO)] + (benzene or water)}/charcoal.[36] In this case the plateau adsorption changes smoothly as the ethylene oxide content of the polymer increases. Plateau adsorption values for (EO–PO–EO) block and (EO–PO) random copolymers in the molecular weight range 2500—11 300 fall on the same curve. When the solvent is benzene the plateau adsorption increases with increasing ethylene oxide content, whereas for aqueous solvent the plateau adsorption decreases. These results reflect the change in solubility of the polymers with composition, the greater the solubility the lower the adsorption. The similarity in the results for the random and block copolymers is explained by the small difference in adsorption energy between the propylene oxide and ethylene oxide monomer units.

Influence of Tacticity.—In a study of the adsorption of conventional and isotactic PMMA with non-porous silica and silicic acid,[26] no dependence of plateau adsorption on tacticity was observed, in general. However, when the solvent was acetonitrile, PMMA prepared by conventional methods (which has no well-defined tacticity) showed no adsorption whereas isotactic PMMA was slightly adsorbed. This suggests that isotactic chains can adopt a lower energy configuration at the surface than those of conventionally prepared polymers.

Influence of Solvent.—In polymer adsorption from solution, polymer segments compete with solvent molecules for surface sites. The interactions of polymer and solvent with the adsorbent are quantified in terms of the differential energy of adsorption of polymer segments over solvent molecules. Consider the adsorption process illustrated for one polymer segment in Figure 1.

The energy difference between the two states is the sum of two parts: (i) the difference in the energies of interaction of a solvent molecule and of a polymer segment with the adsorbent; (ii) the change in energy resulting from a change in the number of solvent–polymer and solvent–solvent contacts for non-athermal solutions. In dilute solution, which is usually the situation in practice, adsorption of a polymer segment reduces the number of polymer–solvent contacts and increases the number of solvent–solvent contacts. Thus changing the solvent to a better solvent (decreasing the value of the Flory parameter χ) makes polymer–solvent contacts energetically more favourable

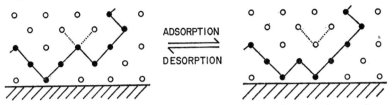

Figure 1 *Adsorption of polymer segment increases number of solvent–solvent contacts and decreases number of polymer–solvent contacts, as shown by dotted lines*

relative to polymer–polymer or solvent–solvent contacts and therefore favours a decrease in the extent of adsorption. A poorer solvent tends to increase adsorption and if the solvent is sufficiently poor, phase separation of the polymer may be nucleated at the surface, leading to multilayer adsorption.

In addition to the above two factors, which are independent of conformation, there will be conformational changes in both the polymer molecules in solution and those in the adsorbed state resulting from the change in polymer–solvent, polymer–polymer, and solvent–solvent interactions. Polymer molecules in solution are more expanded in better solvents, and a similar effect is expected to be found with adsorbed polymers. The net result of these changes, together with (i) and (ii) above, will depend in a complicated way on the signs and magnitudes of the composite effects.

It has been suggested[26] that on increasing the net energy of adsorption the amount of polymer adsorbed should pass through a maximum, since at high energies the polymer molecules lie flat on the surface in a thin film. An increased energy of adsorption is expected to give a steeper initial slope to the adsorption isotherm, but unfortunately this region is experimentally difficult to study and experimental results published during the period of this review have been concerned mainly with plateau adsorption values.

The extent of adsorption of poly(ethylene oxide) on to flame-hydrolysed silica (Aerosil) was found to depend on solvent according to the order:[30] benzene > chloroform > water ≫ dioxan ≈ dimethylformamide. The position of methanol in this list depends on the polymer molecular weight (end-group effect), but lies below water for poly(ethylene oxide) up to 10 000 molecular weight. The solubility of the polymer in the solvents follows the order: chloroform > benzene ≈ dioxan ≈ dimethylformamide ≈ water > methanol. Clearly the change in adsorption with solvent cannot be explained solely in terms of changing solubility. The relative affinities of the solvent for the silica surface, as determined by vapour adsorption, are: water ≈ dimethylformamide > methanol > dioxan > benzene > chloroform. The further down the solvency or affinity lists a solvent lies, the greater would be the extent of polymer adsorption to be expected on the basis of the discussion at the beginning of this section. In fact the two lists are in inverse order so that one effect tends to cancel the other, and an explanation of the solvent dependency of adsorption has to be in terms of specific molecular effects. This perhaps emphasizes the important role of chemistry as distinct from the physical approach adopted in most theories of adsorption.

For the system poly(ethylene oxide) adsorbed on to charcoal[36] the following orders were found: adsorption of polymer from, water > methanol > benzene > dioxan > chloroform > dimethylformamide; affinity of solvent for surface, water < methanol < benzene < chloroform < dioxan < dimethylformamide; solvency, methanol < water < dimethylformamide ≈ dioxan < benzene < chloroform. In this case the adsorption and affinity series correlate well, with the exception of chloroform. The lower adsorption with chloroform than would be expected from the affinity list can be explained by the excellent solvent power of chloroform for the poly(ethylene oxide).

There is further evidence of the combined effects of changing solvency and affinity for the surface. For the system poly(methyl methacrylate) adsorbed on to silica[26] the order of adsorption toluene > 1,2 dichloroethane ≫ acetonitrile ≈ acetone was found. For the solvents acetonitrile and acetone, preferential adsorption of solvent (negative polymer adsorption) was observed. For poly(dimethylsiloxane) adsorbed on to glass[33] the adsorption order was n-hexane > carbon tetrachloride > benzene > methyl ethyl ketone, whereas n-hexane is a better solvent for poly(dimethylsiloxane) than benzene. However, some authors[31,34] have attempted to explain the adsorption results solely in terms of the influence of solvent on the size of the random polymer coils in solution.

Considering the difficulties involved in interpreting the effect of different solvents on polymer adsorption, the use of mixed solvents would appear at first sight to be a further complicating factor. However, if one component of a binary solvent mixture is strongly adsorbed relative to the other component, then over a large mixed solvent composition range most of the surface sites not occupied by polymer segments will be occupied by the more strongly adsorbed solvent. Thus, by suitable choice of solvents, it is possible to change the polymer solubility by varying solvent composition while the polymer molecules effectively compete with only the more strongly adsorbed solvent for the surface sites. In other words, the polymer–surface interactions can be held constant while varying the polymer–solvent interactions. Clearly this argument requires a more rigorous analysis, since the use of mixed solvents could be a powerful technique.

For example, in the system [poly(ethylene oxide) + benzene + n-heptane]/charcoal,[36] the benzene is preferentially adsorbed from n-heptane. Increasing the solvent fraction of n-heptane from zero to 20% decreased the solvent power and increased the plateau adsorption. However, a more stringent test of the technique would require the poorer solvent to be the more strongly adsorbed solvent. For the system [poly(β-cyanoethyl vinyl ether) + acetone]/channel black, the addition of either n-heptane or water increased adsorption.[37] A plot of plateau adsorption against intrinsic viscosity in the mixed

³⁷ T. M. Polonskii and V. P. Zakordonskii, *Kolloid. Zhur.*, 1971, **33**, 721 [*Colloid J. (U.S.S.R.)* 1971, **32**, 601.]

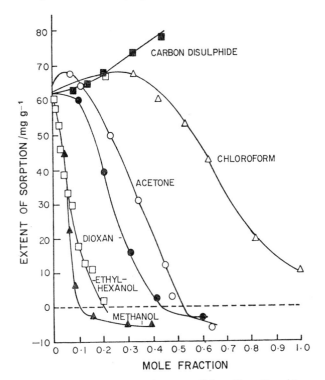

Figure 2 *Sorption of poly(vinyl acetate) on to cellulose fibres from binary solvents based on benzene. The abscissa shows the mole fraction of the second solvent components; the ordinate, the sorption at an equilibrium concentration of $4\,g\,l^{-1}$*

(Reproduced by permission from *J. Colloid Interface Sci.*, 1970, **33**, 586)

solvents was the same for both diluents. From this it is argued that polymer–solvent interactions are the dominant feature.

Some results of an extensive examination of solvent effects on the adsorption of poly(vinyl acetate) and poly(methyl methacrylate) on to porous cellulose from binary mixtures containing benzene as one component are shown in Figure 2.[38,39] The authors conclude that the change in configuration on adding the second solvent to benzene is the predominant factor for the solvents carbon disulphide and chloroform, while the competition for surface sites is predominant for acetone, dioxan, methanol, and ethylhexanol. For the alcohols, configuration changes are also important. The authors advocate the point of zero adsorption, *i.e.* the solvent composition at which the polymer adsorption is zero, as a useful parameter which would be independent of

[38] F. S. Chan, P. S. Minhas, and A. A. Robertson, *J. Colloid Interface Sci.*, 1970, **33**, 586.
[39] F. S. Chan, P. S. Minhas, and A. A. Robertson, *J. Colloid Interface Sci.*, 1970, **33**, 598.

polymer molecular weight, adsorbent surface area, polymer branching, or tacticity, and would depend only on the nature of the polymer functional groups and adsorbent. However, not all solvent pairs exhibit such a point.

The change in solution configuration with change in solvent composition is particularly important in adsorption by porous materials. Thus poly-(ethylenimine) was found to adsorb in the micropores of a cellulose pulp from aqueous solvent at pH 12.5.[40] On changing the acidity to pH 3.5 the volume of the polymer molecule approximately doubled, and those within the micropores became trapped.

Influence of Adsorbent.—An alternative way of discriminating between the polymer–solvent and polymer–adsorbent interactions is to vary the nature of the adsorbent surface without changing the solvent.

The plateau adsorption of poly(ethylene oxide) adsorbed from methanol, benzene, and chloroform on to aerosil silicas is reduced when the surface hydroxy-groups are methylated,[30] because the attraction of the ether linkage to hydroxy-groups by hydrogen bonding is greater than that to methoxy-groups. Heat treatment of the silica to reduce the number of surface hydroxy-groups also decreased the adsorption of poly(ethylene oxide), except when the solvent was methanol.

Plateau adsorption and i.r. bound fractions have been measured for the adsorption of poly(methyl methacrylate) from trichloroethylene on to flame-hydrolysed γ-Al$_2$O$_3$, flame-hydrolysed silica, flame-hydrolysed silica treated with dimethyldichlorosilane, and a re-precipitated silica.[41] The values were found to vary significantly from adsorbent to adsorbent. Plateau adsorption values (weight per unit area) were in the order: flame-hydrolysed SiO$_2$ > γ-Al$_2$O$_3$ = re-precipitated SiO$_2$ > treated SiO$_2$, and i.r. bound fractions in the order: re-precipitated SiO$_2$ > flame-hydrolysed SiO$_2$ > γ-Al$_2$O$_3$ > treated SiO$_2$. From i.r. data, γ-Al$_2$O$_3$ and re-precipitated silica are fully hydroxylated and contain physically bound water. Flame-hydrolysed silica contains less bound water and treated silica fewer hydroxy-groups. Thus the order of p values for the silicas correlates with degree of surface hydroxylation. γ-Al$_2$O$_3$ contains more hydroxy-groups per unit area than fully hydroxylated silica, but the lower polarity of the γ-Al$_2$O$_3$ hydroxy-groups reduces their effectiveness as adsorbing sites. The low position of re-precipitated silica in the plateau adsorption list can be explained in terms of the porous nature of this adsorbent. Calculation showed that twice as many sites were occupied by polymer segments on flame-hydrolysed silica as on re-precipitated silica or treated silica and eight times as many as on γ-Al$_2$O$_3$.

The influence of porosity on adsorption has been studied using the system [poly(vinyl acetate)+benzene]/cellulose.[35] Excellent agreement was found between the plateau adsorption values and the area available for adsorption based on pore size analysis of several swelled cellulose samples.

[40] G. G. Allan, K. Akagane, A. N. Neogi, W. M. Reif, and T. Mattila, *Nature*, 1970, **225**, 175.
[41] C. Thies, *J. Polymer Sci., Part C, Polymer Symposia*, 1971, **34**, 201.

Competition between Polymers.—Competition both between polymers of different structure and between polymers of the same type but of different molecular weight have been studied.

In adsorption on to a non-porous silica from trichloroethylene,[17] subsequent addition of a second polymer displaced a previously adsorbed polymer according to the general orders: poly(vinyl acetate) > ethyl cellulose ≥ poly-(methyl methacrylate) > (?) poly(ethylene–vinyl acetate) > polystyrene, and poly(ethylene–vinyl acetate) (28.8% vinyl acetate) and PEVA (44.2% VA) > PEVA (20.4% VA). The order for poly(ethylene–vinyl acetate) copolymers correlates with i.r. bound fraction values which were 0.089—0.058, 0.18—0.11, and 0.060—0.041, respectively. The orders depend to some extent on the solution concentrations and molecular weights of the polymers. A second part of this paper concerns the adsorption of mixtures of incompatible polymers in the region of phase separation. Thus the addition of polystyrene or polydimethylsiloxane increased substantially the adsorption of poly(methyl methacrylate) and poly(vinyl acetate). An increase in the molecular weight of the polydimethylsiloxane increased the adsorption of poly(methyl methacrylate). Phase separation most probably occurs in the interfacial region of high polymer concentration, leading to a thick layer of the preferentially adsorbed polymer. I.r. bound fraction data for this system substantiate this conclusion.

The competitive adsorption between two polymers of poly(vinyl acetate) of different molecular weights ($M_v = 470\,000$ and $15\,600$) from benzene on to a glass surface has been studied by measuring the change in solution intrinsic viscosity on adsorption.[42] For equi-weight fraction solutions the intrinsic viscosity decreased on adsorption, indicating that a greater weight of higher molecular weight polymer is adsorbed than of lower molecular weight polymer, although a greater number of lower molecular weight polymer molecules were adsorbed. For adsorption from equi-molar solution the intrinsic viscosity stayed constant or increased at higher surface coverages. At low coverages equal numbers of molecules of both polymers were adsorbed, whereas at higher coverages more of the lower molecular weight polymer molecules but a greater weight of the larger molecules was adsorbed.

Preferential adsorption by weight of higher molecular weight polymer was observed also by Felter,[43,44] who analysed the change in the molecular weight distribution of commercial poly(vinyl chloride) adsorbed on to non-porous calcium carbonate from solution in chlorobenzene. Results for three different polymers with $M_n = 20\,200$, $38\,000$, and $52\,900$ are presented. Typical results for the molecular weight distribution in solution and in the adsorbed layer in the region of high surface coverage (Figure 3) show that the higher molecular weight polymers are preferentially adsorbed and the molecular weight distribution in the adsorbed layer is narrower than in solution.

[42] K. Mizuhara, K. Hara, and T. Imoto, *Kolloid-Z.*, 1970, **238**, 442.
[43] R. E. Felter and L. N. Ray, *J. Colloid Interface Sci.*, 1970, **32**, 349.
[44] R. E. Felter, *J. Polymer Sci.*, Part C, *Polymer Symposia*, 1971, **34**, 227.

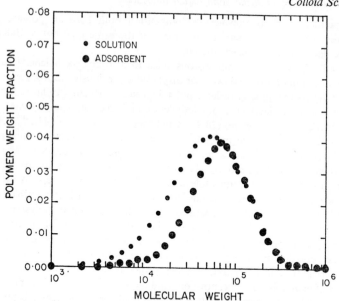

Figure 3 *Molecular weight distribution of poly(vinyl chloride) in solution and in the adsorbed layer at the limiting adsorption plateau*
(Reproduced by permission from *J. Colloid Interface Sci.*, 1970, **32**, 349)

An adsorption efficiency factor, a_i/o_i, for the ith molecular weight fraction is defined, where a_i and o_i are the ith molecular weight masses in the adsorbed layer and in the initial solution respectively. In the molecular weight range up to 10^5 the value of a_i/o_i rises from a low value to a limiting value according to an equation of the form: $a_i/o_i = A + B \log M_i$, where A and B are constants. At molecular weights greater than 10^5 the value of a_i/o_i is essentially constant.

If the adsorbent is porous, then superimposed on the above effect, where the higher molecular weight polymer is preferentially adsorbed, will be a molecular sieve effect favouring preferential adsorption of the lower molecular weight polymer. Thus for poly(vinyl acetate) ($M_w = 79\,000$) adsorbed from benzene on to swollen cellulose fibre[38] the average molecular weights of the adsorbed polymer and equilibrium solution polymers were 22 000 and 98 000 respectively, indicating preferential adsorption of the lower molecular weight polymer. A similar fractionation was observed in the system [poly(methyl methacrylate) + benzene]/porous aluminium silicate.[45]

Adsorption Kinetics.—Adsorption of poly(ethylene glycol) ($M_w = 40\,000$) on to a chrome plate from aqueous solution has been followed by measuring film thickness using ellipsometry and adsorption isotherms.[22] At low solution concentrations (< 10 mg ml^{-1}) the adsorption and thickness increase from

[45] V. M. Patel, C. K. Patel, and R. D. Patel, *Angew. Makromol. Chem.*, 1970, **13**, 195.

zero initially and reach limiting values in less than 10 min. At higher concentrations, although the adsorption appears complete after 10 min, the layer thickness continues to increase, reaching a limiting value after 40 min. This implies that the density of the polymer in the adsorbed layer decreases during the time interval between 10 and 40 min, resulting from molecular rearrangement at the surface.

In another ellipsometric study[23] in which the adsorption was calculated directly from the film thickness and refractive index, adsorption and film thickness reached plateau values at similar times for the systems (polystyrene + cyclohexane)/solid metal surfaces. For polystyrene adsorbed on to mercury, and low molecular weight polyesters on to steel and chrome, film thickness remained constant with time, despite increasing adsorption. In these cases the conformations remain relatively flat and do not extend as surface coverage increases with time.

Also on a non-porous surface, Botham and Thies[17] observed displacement rates of one polymer by another more strongly adsorbed polymer. For systems showing strong displacement, equilibrium is reached within one hour, whereas for less strongly displacing systems several hours may be needed to complete the displacement process.

Even longer times to equilibrium, from 50 to 100 h, were observed for the system [sodium poly-(2-sulphoethyl methacrylate) ($M_v = 2\ 400\ 000$) + aqueous NaCl solutions]/non-porous polyethylene powder.[27] For polymers of lower molecular weights the times to reach equilibrium were: $M_v = 400\ 000$, $t \approx 25$ h; $M_v = 35\ 000$, $t \approx 4$ h. Possibly electrostatic interactions are important in determining the rate in this case.

The initial slope of the change in solution composition with time has been correlated with diffusion coefficients for adsorption into a porous substrate.[46] For the system [poly(ethylenimine) + water]/cellulose fibres, good agreement was found with the theoretical equation:

$$(dc^*/dt)_{t \to 0} = (g_2/g_1)(D^{0.66} S_e W_f C_0/V) \tag{6}$$

where $(dc^*/dt)_{t \to 0}$ is the initial rate, g_1 and g_2 are constants, D is the diffusion coefficient, S_e is the effective specific surface area, calculated from the equilibrium isotherm and assumed polymer dimension, W_f is the weight of fibre, C_0 is the initial solution concentration, and V is the total volume of the system. For several molecular weights, solution concentrations, and salt concentrations a single value of g_2/g_1 was calculated and, from this, theoretical adsorption–time curves were constructed. These agreed well with the experimental curves. Deviations were explained in terms of the neglect of electrostatic interactions. However, the calculated barrier length to diffusion was excessively high, indicating that the model assumptions were too simple. First-order kinetics have been observed also in the system [poly(methyl methacrylate) + benzene]/aluminium silicate.[45]

[46] W. A. Kindler and J. W. Swanson, *J. Polymer Sci., Part A-2, Polymer Phys.*, 1971, **9**, 853.

Evidence that systems may never reach time equilibrium states comes from a study of the adsorption of gelatin on to the surfaces of silver halide crystals.[47] The extent of adsorption depends on such experimental conditions as order and speed of mixing. Slow mixing enabled the gelatin macromolecules to spread on the adsorbent surface.

[47] J. Pouradier and A. Rondeau, *J. Chim. phys.*, 1971, **68**, 1108.

4
Capillarity and Porous Materials: Equilibrium Properties

By D. H. EVERETT and J. M. HAYNES

1 Introduction

Capillarity is a branch of science with a long history, whose fundamental phenomenological laws were established over a century ago. It is, however, still a topic which continues to attract considerable attention. There are many reasons for this, among the more important of which is the fact that capillary phenomena play a dominant role in determining the properties of porous media in relation to their interaction with fluids. The application of the theory of capillarity to porous media has to overcome the complications inevitably introduced by the generally complex structure of real porous materials. In practice these complications have, in the past, been minimized by the adoption of simple geometrical models of pore structures. But even regularly packed spheres pose many problems for which there are still only incomplete and approximate solutions. The new impetus to studies of this kind has been stimulated partly by the relevance of the work to practical problems, partly by a wish to complete the theoretical structure of the subject, and partly by the advent of computing techniques which make possible studies which would hitherto have been prohibitively difficult.

A number of general publications in this field have appeared recently. Reference should first be made to the American Chemical Society Symposium on Flow through Porous Media, published in *Industrial and Engineering Chemistry*, which includes two extensive surveys, one by Dullien and Batra[1] on the determination of the structure of porous media, the other by Morrow[2] on the physics and thermodynamics of capillary action in porous media.

Among the other publications which appeared during the period under review and which contain important material relevant to the subject of this Chapter are Volume 3 of 'Surface and Colloid Science' (edited by Egon Matijevic),[3] the volume dedicated to J. H. de Boer (edited by B. G. Linsen),[4]

[1] F. A. L. Dullien and V. K. Batra, *Ind. and Eng. Chem.*, 1970, **62**, 25.
[2] N. R. Morrow, *Ind. and Eng. Chem.*, 1970, **62**, 33.
[3] 'Surface and Colloid Science', ed. E. Matijevic, Wiley–Interscience, New York, 1971, vol. 3.
[4] 'Physical and Chemical Aspects of Adsorbents and Catalysts', ed. B. G. Linsen, dedicated to J. H. de Boer, Academic Press, London and New York, 1970.

123

and the proceedings of the IUPAC/SCI Symposium on Surface Area Determination.[5]

Among recent contributions to work on the theory of interfacial regions we mention in particular the review by Goodrich on the statistical mechanics of the capillary layer,[6] and Melrose's contributions on the thermodynamics of solid/liquid interfacial regions, and of surface phenomena.[7,8] Between them these papers provide a modern introduction to many of the ideas which will necessarily form the basis of further advances in this field.

Most of the present review will be concerned with studies of the relationship between the adsorbent properties and pore structure of solids, having as their main eventual practical aim the characterization of porous media by adsorption techniques. We shall limit consideration to solids whose pores are sufficiently large to allow heterogeneous phase equilibria to be established within the pore space. We thus exclude those very small pores (micropores) in which the adsorption fields arising from opposite walls of the pore overlap; pores of this kind are treated in Chapter 1, on the basis of the assumption that adsorption in them occurs by a volume-filling, or homogeneous process. In larger pores, on the other hand, it is supposed that a condensed phase (liquid or solid) can exist in equilibrium with a vapour phase whose local concentration in the pore is the same as that of the bulk vapour phase external to the pore system. Thus *two* (or more) phases are supposed to exist in heterogeneous equilibrium within the pores, separated by phase boundaries whose geometrical properties (curvature, area, and angles formed at lines of intersection with other boundaries) can be specified. These larger pores are defined as falling within the two classes of meso- and macro-pores, and comprise, by conventional definition, all pores of widths greater than about 2.0 nm (20 Å). In practical terms, the distinction between micropores and these larger pores is by no means as clear as this statement implies: on the one hand such a pore size is not susceptible to any form of precise measurement, and on the other the functional differentiation between volume-filling and capillary condensation is based on a qualitative definition and is likely to depend both on specific geometrical and energetic factors characteristic of a given adsorbate–adsorbent system, and upon temperature.

In discussing the relationship between structure and properties, several major fundamental problems have to be solved, among which the more important to have received recent attention are the following:

(i) the relationship between the configuration and size of pores and the potential energies of adsorbate molecules within the pores;

(ii) the fundamental theory of capillary condensation, including an understanding of the properties of the condensed phase;

[5] 'Surface Area Determination', ed. D. H. Everett and R. H. Ottewill, Butterworths, London, 1970.
[6] F. C. Goodrich, Chapter 1 of ref. 3.
[7] J. C. Melrose, *Ind. and Eng. Chem.*, 1968, **60**, 53.
[8] J. C. Melrose, *Pure Appl. Chem.*, 1970, **22**, 273.

(iii) methods of deriving 'pore-size distributions' from experimental data, including mercury porosimetry;

(iv) further understanding of adsorption hysteresis; and

(v) phase changes of material adsorbed in pores.

A summary is also included of the many studies which have appeared concerning the preparation and pore structure of individual materials.

Among a number of related topics which are omitted from the present review, but which will be covered in subsequent reports, are the following:

(a) drainage of liquids from, and imbibition of liquids by, porous solids;

(b) transport processes involving vapour and capillary-condensed phases.

2 Adsorption Energies in Pores

The enhancement of the potential energy of adsorption within a pore, compared with that on a plane surface, is of major importance only when the dimensions of the pore are less than two or three times the diameter of the adsorbed molecule. Its main significance is thus confined to the phenomenon of micropore filling, although it cannot be entirely ignored when one considers capillary condensation in pores in the size range lying between meso- and micro-pores. Furthermore, adsorption in cracks of molecular dimensions may well, in many instances, provide the necessary nucleation centres for the propagation of capillary condensation in coarser pore structures.

Recent work in this field has therefore been reviewed in Chapter 1 and its relevance to an understanding of the transition from adsorption to capillary condensation will be discussed later in the present Chapter (Section 4).

3 Fundamental Theory of Capillary Condensation

The fundamental thermodynamic equation governing capillary condensation equilibrium is that derived by Kelvin[9] just over a century ago.[10] The exact differential form of the equation[11] reads:

$$\delta\left(\frac{2\sigma}{r_m}\right) = \left(\frac{v^g - v^l}{v^l}\right)\delta p, \tag{1}$$

where σ is the surface tension of a surface of mean radius of curvature r_m [defined by $1/r_m = \frac{1}{2}(1/R_1 + 1/R_2)$, where R_1 and R_2 are the principal radii of curvature of the surface]*; v^g and v^l are the molar volumes of vapour and

---

* We use for liquid/vapour interfaces the convention that R is taken as positive if the corresponding centre of curvature lies on the liquid side of the interface.

[9] Kelvin, Lord (W. Thomson), *Phil. Mag.*, (4), 1871, **42**, 448.
[10] The history of this equation has been reviewed by L. M. Skinner and J. R. Sambles, *Aerosol Sci.*, 1972, **3**, 199.
[11] R. Defay, I. Prigogine, A. Bellemans, and D. H. Everett, 'Surface Tension and Adsorption', Longmans, London, 1966, pp. 218 *et seq.*

liquid respectively, and p is the equilibrium vapour pressure over the surface. This equation was derived and used originally in the approximate difference form:

$$\Delta p = p - p^0 = \left(\frac{v^l}{v^g - v^l}\right) 2\sigma\left(\frac{1}{r_m}\right), \tag{2}$$

where p^0 is the vapour pressure over a plane surface, for which $1/r_m = 0$; and Kelvin himself considered that the macroscopic thermodynamic treatment of capillary phenomena would break down close to the limits of the acceptability of the mathematical approximation implicit in equation (2), *i.e.*, for p/p^0 less than about 0.99 or for r_m less than about 100 nm (1000 Å). However, it has become customary to attempt to use the equation over a wider range of conditions, and the integration of equation (1) is usually carried out subject to the assumptions (i) that the liquid is incompressible, so that v^l is independent of r_m, (ii) that v^l in the numerator of equation (1) can be neglected in comparison with v^g, and (iii) that the vapour behaves as a perfect gas. These lead to the usual form of the equation:

$$\ln p/p^0 = 2\sigma v^l/(r_m RT), \tag{3}$$

where R is the gas constant. It is to be noted that it is not necessary to assume that σ is independent of r_m, provided that the value of σ inserted in equation (3) is that corresponding to a surface of mean radius r_m. More exact equations which take account of the compressibility of the liquid, and in which v^l is not neglected, are needed when p is high; they are readily derived.[12] Such refinements become significant only for very small radii of curvature or for systems close to the critical state.

Discussions of the applicability of Kelvin's equation to porous systems have had as one main theme the question of the validity of attributing to the capillary-condensed liquid the values of σ and v^l appropriate to the bulk liquid.

The thermodynamic analysis of the effects of curvature of a liquid/gas interface on the surface tension goes back to Gibbs,[13] and has been studied more recently by Defay,[14] Hill,[15] Tolman,[16] and Koenig.[17] The problem has been expressed in statistical terms by Kirkwood and Buff,[18] and reviews of this topic and clear expositions of some of the conceptual problems involved have been given recently in the papers by Goodrich[6] and Melrose[7, 8] already mentioned. According to calculations of this kind, the surface tension should be essentially independent of curvature down to a radius of about 100 nm, and should not depart by more than 1 % or so from the bulk value until radii

[12] *e.g.* ref. 11, p. 222, and J. C. Melrose, *Amer. Inst. Chem. Engineers J.*, 1966, **12**, 986.
[13] J. W. Gibbs, 'Scientific Papers', Longmans Green, London, 1906, Vol. 1, p. 232; Dover reprint, New York, 1961.
[14] R. Defay, *Bull. Classe Sci. Acad. roy. Belg.*, 1930, **16**, 1249.
[15] T. L. Hill, *J. Phys. Chem.*, 1952, **56**, 526.
[16] R. C. Tolman, *J. Chem. Phys.*, 1949, **17**, 333.
[17] F. O. Koenig, *J. Chem. Phys.*, 1950, **18**, 449.
[18] J. C. Kirkwood and F. P. Buff, *J. Chem. Phys.*, 1949, **17**, 338; F. P. Buff, *ibid*, 1955, **23**, 419.

of less than 10 nm (100 Å) are involved: if the liquid surface is convex (droplet), σ is expected to fall when r is small; if concave (bubble or capillary condensate in a pore wetted by liquid), an increase in σ is expected. However, the validity of this conclusion (like that of many others in this field) depends on the justification for the application of thermodynamic and conventional statistical-mechanical theory to those systems that contain a relatively small number of molecules, and whose overall properties exhibit wide fluctuations from their mean value. The theory of the thermodynamics of small systems[19] has yet to be developed into a useful tool. The problem is complicated by the fact that in interfacial regions the state of stress in the surface is no longer isotropic, but has to be represented by a stress tensor: in very small systems the interfacial regions overlap so that in no part of the system can one assume the pressure to be hydrostatic. Earlier discussions of the problem have been taken up in the past few years by Rusanov and his co-workers in Leningrad. The stress tensor depends, among other things, on the molecular distribution functions in the system.[20] Asymptotic equations for these have been developed[21] and used to calculate the stress in flat boundary layers[22] and in cylindrically symmetrical liquid systems[23] (*e.g.*, liquid in a capillary, a liquid filament surrounded by gas or by a second immiscible liquid). The final equations obtained are complex and no numerical results are given to enable one to get a feel for the magnitude of the effects or their range in space; they do show, however, both the variation of pressure within a small system, and its dependence upon the curvature of the boundary of the system. Although this work is undoubtedly of importance, and one will look forward to further papers, it is still constrained by the assumption (whose seriousness is difficult to assess) that the radius of the boundary is large in comparison with molecular dimensions. Unfortunately, many experimentally interesting situations involve systems where this assumption is unlikely to be justified.

It may also be questioned whether the molar volume of the bulk liquid is appropriately used in equation (3) as noted above. Account is easily taken of the change in volume arising from the pressure changes in the condensed liquid; the remaining assumption is then that the compressibility of the condensed liquid is the same as that of the bulk. No theoretical studies seem to have been directed to this assumption.

Finally, in any application of the Kelvin equation to porous media, it is necessary to relate the mean curvature of the meniscus (r_m) to the pore size. This is commonly done by choosing the model of a cylindrical capillary of radius r, whose walls make a contact angle of θ with the liquid/vapour surface.

[19] T. L. Hill, 'The Thermodynamics of Small Systems', Parts 1 and 2, Benjamin, New York and Amsterdam, 1963.
[20] *e.g.* F. C. Goodrich, Chapter 1 in ref. 3; ref. 11, p. 142.
[21] E. N. Brodskaya, A. I. Rusanov, and F. M. Kuni, *Kolloid. Zhur.*, 1970, **31**, 860 [*Colloid J. (U.S.S.R.)*, 1970, **31**, 691].
[22] F. M. Kuni, A. I. Rusanov, and E. N. Brodskaya, *Zhur. fiz. Khim.*, 1970, **44**, 756 (*Russ. J. Phys. Chem.*, 1970, **44**, 419).
[23] E. N. Brodskaya, F. M. Kuni, and A. I. Rusanov, *Kolloid. Zhur.*, 1971, **33**, 23 [*Colloid J. (U.S.S.R.)*, 1971, **33**, 17].

In this simple case,

$$r_{\mathrm{m}} = -r/\cos\theta. \tag{4}$$

Some shortcomings of this simplification are discussed later (Section 6).

That earlier estimates of the effect of curvature on the properties of capillary-condensed liquids might be in error was suggested by the work of Shereshefsky[24] and Derjaguin,[25] which indicated that substantial effects occurred in capillaries 10—100μm in diameter. This work has led to the controversy on 'anomalous water' which has centred round the contention that not only are the properties of liquids modified over large distances from the surface, but also that these modifications are irreversible in the sense that they persist when the liquid is removed from the capillary. A full discussion of this problem lies outside the scope of this survey, but a review of the situation up to early 1971 has been published,[26] as have the proceedings of an international conference on the subject.[27]

A much more controversial theory has been proposed by O'Brien,[28] who claims to show that increases of surface tension by a factor of 10 to 20 can occur in isolated capillaries in the micron size range. The effect is claimed to exhibit the unexpected and curious behaviour of increasing in magnitude as the extent of filling of the capillary increases. The theory is based on the idea that a liquid can be regarded as a two-component system consisting of 'pure liquid' and 'vacancies', and that when a liquid is brought into contact with a solid, vacancies are desorbed (*i.e.*, the density of the liquid near the surface increases). These vacancies migrate to the liquid/vapour interface, which therefore has a decreased density. While these basic concepts are readily acceptable and in accordance with current views, the development of the theory is expressed in ill-defined terms. It seems to involve the assumption that vacancies are conserved, so that desorption of vacancies at the solid/liquid interface must be compensated for by their adsorption at the liquid/vapour interface: this appears to be the origin of the curious behaviour noted above, since the larger the area of solid/liquid interface relative to that of the liquid/vapour interface, the larger the decrease in density at the liquid/vapour interface and the larger the surface tension change. It is also assumed without comment that the enormous changes in liquid/vapour surface tension which are predicted occur without affecting the contact angle, and that changes in the area of solid wetted by liquid occur at constant area of liquid/vapour interface even in pores of variable cross-section. Furthermore, the key equation on which the quantitative development of the theory is based appears to be thermodynamically unsound, although in the first paper it is implied that it has a thermodynamic basis: in a later paper,[29] however, it is called a postulate.

[24] J. L. Shereshefsky, *J. Amer. Chem. Soc.*, 1928, **50**, 2966; J. L. Shereshefsky and C. P. Carter, *ibid.*, 1950, **72**, 3682.

[25] See, *e.g.*, B. V. Derjaguin, *Scient. Amer.*, 1970, **223**, 52.

[26] D. H. Everett, J. M. Haynes, and P. J. McElroy, *Sci. Progr.*, 1971, **59**, 279.

[27] Papers in *J. Colloid Interface Sci.*, 1971, **36**, 415—566.

[28] W. J. O'Brien, *Surface Sci.*, 1970, **19**, 387.

[29] W. J. O'Brien, *Surface Sci.*, 1971, **25**, 298.

This work was criticized in some relatively minor details by Leinfelder.[30] The author's reply[31] and subsequent restatement of the theory[29] stress that it gives a satisfactory account of the strength of liquid bridges between spheres in contact and leads to the same equation as that derived by Fisher.[32] Since it is generally accepted[33] that Fisher's equation (which assumes a constant surface tension equal to the bulk value) is based on a correct thermodynamic analysis, it is clear that O'Brien must have arrived at the same equation fortuitously. A careful analysis of O'Brien's assumptions and analysis would no doubt reveal the source of error in this work. His use of the theory to account for anomalous water (by assuming that the enhanced surface tension renders the water in the capillary a much better solvent) and for adsorption–desorption hysteresis (on the assumption that on the desorption branch some of the solid has dissolved in the condensate) are untenable as general explanations.

The curvature dependence of surface tension, but now related to cavities of molecular size, has been invoked by Choi, Jhon, and Eyring[34] to account for the solubilities and heats of solution of simple gaseous molecules in organic liquids. This paper is not directly concerned with colloid science, but some of the ideas it contains may find application in this field.

One may certainly conclude that theoretical studies have still not yet expressed in a quantitative form the objections which can be made in principle to the application of the Kelvin equation to very small systems, or to systems having high surface curvature.

The story is no more conclusive on the experimental side. Attempts to obtain direct experimental evidence of the ability of the Kelvin equation to give an accurate quantitative representation of the vapour pressure of liquids confined in capillaries of known dimensions were, as noted above, the forerunners of the 'anomalous water' problem: there is little doubt that any further direct experiments will have to be carried out under the most rigorously controlled conditions. Insofar as these phenomena have effectively prevented experimental substantiation of the Kelvin equation for negative curvatures, they must also stand as a warning against the literal interpretation of Kelvin effects in terms of the size of pores (see later).

However, an indirect thermodynamic method for studying the applicability of the equation and for identifying the point beyond which it fails (either in itself or because capillary condensation no longer occurs) has been suggested by Dubinin.[35] The *adsorption potential* is equal to the change (ΔG^0) in Gibbs free energy when unit amount of material is transferred from the standard state (here taken as pure liquid) to the adsorbed state: with its sign changed,

[30] K. Leinfelder, *Surface Sci.*, 1970, **23**, 427.
[31] W. J. O'Brien, *Surface Sci.*, 1970, **23**, 428.
[32] R. A. Fisher, *J. Agric. Sci.*, 1926, **16**, 492.
[33] *e.g.* J. C. Melrose and G. C. Wallick, *J. Phys. Chem.*, 1967, **71**, 3676; M. A. Erle, D. C. Dyson, and N. R. Morrow, *Amer. Inst. Chem. Engineers J.*, 1971, **17**, 115.
[34] D. S. Choi, M. S. Jhon, and H. Eyring, *J. Chem. Phys.*, 1970, **53**, 2608.
[35] M. M. Dubinin, *Proc. k. ned. Akad. Wetenschap.*, 1970, **B73**, 33; M. M. Dubinin in ref. 5, p. 131.

it may also be called the *standard affinity of adsorption* (A^0), defined by:

$$A^0 = -\Delta G^0 = RT \ln p^0/p, \tag{5}$$

where p^0 is the vapour pressure of pure bulk liquid. The entropy of adsorption and enthalpy of adsorption are obtained by standard formulae from the temperature variation of A^0 at constant amount of adsorbed material, n^a. In porous materials the maximum amount adsorbed at the saturation vapour pressure ($n^{a,0}$) varies with temperature; consequently the variation of A^0 with temperature at constant n^a involves a variation in the degree of pore filling $\theta = n^a/n^{a,0}$. It is easily shown that:

$$\left(\frac{\partial A^0}{\partial T}\right)_\theta = \left(\frac{\partial A^0}{\partial T}\right)_{n^a} - \alpha\left(\frac{\partial A^0}{\partial \ln n^a}\right)_T$$

$$= \overline{\Delta S^0} - \alpha\left(\frac{\partial A^0}{\partial \ln n^a}\right)_T, \tag{6}$$

where $\overline{\Delta S^0}$ is the standard differential entropy of adsorption (*i.e.*, from the bulk liquid as standard state to the adsorbed state at a filling θ), and $\alpha = -\mathrm{d}\ln n^{a,0}/\mathrm{d}T$. Dubinin discusses these equations in relation to various types of adsorption [for example, in the region in which the characteristic curve $A^0(\theta)$ is independent of temperature, so that the left hand side of equation (6) is zero]. Of particular interest here is the region of capillary condensation in which, from equation (3),

$$A^0 = -2\sigma v^l/r_m. \tag{7}$$

Now in a porous medium which has itself a negligible coefficient of thermal expansion, constant θ implies that the radius of exposed menisci (r_m) remains constant, so that

$$\left(\frac{\partial A^0}{\partial T}\right)_\theta = A^0 \frac{\mathrm{d}\ln \sigma v^l}{\mathrm{d}T}. \tag{8}$$

According to this equation $(\partial A^0/\partial T)_\theta$ should be a linear function of A^0, of slope $(\mathrm{d}\ln \sigma v^l/\mathrm{d}T)$ wherever equation (7) is valid. If σ and v^l are assumed to have the same values as bulk liquid, the accuracy of (7) can be tested. When applied to the desorption curves for the systems (benzene + ferric oxide gel) and (benzene + silica gel), straight lines of the predicted negative slope are obtained up to values of A^0 of about 2.8 kJ mol^{-1}, above which the curve turns upward (Figure 1). The point at which deviations from equation (7) become appreciable corresponds to values of r_m of *ca.* 1.6nm (16Å), and it is concluded that below this either the Kelvin equation breaks down or capillary condensation ceases to be a stable situation. The literature so far seems only to contain two curves of the type of Figure 1: it would be very interesting to know whether this criterion is of general applicability.

Before leaving the question of the validity of Kelvin's equation it is interesting to note that although attempts at a direct experimental confirmation for

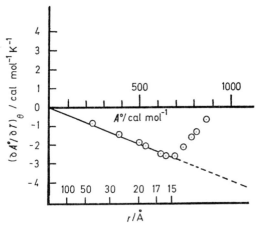

Figure 1 *Variation of the temperature coefficient of the standard affinity of adsorption at constant degree of filling* $(\partial A^0/\partial T)_\theta$, *with* A^0 *for the adsorption of benzene by silica gel*
(Redrawn by permission from 'Surface Area Determination', Butterworths, London, 1971)

concave menisci have been inconclusive or of limited precision,[36] more success has been achieved with convex menisci (droplets). Earlier work by LaMer and Gruen[37] gave some support for the quantitative applicability of the equation, although the precision achieved was not very great. Recently, Sambles, Skinner, and Lisgarten[38] studied the kinetics of evaporation of liquid lead particles in the electron microscope: the values of surface tension required to account for the dependence of rate of evaporation on droplet size were within 5% of that measured by conventional techniques. These authors also report that solid silver particles behaved in a similar fashion, and this enabled an estimate to be made of the surface tension of solid silver. This work was subsequently extended to liquid droplets and solid particles of gold.[39]

4 Range of Stability of Capillary-condensed Liquid

Material adsorbed in a meso- or macro-pore may exist either as a thin adsorbed film on the walls of the pore, or as a thicker layer having essentially liquid-like properties. Adsorbed films may be regarded as being defined by one interfacial region, the solid/adsorbate interface as shown schematically in Figure 2(a), the adsorbate/vapour interface being diffuse. Capillary-condensed liquid, on the other hand, is of such a thickness that it has to be

[36] N. L. Cross and R. G. Picknett, *Trans. Faraday Soc.*, 1963, **59**, 846.
[37] V. K. LaMer and R. Gruen, *Trans. Faraday Soc.*, 1952, **48**, 410.
[38] J. R. Sambles, L. M. Skinner, and N. D. Lisgarten, *Proc. Roy. Soc.*, 1970, **A318**, 507.
[39] J. R. Sambles, *Proc. Roy. Soc.*, 1971, **A324**, 339.

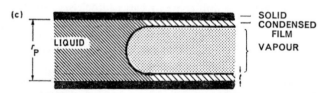

Figure 2 *Transition from adsorption to capillary condensation (schematic)*
(a) *gas adsorbed at solid surface characterized by gas/solid interface;*
(b) *condensed film on inside of cylindrical capillary characterized by a condensed film/solid and a condensed film/gas interface;*
(c) *as* (b) *but with capillary partially filled with condensate with a liquid/gas interface*

defined in terms of two interfaces, the solid/condensed liquid interface and the condensed liquid/vapour interface [Figure 2(b)]; the latter is assumed to exhibit the surface tension of the pure bulk liquid. These two interfaces are not necessarily parallel [Figure 2(c)].

We consider first the transition 2(a)→2(b)→2(c). No final and satisfying answer has yet been given to the problem of identifying the origin and nature of the transition from 2(a) to 2(b), although the change from 2(b)→2(c) at constant vapour pressure, once the cylindrical film has achieved a sufficient thickness and developed a surface tension, has long been recognized as a spontaneous pore-filling process.

Broekhoff and Linsen, writing in 'Physical and Chemical Aspects of Adsorbents and Catalysts',[40] outline the views of the de Boer School on this transition from adsorption to capillary condensation. These are a quantitative development of earlier formulations of Foster[41] and Everett,[42] in which it is assumed that the state of a thin adsorbed layer on the inside of a cylindrical pore is influenced simultaneously by the adsorption potential and by the curvature of the cylindrical surface. To put these ideas into a quantitative form, Broekhoff and de Boer assumed that the adsorption potential could be calculated from the *t*-curve (see Chapter 1), and that the effect of curvature could be added to this by using the Kelvin equation, together with the assumption that the cylindrical adsorbed layer had already developed a surface tension, at the adsorbed layer/vapour boundary, equal to that of bulk liquid. Thus, although a cylindrical liquid/vapour interface in the absence of adsorption effects is unstable with respect to the formation of a hemispherical meniscus, it can be stabilized by the presence of an adsorption potential. However, because the adsorption potential falls off with distance from the surface, this stabilizing effect decreases as the thickness of the film increases, and at a critical thickness the transition 2(b)→ 2(c) occurs spontaneously. On this basis the layer thickness and relative pressure, p/p^0, at which an adsorbed film of nitrogen becomes unstable and passes over spontaneously to a filled pore can be calculated. For example, for a cylindrical pore 2.58 nm (25.8 Å) in radius, these critical values are found to be 1.1 nm and 0.6 respectively. The status of this theory will be discussed later (Section 6); meanwhile two possible criticisms of its validity are, first, that simple additivity of the two effects may not be justified and, secondly, that one must be cautious of applying the equations quantitatively to such thin layers in small capillaries where it seems doubtful whether the bulk surface tension can have been established.

The reverse process 2(c)→ 2(b) may be analysed similarly: it is found that the forward and reverse transitions do not occur at the same values of the relative pressure. This was first appreciated clearly by Cohan[43] and by Foster[44] and formed the basis of their theories of adsorption hysteresis.

Quite apart from the above considerations, there is another factor which may be of crucial importance in deciding whether a configuration such as 2(c) is stable. Ignoring for the moment the effect of adsorption forces, the lowering of the vapour pressure of a pure liquid condensed in a pore is a consequence of the pressure difference across a liquid/vapour meniscus given by the Laplace equation:

$$p^l - p^g = \frac{2\sigma}{r_m}. \tag{9}$$

[40] J. C. P. Broekhoff and B. G. Linsen in ref. 4, Chapter 1.
[41] A. G. Foster, *J. Chem. Soc.*, 1952, 1806.
[42] D. H. Everett in 'Colston Papers, Vol. X, Structure and Properties of Porous Materials' ed. D. H. Everett and F. S. Stone, Butterworths, London, 1958, p. 95.
[43] L. H. Cohan, *J. Amer. Chem. Soc.*, 1944, **66**, 98.
[44] A. G. Foster, *Trans. Faraday Soc.*, 1932, **28**, 645.

On the liquid side of a concave meniscus the hydrostatic pressure, and hence the chemical potential of the liquid, is lowered. The magnitude of this effect is such that even when the curvature of the interface reaches about $1 \mu m$ the pressure decrease in the liquid has exceeded its own vapour pressure, so that the liquid is subjected to a negative pressure (*i.e.*, is under tension). Most simple equations of state predict that a liquid can be maintained in metastable equilibrium under a negative pressure. Experimental observation of this state in macroscopic systems is very difficult since it is almost impossible to exclude completely dust or surface asperities which nucleate the breakdown of the metastable state. Consequently, although negative pressures in liquids have been observed,[45] the ultimate tensile strengths which have been determined probably do not have any absolute significance. Hayward[46] has recently reviewed the history of negative pressure experiments in liquids. However, it appears that, when confined in a small pore, the metastable state is less subject to breakdown than when present in bulk. A possible reason for this is that the pore space is too small to accommodate a bubble of vapour large enough to act as a critical nucleus for evaporation.[47] However, as the relative pressure of vapour in equilibrium with capillary-adsorbed liquid is progressively lowered, the menisci retreat into pores, or pore constrictions, of decreasing size and the liquid is subjected to an ever-increasing tension. If this tension exceeds the ultimate tensile strength of the liquid, then the liquid state becomes unstable and spontaneous evaporation occurs. That the range of existence of capillary-condensed liquid might be limited in this way was envisaged by Schofield;[48] he therefore concluded that insofar as adsorption–desorption hysteresis was a consequence of capillary condensation, there should be a lower limit of relative pressure, h_0, below which hysteresis should be absent. This idea was developed by Flood[49] and by Everett,[50] and has recently been subjected to a critical examination in relation to experimental data by Kadlec and Dubinin[51] and by Burgess and Everett.[52] It has long been recognized that the relative pressure, h_0, at which the hysteresis loop closes is usually characteristic of the adsorptive rather than the adsorbent.[53] It may be recalled also that Harris[54] illustrated this point by comparing the pore radius calculated from the Kelvin equation with the 'hydraulic radius' calculated according to the equation

[45] See, *e.g.*, H. N. V. Temperley, *Proc. Phys. Soc.*, 1947, **59**, 199; H. N. V. Temperley and L. G. Chambers, *ibid.*, 1946, **58**, 420, where earlier work is discussed.
[46] A. T. J. Hayward, *Amer. Scientist*, 1971, **59**, 434.
[47] D. H. Everett in 'The Solid–Gas Interface', ed. E. A. Flood, Dekker, New York, 1967, Vol. 2, Chapter 36.
[48] R. K. Schofield, *Discuss. Faraday Soc.*, 1948, No. 3, p. 105.
[49] E. A. Flood in 'The Solid–Gas Interface', ed. E. A. Flood, Dekker, New York, 1967, Vol. 1, Chapter 1.
[50] D. H. Everett, ref. 47, p. 1086.
[51] O. Kadlec and M. M. Dubinin, *J. Colloid Interface Sci.*, 1969, **31**, 479.
[52] D. H. Everett, ref. 5, p. 138; C. G. V. Burgess and D. H. Everett, *J. Colloid Interface Sci.*, 1970, **33**, 611.
[53] *e.g.* D. H. Everett, ref. 47.
[54] M. R. Harris, *Chem. and Ind.*, 1965, 268.

$r_h = 2V_p/A$, where V_p is the pore volume and A the surface area.[55] When applied to N_2-desorption curves on porous titania, alumina, and silica, while the two values agreed for larger pore sizes, deviation occurred at 1.8 nm (18 Å), at which point the Kelvin radii remained constant while the hydraulic radii continued to fall. Kadlec and Dubinin produce additional experimental data on the adsorption of argon, benzene, hexane, dimethylformamide, and water by porous carbons, porous glasses, and silica gels to support the contention that h_0 is not sensitive to the nature of the adsorbent. They analyse their data using a recent estimate[56] of the ultimate tensile strength (τ_{max}) of a liquid, based on a relatively crude liquid model:

$$\tau_{max} = \frac{2.06\sigma}{d_0}, \tag{10}$$

where d_0 is the 'distance between neighbouring molecules', and consequently somewhat larger than the molecular diameter. By combining equation (10) with the Laplace equation (9), which for this purpose is approximately ($\tau = -p^l \gg p^g$),

$$\tau = -2\sigma/r_m, \tag{11}$$

they showed that the radius r_0 of the pore in which the meniscus would exert a negative pressure equal to the tensile strength of the liquid is (assuming zero contact angle) given by:

$$r_0 = 0.97 \, d_0, \tag{12}$$

independent of the porous material considered. Values of r_0 calculated by inserting the observed values of h_0 into equation (3) gave figures of about 1.1 nm for Ar, 1.5—1.6 nm for benzene, 1.1—1.5 nm for water: these are at least two to three times larger than estimated values of d_0, so that the quantitative agreement with equation (12) is not particularly good. However, this may well result from the shortcomings of equation (10) rather than failure of the basic theory. Burgess and Everett also assembled data on the closure points of hysteresis loops on porous carbons and glasses, but analysed them in a somewhat different fashion. Combination of the Kelvin equation, the Laplace equation, and equation (4) enables one to derive values of the tensile strength of the liquid:

$$\tau_{max} = \frac{\sigma RT}{v^l \cos \theta} \ln h_0 - p^g, \tag{13}$$

in which p^g is usually negligible and θ is taken as zero.

If τ_{max} is expressed as a fraction of the critical pressure p_c of the adsorptive, then it is found that when τ_{max}/p_c is plotted against T/T_c (where T_c is the critical temperature of the adsorptive), a broad band of points is obtained which lies between the curves which would be predicted on the basis of the

[55] *e.g.* R. G. van Nordstrand, W. E. Kreger, and H. E. Ries, *J. Phys. Chem.*, 1951, **55**, 621.
[56] J. L. Gardon, 'Treatise on Adhesion and Adhesives', Dekker, New York, 1967, Vol. 1, p. 269.

van der Waals and Berthelot equations of state respectively. The modified van der Waals equation suggested by Guggenheim[57] predicts a curve which passes through the centre of this band.

Although further theoretical and quantitative experimental work is needed, the general concept that the lower limit of capillary condensation hysteresis is determined by the tensile properties of the adsorbate seems justified. However it needs to be shown more convincingly that the observed values of τ_{max} are independent of the pore structure of the adsorbent (provided that it contains pores down to r_0 in radius); the theory should also be refined to take account of the compressibility (or here the extensibility) of the liquid, and to include the effect of the adsorption field of the solid, which in very narrow pores may in part offset the tension in the liquid.

5 Model Capillary Systems

An understanding of the capillary properties of pore systems having a simple geometry is a prerequisite of any theory which purports to derive information about the pore structure of a solid from its adsorbent properties. Indeed many of the basic ideas about capillary condensation have been developed in relation to the simplest pore geometry, namely that of uniform circular capillaries. Yet it is only recently that a detailed discussion of the filling and emptying of capillaries has been developed.[58] Most studies so far have been concerned with circular cylinders, but recently Nicholson[59] has examined the properties of cylinders of other cross-sections to establish whether the assumption that the cross-section of a pore is circular imposes serious limitations on the significance of the analysis of data for real pore systems. He has calculated the 'local hydraulic radius', r_h, defined in terms of the local cross-sectional area of the cylinder (A_c) and the perimeter of the cross-section P by the equation:

$$r_h = 2A_c/P. \qquad (14)$$

This has been done for cylindrical pores of polygonal cross-section and for the window spaces between monodisperse spheres in a regular packing. It is assumed that r_h calculated in this way determines the capillary properties of the pore system. If the true geometrical volume and surface area of the pores are defined by the geometrical surfaces of the polygonal cylinders, or of the spheres, then the correction factors required to convert the volumes and areas calculated from the hydraulic radii on the assumption of circular cylinders to the true values can be evaluated, and they are tabulated in the paper. Much of this paper is concerned with pores of rectangular cross-section since this

[57] E. A. Guggenheim, 'Thermodynamics', North Holland, Amsterdam, 5th edn., 1967, p. 142.
[58] *e.g.* R. M. Barrer, N. McKenzie, and J. S. S. Reay, *J. Colloid Sci.*, 1956, **11**, 479; J. M. Haynes, Ph.D. Thesis, Bristol, 1965; D. H. Everett, in ref. 47, p. 1078.
[59] D. Nicholson, *Trans. Faraday Soc.*, 1970, **66**, 1713.

provides a convenient way of studying the effect of changing from a square cross-section (which can be approximated by a circular cylindrical pore) to a narrow rectangle defining a slit-shaped pore. Adsorption and desorption isotherms for an assembly of isolated cylinders are calculated by methods essentially similar to those of de Boer and his co-workers except that the adsorption potential is assumed to follow a simple inverse-cube law. A range of isotherm and hysteresis-loop shapes analogous to many of those observed experimentally are obtained and shown to depend upon the pore dimensions and their ratio, and, for assemblies of pores of different sizes, on the pore-size distribution. It is concluded that when the cross-section of the cylinders is not far from square, and the distribution of sizes is narrow, the circular cylinder assumption is satisfactory, but that for slit-like pores, especially if the distribution is wide, the cumulative surface area (obtained from the desorption isotherm by calculating the areas of surface exposed in successive desorption steps, and summing these contributions)[60] becomes much less than the true surface area and the calculated most probable radius becomes greater than the mean hydraulic radius. Furthermore, because of the different mechanisms for filling and emptying of slit-like pores, the area of the hysteresis loop increases as the pores change from square to slit-like. This paper sounds a warning note – echoed later in this review – concerning the uncritical acceptance of the simpler methods of calculating pore-size distributions.

One must, however, in reviewing this paper, point out some of the dangers in the use of the concept of local hydraulic radius as defined here to describe the capillary properties of a pore. For the particular case of a circular cylindrical capillary already partially filled with liquid [Figure 2(c)], addition of a further volume dV will cause the meniscus to move a distance dl along the capillary, where

$$dV = A_c\, dl = \pi r^2\, dl; \qquad (15)$$

and an extra area dA^{sl} becomes wetted with the liquid, where

$$dA^{sl} = P\, dl = 2\pi r\, dl. \qquad (16)$$

Thus

$$\frac{dA^{sl}}{dV} = \frac{2}{r} = \frac{P}{A_c} = \frac{2}{r_h}, \qquad (17)$$

so that $r_h = r$.

It was shown, however, by Gauss in 1830[61] that, in general, the perturbation of an interface of constant curvature confined by solid boundaries and intersecting them with a constant contact angle, θ, such that an increase in the volume of liquid, dV, is accompanied by changes in the areas of the solid/

[60] J. H. de Boer, in 'Structure and Properties of Porous Materials', ed. D. H. Everett and F. S. Stone, Butterworths, London, 1958, p. 68.
[61] C. F. Gauss, 'Principia generalia theoriae figurae fluidorum in statu aequilibrii', in *Comment. soc. reg. scient. Gottingen. recent.*, Volume 7, 1830 (trans. by R. H. Weber in 'Ostwald's Klassiker der Exakten Wissenschaften', No. 135, Engelmann, Leipzig, 1903).

F

liquid, gas/solid, and liquid/gas interfaces of $dA^{sl} = -dA^{gs}$, and dA^{lg} respectively, is governed by the equation:

$$\frac{d(A^{lg} - A^{sl} \cos \theta)}{dV} = C^{lg} = \frac{2}{r_m}, \tag{18}$$

where C^{lg} is the (constant) curvature of the liquid/gas interface defined by (*cf.* footnote p. 125):

$$C^{lg} = \frac{1}{R_1} + \frac{1}{R_2} = \frac{2}{r_m}. \tag{19}$$

The value of C^{lg}, or r_m, thus defines the capillary properties of the system.

For a circular capillary of constant cross-section containing a thread of liquid, $dA^{lg} = 0$, while dA^{sl} is given by equation (16). The hydraulic radius given by equation (17) is thus numerically equal to r_m, provided that the contact angle is zero. Thus even in this case the hydraulic radius approach is limited to systems in which the liquid wets the solid completely. This is because r_m is an *interfacial* radius of curvature, whilst r_h (and r for cylinders) refers to the *pore* radius. The two are related *via* the contact angle, see equation (4).

In a non-circular capillary the line of contact between solid, liquid, and gas no longer lies in a plane. However, if A_c is taken as the area of the projection of the boundary of contact on a plane normal to the axis of the cylinder, and P is the perimeter of this projection, then, again for zero contact angle, the hydraulic radius and r_m should be the same. However, when one considers pores of non-uniform cross-section, the change in volume dV is accompanied by a change in A^{lg} which is positive or negative depending on the direction of taper of the pore. In this case it is therefore essential to employ equation (18). The problem becomes particularly acute, and has not yet been solved, for the movement of a meniscus through the window spaces between spheres. The calculated values of r_h given for this case by Nicholson cannot be more than very approximate estimates of r_m; and indeed are in disagreement with experimental observations on macroscopic systems.[62]

The only other model pore system to have attracted detailed attention in the past is that of packed spheres, and over the years studies have been devoted, at various levels of sophistication, to the condensation between two spheres in contact, between groups of three, four . . . contacting spheres, and in various lattice arrangements of spheres of equal size. The problem of liquid bridges between spheres in contact (and also for other configurations) has recently been subjected to a detailed mathematical analysis by Erle, Dyson, and Morrow.[33] The results have been applied specifically to the calculation of the force between the spheres. For spheres in contact the theory is in agreement with experiment and earlier calculation; for spheres out of contact, not previously treated, agreement with the only available experimental data is

[61] *e.g.* F. E. Hackett and J. S. Strettan, *J. Agric. Sci.*, 1928, **18**, 671; J. M. Haynes, Ph.D. Thesis, Bristol, 1965.

excellent, except for very small separations. This work provides information to enable the curvature, and hence relative vapour pressure, of the bridge left between two spheres at constant separation to be calculated; and also the limiting volume of the bridge (and its vapour pressure) at which, during evaporation, the bridge collapses. Some simple measurements of the rupture of bridges between 5 mm diameter ball bearings confirmed the theoretical predictions. This work is undoubtedly of potential importance in its application to capillary condensation in random sphere packs in which a proportion of spheres just fail to make contact.

Further work continues on the relation between the geometry and adsorbent properties of regularly packed spherical assemblies. The status of the subject has been summarized by Nagiev and Ibragimov,[63] although this article is unlikely to be widely available. Karnaukhov and Kiselev[64] have now elaborated their earlier work.[65] In previous studies it had been assumed that the thickness of the adsorbed film on the spherical particles was the same continuous function of the relative pressure as that corresponding to adsorption on a planar surface of the same material. As adsorption proceeds, condensation occurs at points of contact of the spheres, and the exposed surface decreases. At a critical point when the advancing menisci meet, the windows formed by groups of spheres fill spontaneously, and finally the remaining voids are filled. The methods that were previously proposed apply only when the sphere diameter is much greater than the molecular diameter. The present work was planned to give a more precise account of the phenomena involved in the earlier stages of adsorption, especially for the case in which the sphere size and molecular size are more nearly equal. Adsorption was supposed to occur layer by layer in a stepwise process, and more accurate equations were used to estimate the area of original surface 'blocked' as a result of adsorption. For systems of the kind under consideration, account must also be taken of the mutual overlapping of blocked zones. Calculations were made for regular arrangements of spheres having 0, 2, 4, 6, 8, and 12 points of contact: they show clearly the importance of the decrease in available surface area caused by adsorption. Thus for a sphere radius/adsorptive molecule radius ratio of about eleven in an eight- or twelve-co-ordinated lattice, the first adsorption layer reduces the surface available for further adsorption by 50%. When account is taken of the enhanced adsorption potential in small pores and the possible intrusion of capillary condensation, this reduction in surface area will be even greater; but these factors also increase the amount adsorbed per unit surface, so that despite the reduction in exposed area the adsorption may exceed that on a plane surface of the same area.

[63] M. F. Nagiev and C. S. Ibragimov, *Issled. Obl. Kinet., Model. Optimizatsii Khim. Protsessov*, 1970, **1**, 105.
[64] A. P. Karnaukhov and A. V. Kiselev, *Zhur. fiz. Khim.*, 1970, **44**, 2354 (*Russ. J. Phys. Chem.*, 1970, **44**, 1332).
[65] A. P. Karnaukhov and A. V. Kiselev, *Zhur. fiz. Khim.*, 1960, **34**, 2146 (*Russ. J. Phys. Chem.*, 1960, **34**, 1019); *cf.* W. H. Wade, *J. Phys. Chem.*, 1964, **68**, 1029; 1965, **69**, 322.

Another approach to the problem of capillary condensation in regularly packed lattices of spheres has been published by Morioka, Kobayashi, and Higuchi.[66] Full details are not available but their calculations, for various regular lattice packings, are said to give a more satisfactory account of Kiselev's experimental data than Kiselev's own theoretical work.

The problem of randomly packed spheres has still to be tackled in detail. A method of describing the geometry of random sphere packing has been described by Mason.[67] The structure is analysed in terms of 'pores', each defined as the free space within a tetrahedral subunit formed by joining together the centres of neighbouring – not necessarily touching – spheres. When the co-ordinates of the centres of the spheres are known, then the geometry of the tetrahedra is defined, together with the network of interconnections between the tetrahedra. The common face of two adjacent tetrahedra defines the interconnection between the two pores to which they correspond. Each pore is thus quadruply connected to its neighbours. By making the crude approximation that for each connection there is a critical associated curvature of the meniscus that will just pass through the common face (by evaporation or drainage) and that similarly there is a critical curvature which will refill the pore (by condensation or imbibition), it is possible to examine the capillary properties of the random sphere pack using either the independent-domain model[68] or an appropriate network model to take some account of interconnectivity.

Various methods of building up model random sphere packs are possible;[69, 70] in ref. 67 Mason starts from the radial distribution function for random sphere packing that was derived experimentally by Scott[70] and then uses an approximate statistical method to derive the distribution of properties of tetrahedra consistent with this function. The capillary properties of this packing are derived and shown to have some of the characteristics of both adsorption–desorption curves and drainage–imbibition curves. This paper represents a useful start to the problem of understanding the complex problem of random sphere packs, whose properties depend to a significant extent on network effects.

6 Calculation of Pore-size Distributions from Experimental Data

The above more sophisticated studies of adsorption and condensation in porous bodies do not yet provide a new basis for the analysis of adsorption isotherms to obtain pore-size distributions. Consequently most current methods are still based on a model of independent circular cylindrical capillaries whose properties can be described in terms of the Kelvin equation,

[66] Y. Morioka, J. Kobayashi, and I. Higuchi, *Nippon Kagaku Zasshi*, 1970, **91**, 603.
[67] G. Mason, *J. Colloid Interface Sci.*, 1971, **35**, 279.
[68] D. H. Everett and F. W. Smith, *Trans. Faraday Soc.*, 1954, **50**, 187; D. H. Everett, *ibid.*, p. 1077; 1955, **51**, 1551; see also ref. 47, pp. 1095–1103.
[69] G. Mason, *Discuss. Faraday Soc.*, 1967, No. 43, p. 67.
[70] G. D. Scott, *Nature*, 1962, **194**, 956; G. Mason, *ibid.*, 1968, **217**, 733.

using the properties of the normal bulk liquid to derive the required parameters. We first consider recent work based on these standard methods before dealing with modified procedures, depending on alternative theoretical principles. The feature of the theory about which there has been greatest discussion is the relationship between the radius r_K, calculated from the Kelvin equation, and the physical size of the pore (r_p). The simplest assumption is that leading to equation (4). However, the retreating meniscus in the capillary leaves behind an adsorbed film of thickness t [Figure 2(c)], which is a function of relative pressure, so that (if $\theta = 0$) the hemispherical meniscus has a radius of $(r_p - t)$. It is usually assumed that t may be estimated from the t-curve (see Chapter 1) measured for a non-porous surface. Account must also be taken, in analysing desorption curves, of the decrease both in the amount of capillary-condensed liquid, and of the thickness of the adsorbed layer. Various methods have been proposed for making the appropriate corrections.[71] The calculations needed are conveniently carried out on a computer, and several rather inaccessible papers[72-75] have been published recently dealing with the computational processing of experimental data using one of the previously established methods.

Discussion still continues on the best method of establishing the t-curve, and Dollimore and Heal[76] have examined in detail (again using a computer) the effect of making various assumptions about the t-curve on the final pore-size distribution curve. Nine different t-curves were applied to experimental data on N_2 adsorption by thirty-six samples of silica and alumina. A large measure of agreement was found between the values of the radius at the peak of the distribution, calculated in various ways; however, the 'cumulative surface area' of the pores which is obtained from this analysis often deviated considerably from the area obtained by the application of the BET equation to the lower-pressure adsorption points on the isotherm. The overall conclusion was that the most acceptable expression for t is that attributed to de Boer,[71] namely, $t/\text{nm} = 0.354\,[-5/\ln(p/p^0)]^{1/3}$.

For application to studies of the structure of clays and certain other materials having a laminar structure, one must replace the model of cylindrical pores by one of slit-shaped pores: Jantti and Penttinen's paper[75] deals with this case. The appropriate equations are also dealt with in detail in ref. 4. A recent paper by Delon[77] describes a similar method applied to a number of clay minerals. It was found that most samples showed a maximum in the

[71] J. C. P. Broekhoff and B. G. Linsen, in ref. 4, Chapter 1; J. H. de Boer, B. C. Lippens, and B. G. Linsen, *J. Catalysis*, 1964, **3**, 36; J. H. de Boer, B. G. Linsen, and Th. J. Osinger, *ibid.*, 1965, **4**, 643.
[72] R. A. Churruarin and S. A. Hillar, *Rev. Fac. Ing. Quim., Univ. Nat. Litoral*, 1968, **37**, 363.
[73] Z. Spitzer, V. Biba, and Z. Kalkant, *Sbornik Pr. UVP* (*Vyzk. ustav vyusili paliv*), 1970, No. 22, p. 152.
[74] R. H. C. Rodarte, *Rev. Inst. Mex. Petrol.*, 1971, **3**, 68.
[75] O. Jantti and M. Penttinen, *Suomen. Kem.*, 1970, **43**, B, 239.
[76] D. Dollimore and G. R. Heal, *J. Colloid Interface Sci.*, 1970, **33**, 508.
[77] J. F. Delon, *Silicates Ind.*, 1971, **36**, 89.

pore-size distribution around 2.7 nm, while a number had two maxima, at 1.4 and 2.7 nm, these being simple multiples of the thickness of clay structures. This is interpreted as meaning that the porosity is that of microagglomerations of clay in which the lamellae are held apart by particles of debris which may consist of one or two crystallite layers. Delon and Cases[78] discuss the determination of pore sizes in layered clay minerals using a modified *n*-layer BET equation.[79] The texture of platelet or laminar solids as revealed by adsorption methods also forms the subject of a paper by Rouquerol, Rouquerol, and Imelik,[80] who studied N_2 adsorption by montmorillonite, beryllium oxide and hydroxide, and aluminium oxide and hydroxide (hydrargillite). Hysteresis loops were observed for all samples, and in the case of glucine treated at 1075 °C and alumina treated at 1092 °C a 'waisted loop' was observed (*i.e.*, at intermediate adsorptions the loop was narrower than at higher and lower adsorptions). Attempts to deduce pore-size distributions from these isotherms led to cumulative surface areas which were up to four times larger than the BET surface areas. Arguments are put forward to support the view that the hysteresis is not caused by porosity of the solids themselves, but rather by interlamellar or interparticle condensation.

Kinloch and Machin[81] have studied the hysteresis in the capillary condensation of tetrachloroethylene and trichlorofluoromethane by anhydrous nickel sulphate at -7.5 °C, 0 °C, and $+20.0$ °C. The two adsorbates exhibited interesting differences. For example, the isosteric enthalpy of adsorption (q^{st}) of C_2Cl_4 at coverages around the monolayer capacity is between 8.5 and 9.0 kcal mol^{-1}, and rises to 9.7 (the enthalpy of vaporization of the pure liquid) as adsorption proceeds; for $CFCl_3$ in contrast, q^{st} *falls* from about 7.5 to 6.4 kcal mol^{-1} as adsorption increases. The temperature dependence of the lower closure point of the hysteresis loop was also regarded as anomalous by the authors. However, the above discussion (Section 4) indicates that this is to be expected, and is indeed observed in other systems. The authors interpret their data on the basis of a parallel-plate pore model which enables them to write the Kelvin equation in a linear form according to which $(-1)/\ln(p/p^0)$ should be proportional to the amount adsorbed. That the lines so obtained did not pass through the origin was taken to mean that only part of the adsorbed material was present as capillary condensate, the rest forming an adsorbed film on the surface. The data suggest that the thickness of this film is independent of partial pressure. The slopes of the lines obtained in this analysis lead to values of the contact angle between condensed liquid and solid. The contact angle is found to vary with temperature and to have different values along the adsorption and desorption branches of the isotherms. This leads the authors to attribute the hysteresis in these systems to contact-angle hysteresis. However, the assumptions of the model employed are not set

[78] J. F. Delon and J.-M. Cases, *J. Chim. phys.*, 1970, **67**, 662.
[79] R. Dellyes, *J. Chim. phys.*, 1963, **60**, 1008.
[80] F. Rouquerol, J. Rouquerol, and B. Imelik, *Bull. Soc. chim. France*, 1970, **11**, 3816.
[81] G. A. Kinloch and W. D. Machin, *Canad. J. Chem.*, 1971, **49**, 1515.

out in sufficient detail to enable one to establish whether the conclusions of this work are necessary or represent only one possible interpretation. It will be interesting to see whether measurements of adsorption on other materials with slit-like pores are amenable to a similar analysis.

While the older, more-conventional methods of estimating pore-size distributions based on cylindrical or slit-shaped pores continue to be used, more-sophisticated methods based on alternative theoretical treatments are being developed.

Brunauer and his collaborators[82] have taken up and elaborated a method first used by Kiselev,[83] and more recently by Bakardjiew.[84] This method is based on the equation:

$$\sigma^{lg} \, dA = -(\mu - \mu^0) \, dn^a, \qquad (20)$$

where A is the 'surface area within the pores', μ the chemical potential within the pores, and μ^0 that of the saturated vapour; n^a is the amount of adsorbate. Introduction of

$$\mu - \mu^0 = RT \ln p/p^0, \qquad (21)$$

where p^0 is the saturated vapour pressure, gives:

$$dA = -\frac{RT}{\sigma^{lg}} \ln p/p^0 \, dn^a. \qquad (22)$$

Kiselev assumed that if this expression is integrated from the point at which multilayer adsorption passes over into capillary condensation (n^a_h) (which he identifies as the lower closure point of the hysteresis loop) to the saturated vapour pressure, where the adsorption is n^a_s, then

$$A = -\frac{RT}{\sigma^{lg}} \int_{n^a_h}^{n^a_s} \ln p/p^0 \, dn^a \qquad (23)$$

gives the surface area of the pores. The area so obtained is, according to Brunauer, more accurately described as the internal area of the adsorbed layer within the pore at the point of inception of condensation, *i.e.*, the 'core' area of the pore [*cf.* Figure 2(b)].

To utilize these equations for the calculation of pore sizes, Brunauer converts dn^a into dV, using the bulk liquid density, and replaces the integration by a summation involving a number of small finite steps:

$$A = \sum \Delta A = -\frac{RT}{\sigma^{lg} v^l} \sum \ln p/p^0 \, \Delta V. \qquad (24)$$

[82] S. Brunauer, ref. 5, p. 63.
[83] A. V. Kiselev, *Uspekhi Khim.*, 1945, **14**, 367; A. V. Kiselev in 'Colston Papers, Vol. X, Structure and Properties of Porous Materials', ed. D. H. Everett and F. S. Stone, Butterworths, London, 1958, p. 128.
[84] I. Bakardjiew, *Z. phys. Chem.* (*Leipzig*), 1964, **225**, 273; *ibid.*, 1966, **231**, 382; *Compt. rend. Acad. bulg. Sci.*, 1965, **18**, 545.

Thus for a desorption step the change in area ΔA accompanying a decrease in volume of adsorbate, ΔV, can be calculated. A correction to ΔV is necessary since the desorption includes two processes – emptying of the pores containing capillary-condensed liquid, ΔV^{cap}, and desorption from the multilayer films on the walls of pores which have emptied, ΔV^{ads}. This correction is made, as in standard methods, by estimating the cumulative area of pore wall exposed at the point concerned and multiplying this by Δt, the change in thickness of the adsorbed layer in the step considered. Having found ΔA and ΔV^{cap}, the hydraulic radius, r_h, can be calculated from equation (17). This radius is taken as the mean radius of pores emptying in the step. Brunauer describes the method as 'model-less', since equation (20) is said not to be limited to cylindrical or slit-like pores, although the adoption of a specific pore model is necessary when the correction to ΔV is made and when the core areas are converted to solid surface areas. Recent applications of this method have been to the pore-structure analysis of several silica gels using water adsorption isotherms[85] and of carbon blacks and activated carbon adsorbents using N_2 adsorption.[86]

This method of analysing the properties of porous materials (leaving aside the problem of microporosity – see Chapter 1) was subjected to critical discussion at the international conference on surface area determination.[5] It was pointed out by de Vleesschauwer[87] that there was some apparent discrepancy in the logic of the technique since each term in equation (24) can be written, using equation (3), as:

$$\Delta A = \frac{2\Delta V^{cap}}{r_K}, \qquad (25)$$

where r_K is the Kelvin radius. Having calculated ΔA from equation (25), the ratio $\Delta A/\Delta V$ is then used to calculate the hydraulic radius, which must then equal the Kelvin radius. Furthermore, de Vleesschauwer concluded that equation (25) would give misleading results in pores of non-uniform cross-section, such as 'ink bottles'. Brunauer[88] was of the opinion that equation (20) is true for pores of all shapes, and consequently the criticism is invalid.

A quite different criticism was made by Haynes[89] on the grounds that equation (20) is valid only for reversible processes* and that it must in principle be inaccurate when applied to isotherms showing hysteresis. To assess the seriousness of this limitation he made use of the data of Ferguson and Wade,[91] who measured the complete isotherm of adsorption of water on Vycor porous glass and also the BET N_2 surface areas of the same sample of glass containing

---

* This point had been made much earlier by Kistler, Fischer, and Freeman[90] in connection with their own use of equation (20); and by Kiselev.[83]

[85] J. Hagymassy and S. Brunauer, *J. Colloid Interface Sci.*, 1970, **33**, 317.
[86] J. Skalny, E. E. Bodor, and S. Brunauer, *J. Colloid Interface Sci.*, 1971, **37**, 476.
[87] W. de Vleesschauwer, in ref. 5, p. 82.
[88] S. Brunauer, in ref. 5, p. 84.
[89] J. M. Haynes, in ref. 5, p. 86; see also *J. Colloid Interface Sci.*, 1967, **24**, 513.
[90] S. S. Kistler, E. A. Fischer, and I. R. Freeman, *J. Amer. Chem. Soc.*, 1943, **65**, 1909.
[91] C. B. Ferguson and W. H. Wade, *J. Colloid Interface Sci.*, 1967, **24**, 366.

various amounts of pre-sorbed water. The areas of exposed surface at various stages in adsorption and desorption were calculated by Brunauer's method, as well as by more conventional methods, and compared with the measured BET areas. It was concluded that the application of equation (24) gave good agreement along the *adsorption* curve, but over-estimated the area on the desorption branch.

Attention needs also to be directed to the status of equation (20). It would appear that the rather imprecise definition of A given by various authors arises from the fact that the equation should strictly be written more generally as [*cf.* equation (18)]

$$\sigma^{\mathrm{lg}}(\mathrm{d}A^{\mathrm{lg}} - \mathrm{d}A^{\mathrm{sl}} \cos \theta) = -(\mu - \mu^0) \, \mathrm{d}n^{\mathrm{a}}. \tag{26}$$

This simplifies in certain special cases. For example, if one considers a film on the inside of a cylindrical capillary [Figure 2(b)], $\mathrm{d}A^{\mathrm{sl}} = 0$ and $\mathrm{d}A$ is to be interpreted as $\mathrm{d}A^{\mathrm{lg}}$; alternatively, for a thread of liquid in a cylindrical capillary [Figure 2(c)], $\mathrm{d}A^{\mathrm{lg}} = 0$ and $\mathrm{d}A = \mathrm{d}A^{\mathrm{sl}} \cos \theta = -\mathrm{d}A^{\mathrm{sg}} \cos \theta$. Thus Brunauer's treatment, which identifies $\mathrm{d}A$ with $\mathrm{d}A^{\mathrm{sl}}$ (or $-\mathrm{d}A^{\mathrm{sg}}$), is strictly applicable only to pores of uniform cross-section which are wetted by the adsorbate: the treatment is therefore not model-less even at its inception. That the use of a model is necessary in the application of this method to actual systems was, of course, appreciated by Brunauer;[82] it has not been made clear, however, that a model is implied in the basic equation. If desorption proceeds through pores of varying cross-section in which the area of the liquid/vapour meniscus varies, equation (26) should be used. However, there seems to be no obvious way of separating $\mathrm{d}A^{\mathrm{lg}}$ from $\mathrm{d}A^{\mathrm{sl}}$. It may be noted, however, that for an adsorbed film immediately prior to capillary condensation $A^{\mathrm{lg}} \approx A^{\mathrm{sl}}$, while at saturation vapour pressure A^{lg} is very small, so that despite the limitations of equation (20), equation (23) may well give (for the case of $\theta = 0$) a reliable estimate of the area of solid A^{sl} wetted at p^0, and hence of the total area of solid.

Equation (20) also features in the method proposed by Broekhoff and de Boer[92] for finding the surface area of mesopores. The method is outlined in the volume dedicated to de Boer[93] and was reviewed and discussed at the Surface Area Symposium.[94] This method differs from Brunauer's in that Broekhoff and de Boer introduce a new feature by taking specific account of the influence of the adsorption potential on the chemical potential (μ^{a}) of liquid near the pore wall: they write:

$$\mu^{\mathrm{a}} = \mu^{\mathrm{l}} - \mathrm{F}(t), \tag{27}$$

where $\mathrm{F}(t)$ is a function of the film thickness, and μ^{l} is the chemical potential of liquid far from the wall. Equation (27) is then substituted into (20) to obtain a so-called 'corrected' Kelvin equation. Thereafter the procedure for

[82] J. C. P. Broekhoff and J. H. de Boer, *J. Catalysis*, 1968, **10**, 368.
[83] J. C. P. Broekhoff and B. G. Linsen, in ref. 4, Chapter 1.
[84] J. C. P. Broekhoff and J. H. de Boer, in ref. 5, p. 97.

calculating the cumulative areas and volumes from desorption curves is essentially similar to Brunauer's, the curve being divided into a suitable number of intervals and the corrected increments summed to find the cumulative area of surface exposed. A number of examples of the use of this method of determining surface area and pore-size distribution have been given which show that it is essential to use equation (27) if the areas of mesopores calculated in this way are to agree with those obtained by the *t*-plot method (see Chapter 1). In applying the Broekhoff–de Boer method to solids having cylindrical or spherical pores, the correction for the volume change of the adsorbed layer is more complicated. For the calculations to remain tractable it is necessary to assume that $F(t)$ is independent of the curvature of the pore walls. As noted in Section 2 above, this is reasonable unless the pores approach the micropore range. By taking account of the different configurations of the adsorbed phase in cylindrical capillaries during adsorption and desorption, it is possible to adapt this procedure to the analysis of both branches of the hysteresis loop: it is found that for a number of porous oxides the cumulative areas calculated from the two branches are in good agreement with one another, and with the conventional BET N_2 area.

The fundamental status of the Broekhoff–de Boer theory is also open to discussion. The use of equation (20) has already been commented upon, but the way in which the adsorption potential term in equation (27) was introduced in the earlier papers was somewhat misleading. Equation (27) implies that the chemical potential of liquid close to the surface varies with the distance from the surface, whereas the condition for diffusional equilibrium in the liquid phase is that the chemical potential should remain constant throughout the phase. Indeed it is puzzling (assuming that they employ the usual definition of chemical potential) that Broekhoff and de Boer should say:[95] 'It is, however, impossible to assume that the chemical potential of the adsorbed layer μ^a is equal to that of the bulk liquid, as this would lead to the consequence that no adsorbed layer may ever be stable!'.

This apparent conflict can be resolved by the following reasoning, which, while not offered as a rigorous development, does nevertheless help to understand the way in which the work of the de Boer School is related to earlier somewhat similar arguments of Derjaguin.[96]

Consider first Figure 3(a), which represents the energy relationships of a liquid in contact with a plane solid surface. The line OA represents the chemical potential of the system, which at equilibrium is equal to that of pure bulk liquid and of the saturated vapour. The curve ABCD is the change of free energy of an element of liquid as it is brought up to the surface *at constant temperature and pressure* and becomes increasingly influenced by adsorption forces which give rise to the potential $\varepsilon(z)$. An element of liquid in this state at C, say, would have a lower chemical potential than the bulk liquid far from

[95] Ref. 5, p. 101.
[96] B. V. Derjaguin, *Acta phys. chim. U.R.S.S.*, 1940, **12**, 181; *Proc. 2nd Internat. Cong. Surf. Act.*, Butterworths, London, 1957, Vol. II, p. 153.

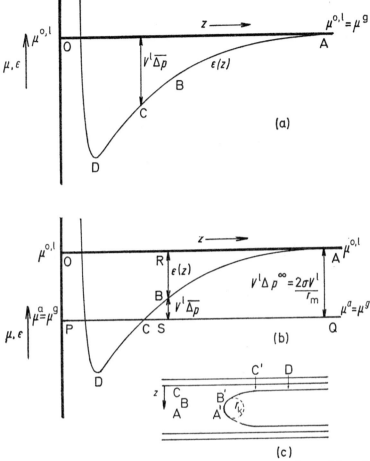

Figure 3 (a, b) *Chemical potential in a liquid in contact with a solid surface as a function of distance from that surface: curves ABCD, for element of liquid at same temperature and state of stress (pressure) as the bulk liquid in equilibrium with saturated vapour pressure:*
(a) *for liquid close to plane surface in equilibrium with vapour at saturation vapour pressure; vertical arrow shows the amount by which the chemical potential must be increased by changes in local stress at C to bring the liquid into equilibrium with bulk liquid;*
(b) *for capillary-condensed liquid in equilibrium with vapour at reduced vapour pressure: vertical arrow at right shows amount by which the chemical potential of liquid far from the wall is decreased by curvature of the meniscus. Because of the contribution from surface forces the reduction in pressure needed at B close to the wall is much less than at A*
(c) *Modification to profile of meniscus resulting from decreased curvature near the walls: the curvature at A′ is equal to the Kelvin radius as measured by the lowering of vapour pressure*

the surface – it is *this* chemical potential to which equation (27) applies. But this is a non-equilibrium situation: to achieve equilibrium the state of stress at C must be changed by an increase in the 'volumetric mean pressure' of $\overline{\Delta p} = \overline{p} - p^{0,1}$ such that:

$$\varepsilon(z) = -v^{\mathrm{l}}\,\overline{\Delta p}. \tag{28}$$

Strictly, $\overline{p}$ has to be defined in terms of the stress tensor at C.[97] (We note in passing that, physically, this means an increase in density at C by movement of molecules from the bulk liquid towards the surface, *i.e.* adsorption at the surface.) We conclude that the adsorption potential $\varepsilon(z)$ produces an excess pressure in the neighbourhood of the surface, and that this is closely related to Derjaguin's 'disjoining pressure'.[98]

Now consider Figure 3(b), where the liquid is supposed to be confined in a capillary the centre of which is outside the range of surface forces. The condensed material is now in equilibrium with vapour at a chemical potential μ^{g} that is lower than $\mu^{0,1}$. At the centre of the capillary the lowering of chemical potential is caused entirely by the pressure drop $\Delta p^{\infty} = 2\sigma/r_{\mathrm{m}}$ across the meniscus; closer to the walls this reduction in pressure is offset by the increase in pressure resulting from the change in adsorption potential:

$$\mu^{\mathrm{a}} - \mu^{0,1} = v^{\mathrm{l}}\,\Delta p^{\infty} = \varepsilon(z) + v^{\mathrm{l}}\,\overline{\Delta p} \tag{29}$$

or

$$\overline{\Delta p} = \Delta p^{\infty} - \frac{\varepsilon(z)}{v^{\mathrm{l}}}. \tag{30}$$

Thus to maintain a constant chemical potential throughout the adsorbed phase the local state of stress must change from a lowering of pressure near the centre of the capillary, through zero mean stress at C, to a compression from C to D. If we now apply similar arguments to elements of liquid behind the meniscus such as A', B', C' ..., and assume that the pressure drop across each element of the interface is given by Laplace's equation, then we must conclude that, assuming the surface tension is uniform and equal to that of the bulk liquid, the mean curvature of the meniscus decreases as the wall (and the adsorbed film) is approached [Figure 3(c)]. This effect was first pointed out by Derjaguin[96] and forms the basis both of his and of Broekhoff and de Boer's 'corrected' Kelvin equations. Incidentally, a change in curvature of an interface arising from a varying pressure in one of the phases is no new concept: this is exactly the situation in a pendant or sessile drop under the influence of gravity. Provided the centre of the capillary (A') is beyond the range of surface forces [$\varepsilon(z) = 0$], the radius of curvature at the centre is equal to the Kelvin radius (r_{K}), but this radius is less than the radius ($r - t$) of the 'core' of the capillary covered with adsorbed material [see Figure 3(c)]. The relationship between r_{K} and ($r - t$) must be calculated by integration

[97] A. G. McLennan, *Proc. Roy. Soc.*, 1968, **A307**, 1.
[98] B. V. Derjaguin, *Colloid J.* (*U.S.S.R.*), 1955, **17**, 1.

round the meniscus, using essentially the same mathematical techniques as those used for calculating the profile of pendant or sessile drops.[99] The equation so obtained, relating $(r-t)$ to $\ln p/p^0$, is what is loosely called the 'corrected' Kelvin equation: strictly speaking the Kelvin equation (3) as such is unaffected by these corrections; what is really calculated is the more sophisticated form of equation (4).

The introduction of the effect of adsorption forces, while taking account of an additional factor, should, however, be carried out with due consideration of its effect on the basic equation (20) (for capillaries of uniform cross-section) or the more general equation (26). The latter is derived from (18), which applies strictly only to surfaces of constant curvature; so it will not, in general, hold for interfaces whose shape is affected by adsorption forces. At first sight, however, equation (20) requires only that the meniscus be of constant area so that, provided this requirement is satisfied, its validity may be unaffected by the introduction of the adsorption potential.

Although the probability that pore-blocking effects will affect the validity of the above methods of finding pore sizes was recognized some time ago,[47] no further critical assessment of this effect on capillary-condensation curves has appeared. However, recent examinations of the effect of pore blocking on the related method of mercury porosimetry will be discussed in the following section.

Clearly there are still many fundamental questions to be answered before we can feel confident of the reliability of these methods of pore-size determination, and of the related methods for estimating the surface areas of porous materials.

7 Mercury Porosimetry

One of the most widely-used practical methods of pore size analysis is that of mercury porosimetry, in which mercury is forced to enter the pores of a previously evacuated sample under successive increments of applied hydrostatic pressure. When equilibrium is reached at each pressure, the curvature of the mercury liquid/vapour interface within the pores is given by equation (9); the volume of mercury intruded at any given pressure is equal to the volume of the pore space which is accessible to an interface of that, or lesser, curvature. Thus, any element of the pore space is characterized by an *entry pressure* which corresponds to the curvature at which mercury enters it.

Since the Laplace equation (9) is a general condition of fluid/fluid equilibrium, it will be seen at once that there is a formal similarity between mercury penetration and the capillary condensation process described earlier, which can be further extended to include the principle of the various suction porosi-

[99] F. Bashforth and J. C. Adams, 'An Attempt to Test the Theory of Capillary Action', Cambridge University Press and Deighton Bell and Company, Cambridge, 1892; see, *e.g.*, J. F. Padday, in 'Surface and Colloid Science', ed. E. Matijevic and F. R. Eirich, Wiley–Interscience, New York, 1969, vol. 1, p. 101.

metry methods. Despite this superficial resemblance, there are important distinctions, which will be discussed later.

The review by Dullien and Batra,[1] quoted earlier, covers the literature up to 1970, and others, by Spencer[100] and by Rootare,[101] deal more specifically with mercury porosimetry up to 1969 and 1968, respectively. Rootare and Nyce[102] have described applications of porosimetry in metallurgy, and a collection of papers has appeared on 'Porosimetry and its Application'[103] (in Czech with English and German summaries), being the published proceedings of a seminar.

Instrumental developments have been concerned mainly with extending the range of the method, to both lower[104] and higher[105] pressures.

The application of pressure porosimetry methods depends on the existence of a contact angle greater than $90°$ between the porosimeter fluid and the solid being examined, and values of the contact angle of mercury against various solids have been reported in the past.[106] In the absence of measurements, a value of $140°$ is generally assumed (see, *e.g.*, ref. 101).

Solids which are wetted by, or which amalgamate with, mercury present special problems, and Svata[107] has described methods which enable porosimetric methods to be applied to porous silver electrodes. The solid may either be protected with a non-reactive and non-wetted sulphide film, or may be rendered hydrophobic by appropriate surface treatment, glycerine being substituted for mercury as porosimeter fluid in the latter case. The results suggest that, for relatively coarse pores at least, these treatments do not cause significant blocking of the pores, or change in their size. Another approach suggested by Svata, described as gas porosimetry, is based on displacement of a wetting liquid from a saturated pore system by application of pressure to a surrounding gaseous medium. It should be noted that the incompressibility of the displaced phase introduces an important difference in principle in this method, since displacement has to be carried out unidirectionally in order that the displaced fluid can be collected. Thus, unlike mercury porosimetry, in which the displaced phase (vacuum) is infinitely compressible, gas porosimetry is restricted to the measurement of pores larger than the largest pore in the supporting material used to maintain contact with bulk wetting liquid. This restriction arises simply from the mechanism by which the wetting phase is transferred; if this took place through the vapour, as in the desorption process in capillary condensation, then the results should more closely

[100] D. H. T. Spencer, *BCURA Monthly Bulletin*, 1969, **33**(10), 228.
[101] H. M. Rootare, *Aminco Lab. News*, 1968, **24**(3).
[102] H. M. Rootare and A. C. Nyce, *Powder Met.*, 1971, **7**, 3.
[103] 'Porozimetrie a Jeji Pouziti' ed. S. Modry, Ceskoslovenska silikatova spolecnost, Prague, 1969.
[104] D. J. Baker, *J. Phys. (E)*, 1971, **4**, 388.
[105] Anon., *Aminco Bull.*, January 1970, 2410A.
[106] A. H. Ellison, R. B. Klemm, A. M. Schwartz, L. S. Grubb, and D. A. Petrash, *J. Chem. and Eng. Data*, 1967, **12**, 607; S. Kruyer and D. W. van Krevelen, *Fuel*, 1958, **37**, 118; P. G. Sevenster, *ibid.*, p. 506.
[107] M. Svata, in ref. 103, pp. 94–121.

resemble those of mercury porosimetry. Thus, gas porosimetry is identical in principle with the pressure-plate method used in suction porosimetry.[108]

An important distinction between mercury penetration and desorption of capillary condensate is that in the latter case (see Section 6) a correction has to be applied to include changes in the volume of the multilayer film which exists, at equilibrium, on the walls of vapour-filled pores. An analogous correction in mercury penetration would take into account the possibility that the density of mercury close to the walls of mercury-filled pores might differ from the bulk density, in a way which depends both on distance from the wall and on applied pressure. No such effect has ever been considered, apparently.

Another distinction arises from the fact that whereas, in capillary condensation, Kelvin equilibrium is maintained by means of evaporation and condensation processes, such processes are altogether ignored in mercury penetration. This is permissible, within the time-scale of the experiment, because of the low vapour pressure of mercury. The principal result is that whilst capillary condensation generally leads to complete filling of the pore space with wetting phase at saturation vapour pressure, the corresponding process of withdrawal of mercury after penetration rarely, if ever, reaches completion; some residual mercury persists, despite its positive interfacial curvature.

At the high pressures needed to force mercury into relatively fine pores, deformable materials may suffer structural damage, and Rossi and Usai[109] have observed irreversible changes in the structure of fragile materials. Ione and Karnaukhov[110] have considered the effects of dimensional changes at high pressure during porosimetric examination of catalyst structures, and have also discussed the interpretation of measurements in pore systems of irregular geometry, and the effects of changes in the contact angle and surface tension of mercury.

Mercury intrusion–withdrawal curves, like adsorption–desorption curves of capillary condensation, generally show marked hysteresis, in addition to the incomplete withdrawal effect noted above. The irreversible process underlying such hysteresis is usually identified as the spontaneous passage of the liquid/vapour interface through regions of non-uniformity in the pore radius, that is, 'ink-bottle' pores, which fill and empty at pressures corresponding to the radii of their 'throats' and 'cavities'. Svata[111] has recently applied an analysis due to Reverberi *et al.*[112] to the investigation of the distribution of throat and cavity radii in powder metallurgical compacts. The method depends on the examination of scanning curves within the main hysteresis loop. For two such re-pressurization curves as those beginning at A and B in Figure 4,

[108] See, *e.g.*, D. Croney, J. D. Coleman, and P. M. Bridge, 'The Suction of Moisture Held in Soil and other Porous Materials', D.S.I.R., Road Research Techn. Paper, No. 24, 1952.
[109] G. Rossi and G. Usai, *Quad. Ing. Chim. Ital.*, 1970, **6**, 134.
[110] K. G. Ione and A. P. Karnaukhov, *Kinetika i Kataliz*, 1971, **12**, 212.
[111] M. Svata, *Powder Technol.*, 1971–72, **5**, 345.
[112] A. Reverberi, G. Feraiolo, and A. Peloso, *Ann. Chim. (Italy)*, 1966, **56**, 1552.

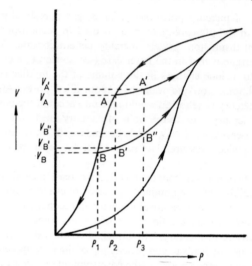

Figure 4 *Illustrating the method proposed by Svata[111] for calculation of distribution of throat and cavity sizes in 'ink-bottle' pores*

for example, the difference in the volumes intruded between two fixed pressures, *e.g.* $[(V_B'' - V_B') - (V_A' - V_A)]$, is identified as the volume of 'ink-bottle' pores having neck radii corresponding to the pressure range p_2 to p_3.

This interpretation can be clarified if we plot 'domain complexion diagrams' after Everett *et al.*[68] (see Figure 5). In such diagrams, orthogonal axes indicate the intrusion pressure and the withdrawal pressure characteristic of each 'domain' of the pore structure; the volume of pores in any domain class may be represented by the height of a surface lying above the plane of the diagram.

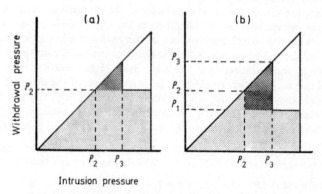

Figure 5 *Use of domain complexion diagrams[68] to examine the validity of Svata's method (see Figure 4)*

Figure 5(a) represents the domain state at the point A′ of Figure 4. The volume of filled pores is given by the volume overlying the shaded area, the more heavily shaded triangle representing that part which is filled in going from A (intrusion pressure p_2) to A′ (intrusion pressure p_3). Figure 5(b) shows the domain state at the point B″; again, the more heavily shaded areas show the result of increasing the pressure from p_2 to p_3, this time passing from B′ to B″. On this basis the difference $[(V_B″ - V_B′) - (V_A′ - V_A)]$ is the volume of only those pores having both an intrusion pressure between p_2 and p_3 and a withdrawal pressure between p_1 and p_2.

It may be noted that the 'ink-bottle' pore model is associated with the assumption, implied in the independent-domain treatment, that a given domain can be identified physically with a particular section of the pore space which is the same for both the filling and emptying processes. In a real porous medium, however, in which pores of various shapes and sizes form a three-dimensional interconnected network, there is a so-called 'pore-blocking' effect: for example, a group of pores having a given entry pressure for mercury intrusion may be accessible to mercury only through pores of higher entry pressure. In one sense, such a group of pores might be said to constitute a single 'ink-bottle'. However, the corresponding pore-blocking effect during mercury withdrawal will in general be associated with a different group of pores. For systems exhibiting blocking effects it is thus incorrect to identify, in any physical sense, a volume element in the domain complexion diagram with a volume element in the pore system. A generalized extension of the independent-domain model in this direction presents difficulties; these can be overcome if it is permissible to assume that pore-blocking effects occur in one direction only, as in the case of capillary condensation, where vapour transport during pore filling renders all pores equally accessible.[47]

One immediate result of the pore blocking that occurs during mercury penetration is that the apparent pore-size distribution will be biased in favour of smaller pores, which will include in their volume that of any larger pores they have concealed. Sneck and Oinonen,[113] comparing the results of mercury penetration and nitrogen adsorption on 26 different building materials, found that in the pore-size range from 18—30 nm (180—300 Å), where both methods are applicable, mercury porosimetry yielded a much larger pore volume, by a factor of up to 10. This they ascribed to the presence of 'ink-bottle' pores. A similar explanation of a comparable observation had earlier been offered by Dubinin,[114] who suggested that information on pore shape might be gained in this way.

Orr[115] attributed hysteresis in mercury penetration to the effect of 'ink-bottle' pores and, further, identified the diameters of the throats and cavities with the peaks of the two pore-size distributions obtained from penetration

[113] T. Sneck and H. Oinonen, 'Measurements of Pore Size Distribution of Porous Materials', Publication 155, State Institute for Technical Research, Finland, 1970.
[114] M. M. Dubinin, in 'Chemistry and Physics of Carbon', ed. P. L. Walker, jun., Marcel Dekker, New York, 1966, Vol. 2, pp. 51–120.
[115] C. Orr, *Powder Technol.*, 1969–70, **3**, 117.

and withdrawal, respectively. This is clearly a much cruder interpretation than that of Svata.[111]

In the same paper, Orr asserts that the specific surface of a porous material can be obtained from an appropriate integration over the mercury penetration curve:

$$A = -\frac{1}{\sigma \cos \theta} \int_0^{p_{max}} p \, dV. \tag{31}$$

The origin of this approach (attributed in this connection by Orr to Rootare and Prenzlow[116]) is clearly the same as that of the so-called Kiselev equation (20), discussed earlier in its application to capillary condensation, and the same reservations apply. Orr restricts its use to cases where 'there are no pores with entrance regions smaller than their dimensions elsewhere' which seems, on the basis of his earlier remark concerning 'ink-bottle' pores, to rule out most cases in which hysteresis occurs.

It is worth noting that an analogous equation has been used to derive specific surface from moisture characteristic, or suction, curves.[117]

Svata and Zabransky[118] have derived effective particle sizes for various powdered materials from measurements of mercury penetration pressures in unconsolidated packed beds. They used a theory developed by Mayer and Stowe[119] to describe mercury penetration between regular packings of equal spheres, a feature which might be expected to limit its applicability. Svata and Zabransky did indeed find better agreement with optically determined particle size for the more uniform of their powders, although departures from sphericity had little effect. It should be pointed out that the success of Mayer and Stowe's theory, even at this level, must be largely accidental, since their analysis of interfacial curvature within a lattice packing of spheres is incorrect. The theory depends on geometrical calculations of a one-dimensional (line) analogue of the 'effective area' [defined, see equation (18), by $A^{lg} - A^{sl} \cos \theta$], which is projected on to the plane of minimum cross-section of the pore. They did not, however, take into account the fact that the intersection of the mercury surface with the solid spheres does not lie wholly within this plane, but extends into regions where the pore has a different cross-section. The correct calculation of the length of the interfacial perimeter thus becomes a matter of very considerable difficulty.

Svata and Zabransky further calculated distributions of particle sizes, as well as the mean sizes, from porosimetry curves. Here they found significant disagreements with the direct measurements, which they ascribed to the effects of pore blocking. (The bias in pore-size distributions arising from this cause was mentioned above.) In general, it appears difficult to formulate any

[116] H. M. Rootare and C. F. Prenzlow, *J. Phys. Chem.*, 1967, **71**, 2733.
[117] D. Payne, *Nature*, 1953, **172**, 261; G. Ringqvist, *Handlingar Svenska Forsknings-institutet for Cement och Betong*, 1955, No. 28.
[118] M. Svata and Z. Zabransky, *Powder Technol.*, 1969–70, **3**, 296.
[119] R. P. Mayer and R. A. Stowe, *J. Colloid Sci.*, 1965, **20**, 893.

relationship between the particle size distribution of a powder and the size distribution of the pores between the particles packed to a given density, except on the basis of strongly restrictive assumptions. The use of the Mayer–Stowe theory has nevertheless been recommended by Orr.[115]

8 Phase Changes in Adsorbed Materials

We shall limit consideration here to phase transformations in capillary-condensed phases. Since the chemical potentials of *bulk* gas, liquid, and solid phases at a given temperature and pressure are independent of the amount of phase present, co-existence of all three phases is possible at only one point, the triple point. However, when one or more of the phases is present in small aggregates, the chemical potentials are dependent upon the surface to volume ratio of the aggregates, so that co-existence of three phases, one or more of which is finely divided, is no longer confined to the triple point but can occur under a variety of conditions.[120] Theoretical analysis of the problem indicates that, depending upon the types of interface present and the sign and magnitude of their curvatures, the triple point may be raised or lowered in capillary systems. All experimental evidence, however, indicates that the freezing point of capillary-condensed liquid is lower than the normal freezing point, so that the number of possible surface configurations to be considered may be reduced.

The problem is most simply discussed in terms of the changes in hydrostatic pressure in the co-existing phases arising from the curvatures of the interfaces (for simplicity it is supposed that the solid phase may be discussed, at least semi-quantitatively, as though it is subjected to an isotropic pressure). We may first dismiss the possibility that the co-existing liquid and solid are subjected to the same pressure, for, as is well known, this will increase or decrease the freezing point depending on whether the molar volume of the solid is less than or greater than that of the liquid. The configuration of interfaces which is most likely to account for the observed behaviour is that shown in Figure 6(a), in which the inability of the crystals of solid to grow to bulk size within the porous medium leads to a solid/liquid interface of positive curvature. This increases the pressure in the solid and hence its chemical potential, while the capillary-condensed liquid is subjected to a lower hydrostatic pressure than the vapour phase and has a chemical potential lower than bulk liquid. As indicated in Figure 6(b), these two effects reinforce one another in lowering the equilibrium temperature. The magnitude of the freezing-point lowering depends on the curvatures of both liquid/vapour and solid/liquid interfaces and increases with decrease in the pore sizes in the neighbourhood of either of these interfaces. Furthermore, since the vapour, liquid, and solid phases exert differing pressures on the porous matrix, swelling and shrinking phenomena will be expected; the nature of these effects will depend in a complex fashion on the relative amounts of these phases present under given

[120] See, *e.g.* ref. 11, Chapter XV, §9.

(a)

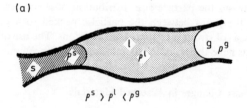

(b)

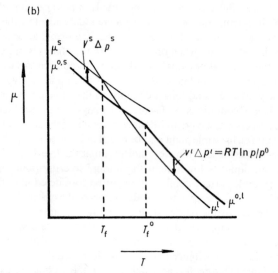

Figure 6 (a) *Probable configuration of interfaces leading to lowering of freezing point in capillary-condensed liquid (schematic)*
(b) *Effect of capillary forces arising from (a) on chemical potential curves of solid and liquid phases leading to reduction of freezing point from T_f^0 to T_f.*

conditions of temperature and of vapour pressure. Complex hysteresis behaviour is also to be expected. These phenomena are important in a number of practical applications, especially those concerned with the influence of adsorption phenomena and phase changes in adsorbed materials on the durability of and damage to building materials and roads: consequently work dealing with these problems is widely dispersed in the literature on soil science, building research, and highway research. In this present short note we deal for the moment only with the few recent papers falling within the ambit of *Chemical Abstracts*: a more comprehensive survey is planned.

According to the above interpretation of phase-change equilibria, the freezing point will be a function of the pore structure of the material so that, since most real adsorbents exhibit a range of pore sizes, the freezing and melting processes will be spread over a range of temperature. The enthalpy of the phase change will also be spread out and will appear as a heat-capacity anomaly rather than as a latent enthalpy. Among the earlier work which demonstrated the existence of heat-capacity maxima associated with the melting of thick adsorbed layers at temperatures below the bulk melting point was that of Patrick and Kemper[121] for water, naphthalene, benzene, and p-nitrotoluene, each adsorbed on silica gel. One recently reported example of this type of behaviour is the work of Berezin, Kiselev, Kozlov, Kuznetsova, and Firsova,[122] in which the adsorption of benzene by silica gel was studied at temperatures well below its melting point, and heat-capacity measurements were made on samples of gel containing known amounts of benzene from − 130 °C to room temperature. This paper is published only as an extended abstract, the full text having been deposited at VINITI, and not all the details of the work are clear. The experimentally observed variation of the heat capacity with temperature for various amounts of adsorbed benzene shows that the freezing point of adsorbed benzene varies from about − 50 °C to 0 °C, with a maximum contribution at temperatures varying from − 30 °C for small amounts adsorbed to − 5 °C near saturation. The paper claims to correlate this with the variation with temperature of the maximum amount of benzene adsorbed (see below). In a related study, Berezin, Kiselev, Kozlov, and Kuznetsova[123] have studied the heat capacity of water adsorbed by silica gel over the temperature range − 150 to + 20 °C. Again a maximum in the heat-capacity curve was observed at adsorptions corresponding to more than one monolayer of adsorbed molecules. The position of the maximum depended on the degree of adsorption.

While it is tempting to interpret these data in terms of the effect of capillary condensation, one must remember that adsorption on non-porous solids can lead to somewhat similar phenomena. This was first convincingly shown by Morrison, Drain, and Dugdale,[124] who found that nitrogen adsorbed on rutile exhibited a heat-capacity maximum about 7 K below the normal melting point when the surface was covered by about two layers of adsorbed material; as the coverage increased the maximum became higher and sharper and moved towards the normal melting point. Similar results have been obtained in the intervening period with other non-porous solids and other adsorptives. A recent paper by Plooster and Gitlin[125] reports work on water adsorbed on two non-porous silica surfaces: heat-capacity maxima exhibiting an increase in height and appearing at increasing temperatures as the thickness of the film increased were observed. At sufficiently high adsorptions a second sharp heat-capacity peak at 0 °C, corresponding to the presence of liquid-like water,

[121] W. A. Patrick and W. A. Kemper, *J. Phys. Chem.*, 1938, **42**, 369.
[122] G. I. Berezin, A. V. Kiselev, A. A. Kozlov, L. V. Kuznetsova, and A. A. Firsova, *Zhur. fiz. Khim.*, 1970, **44**, 541 (*Russ. J. Phys. Chem.*, 1970, **44**, 305).
[123] G. I. Berezin, A. V. Kiselev, A. A. Kozlov, and L. V. Kuznetsova, *Zhur. fiz. Khim.*, 1970, **44**, 1569 (*Russ. J. Phys. Chem.*, 1970, **44**, 879).
[124] J. A. Morrison, L. E. Drain, and J. S. Dugdale, *Canad. J. Chem.*, 1952, **30**, 890.
[125] M. N. Plooster and S. N. Gitlin, *J. Phys. Chem.*, 1971, **75**, 3322.

was found. These results were attributed not to capillary condensation effects, but to the alteration in the structure and properties of the adsorbed water layer by silica surfaces, the magnitude of the effect decreasing with increasing distance from the surface. These effects were found to be greater the more hydrophilic the surface. Conversely, Serpinet[126] claims to have obtained evidence from gas-chromatographic studies over a range of temperatures that a thin film of organic material coated on a macroporous, high-surface-energy support melts at a *higher* temperature than the bulk solid, the effect being more marked for fluorocarbons than for hydrocarbons. No explanation of these observations is put forward.

Further work on the adsorption of water by porous glass at sub-zero temperatures has been carried out by Sidebottom and Litvan.[127] Both adsorption and length-change isotherms were determined, but no evidence of solidification was detected down to -35 °C. The freezing of water in cement has been studied by Bykov and Mikhailov,[129] who also used a dilatometric technique. Expansion of the samples between -10 and -50 °C is attributed to freezing of water in macropores, while the subsequent contraction is said to indicate that all the 'sorbed' water had frozen. It is claimed (although the experimental evidence is far from clear) that adsorbed water shows no signs of freezing down to -100 °C. Experiments to investigate the variation of freezing point with amount of sorbed water seem to show (although again the figure caption and text are ambiguous in the English translation) that the freezing point varies from -50 °C for 50% by weight (of maximum uptake?) to 0 °C for 100% by weight of water. (The alternative interpretation, which is what the text seems to imply, that 100% of the water is frozen at 0 °C and only 50% at -50 °C is physically unacceptable.)

Both Berezin *et al.*[122, 123] and Sidebottom and Litvan[127] pay particular attention to the rapid fall-off in the maximum amount adsorbed at the saturation vapour pressure as the temperature is decreased. Their work seems to confirm the explanation previously given for this phenomenon,[129] namely that although the condensate retains liquid-like properties and the adsorption behaviour is related to $p/p^{0,1}$ where $p^{0,1}$ is the saturated vapour pressure of (supercooled) liquid at the temperature in question, the maximum pressure actually attainable experimentally is $p^{0,s}$, the vapour pressure of the bulk solid. As the temperature is lowered, the upper end of the 'true' isotherm thus becomes truncated and the maximum adsorption decreases. According to this explanation, this behaviour does not depend upon the existence of a phase change; indeed, Sidebottom and Litvan use the fact that isotherms at different temperatures superimpose with fair accuracy when plotted in terms of $p/p^{0,1}$ as an

[126] J. Serpinet, *Nature Phys. Sci.*, 1971, **232**, 42.
[127] E. W. Sidebottom and G. G. Litvan, *Trans. Faraday Soc.*, 1971, **67**, 2726.
[128] V. M. Bykov and N. V. Mikhailov, *Kolloid. Zhur.*, 1970, **32**, 342 [*Colloid J. (U.S.S.R.)*, 1970, **32**, 283].
[129] C. H. Amberg, D. H. Everett, L. H. Ruiter, and F. W. Smith, *Proc. 2nd Int. Congr. Surface Activity*, Butterworths, London, 1957, Vol. II, p. 3; A. V. Kiselev and V. V. Kulichenko, *Zhur. fiz. Khim.*, 1955, **29**, 663.

argument that no freezing had occurred in isotherm experiments down to
$-30\ ^{\circ}$C. On the other hand, Berezin *et al.* identify the temperature range in
which there is a sharp drop in the maximum adsorption with the freezing
range of the adsorbed material. It certainly coincides with the range in which
the heat capacity exhibits a maximum, and the relationship between the heat
capacity and temperature, calculated from the slope of the curve of maximum
adsorption against temperature, is in very good agreement with the experimental heat-capacity curve. However, the basis of the theoretical treatment is
not explained in the abstract, so the assumptions which it involves cannot be
assessed.

9 Preparation and Properties of Specific Porous Media

Technological interest in many different porous solids ensures a continuing
interest in their preparation and development for specific purposes. Of the
many publications dealing with such work which have appeared during the
review period, only those relevant to the principles of preparation of particular
materials will be discussed here.

Alumina.—Porous alumina, with a specific surface of the order of hundreds of
square metres per gram and pore radii in the nanometre range, can be prepared by dehydration ('thermal activation'), at over $500\ ^{\circ}$C, of hydrous alumina
gels, obtained either by slightly alkaline precipitation from aluminium salt
solutions or by controlled hydrolysis of aluminium alkoxides. *X*-Ray studies
show that the amorphous hydrous alumina recrystallizes slowly at room
temperature to give a gel structure containing microcrystalline boehmite,[130]
a reaction which is accelerated by autoclaving[131] at 150—$350\ ^{\circ}$C or by stirring
at $95\ ^{\circ}$C at a pH either moderately alkaline (~ 11) or slightly acid (5.3).[132]
When the initial hydrous alumina is prepared by hydrolysis of aluminium
alkoxides in organic media, the subsequent boehmite recrystallization can be
suppressed, yielding a product of high specific surface, if no more than the
stoichiometric quantity of water is used in the hydrolysis stage.[133] A curious
observation is that of the effect on the rate of recrystallization of carrying out
the precipitation in an ultrasonic field;[134] at a frequency of 20 kHz a more
amorphous product is obtained, with increased specific surface and smaller pore
size and crystallite size, whilst at higher frequencies (1000—2100 kHz)
recrystallization is apparently promoted, yielding a product with larger pores,
larger crystallites, and lower specific surface. No effect is observed at inter-

[130] D. Aldcroft, G. C. Bye, and G. O. Chigbo, *Trans. Brit. Ceram. Soc.*, 1971, **70**, 19.
[131] V. M. Chertov, V. I. Zelentsoz, and I. E. Neimark, *Doklady Akad. Nauk S.S.S.R.*, Ser.
Khim., 1971, **196**, 885 (*Doklady Phys. Chem.*, 1971, **196**, 125).
[132] R. L. Dubrovinskii, I. S. Krasotkin, and A. S. Kuz'menko, *Nov. Issled. Tsvet. Met.
Obogashch*, 1969, 23 (*Chem. Abs.*, 1971, **75**, 80 601).
[133] M. A. Vicarini, G. A. Nicolaon, and S. J. Teichner, *Bull. Soc. chim. France*, 1970, 431.
[134] T. Paryjczak, *Zeszyty Nauk. Politech. Lodz.* (*Chem.*), 1969, No. 19, p. 43 (*Chem. Abs.*,
1970, **72**, 16 028).

mediate frequencies. A porous product can also be obtained by suitable calcination of the boehmite produced by ageing $Al(OH)_3$ gels. By carrying out the boehmite recrystallization in an autoclave at a carefully controlled pH just below 7.0, a final product of uniform pore size (~ 65 nm) is obtained.[135] Because the pores are relatively large, the material is readily permeable and does not swell significantly in various solvents; it is therefore recommended for use in gel-permeation chromatography, under the name 'Poramina'. (A systematic investigation of the behaviour of alumina of various pore sizes as a support medium in gel-permeation chromatography was reported in 1969.[136]) It has also been observed that alumina prepared by the gel precipitation method, in which an organic gel-forming agent such as a polyacrylamide is incorporated in the precipitated hydrous alumina and removed by oxidation above 500°C after drying, has a structure containing coarse pores, which render the fine pores more accessible, and reduce swelling and shrinkage.[137] These coarse pores are presumably left behind on removal of the organic polymer. Incorporation of a few mole percent of ZrO_2[138] or Bi_2O_3,[139] followed by heating to 800—1000 °C, produces a γ-Al_2O_3 that is stabilized against transition to α-Al_2O_3, and more resistant to sintering; the pore size is slightly increased, from 4—5 nm to 6—8 nm, at the same time.

Silica.—By suitable adjustment of the pH of a sodium silicate solution, a silica sol is obtained which can be transformed into a gel by appropriate treatment. Variations of this basic recipe are the subject of numerous patents.[140] If the gel is maintained in contact with water, especially at an elevated temperature, it undergoes an ageing process,[141] resulting in a material of greater mechanical stability but lower specific surface. Even atmospheric moisture, after prolonged contact, can cause a loss of specific surface and of pore volume.[142] In contact with dilute HF, the ageing process is accelerated, supposedly because of the differential solubility of particles of different sizes which leads to an increase in average particle size and a decrease in specific

[135] S. Sato and Y. Otaka, *Chem. Econ. Eng. Rev.*, 1971, **3**, 40.
[136] N. Baba, T. Hara, and Y. Ueda, *Shimadzu Hyoron*, 1969, **26**, 279 (*Chem. Abs.*, 1970, **72**, 70 939).
[137] N. S. Repina, N. F. Ermolenko, and M. D. Efros, *Vestsi Akad. Navuk Belarusk. S.S.R., Ser. khim. Navuk.*, 1969 (6), 113 (*Chem. Abs.*, 1970, **72**, 59 493).
[138] N. F. Ermolenko, N. S. Repina, and M. D. Efros, *Vestsi Akad. Navuk Belarusk. S.S.R., Ser. khim. Navuk.*, 1970 (5), 17 (*Chem. Abs.*, 1971, **74**, 35 008).
[139] N. F. Ermolenko and N. S. Repina, *Vestsi Akad. Navuk Belarusk. S.S.R., Ser. khim. Navuk*, 1971 (2), 5 (*Chem. Abs.*, 1971, **75**, 54 545).
[140] T. Hill, Ger. Offen. 1 958 312 (26 Nov. 1970) (*Chem. Abs.*, 1971, **74**, 45 917); E. G. Acker, U.S.P. 3 526 603 (1 Sept. 1970) (*Chem. Abs.*, 1970, **73**, 102 400); E. G. Acker and M. E. Winyall, Fr.P. 1 560 783 (21 Mar. 1969) (*Chem. Abs.*, 1970, **72**, 36 188); H. A. Aboutboul, J. H. Krekeler, and W. Kirch, Ger. Offen. 1 940 093 (24 Sept. 1970) (*Chem. Abs.*, 1970, **73**, 123 923); H. A. Aboutboul and J. H. Krekeler, Ger. Offen. 1 945 422 (16 Apr. 1970) (*Chem. Abs.*, 1970, **72**, 136 825).
[141] L. M. Sharygin, V. G. Chukhlautsov, and G. F. Bakina, *Kinetika i Kataliz*, 1970, **11**, 187.
[142] R. Yu. Sheinfain and O. P. Stas, *Zhur. priklad. Khim.*, 1969, **42**, 2363.

surface, which in this instance was accompanied by an *increase* in porosity.[143] On the other hand, Robinson and Ross[144] concluded that under the less severe conditions of autoclaving for 2 h at 140 °C, the predominant mechanism was one of aggregation of primary particles into more dense 'globules', with consequent reduction in specific surface and increase in mean pore size (the pores involved presumably being those between, rather than within, the 'globules'). They also found some evidence, both from high-resolution electron micrographs and from the shapes of nitrogen adsorption-desorption hysteresis loops, of a concomitant change in pore shape, described as a reduction in the ratio of maximum ('cavity') to minimum ('throat') diameter along a given pore. A Russian paper[145] suggests that as the 'globules' or aggregates of primary particles grow in size, so their packing density increases. Thus, the pores remain of roughly constant dimensions, while the total pore volume decreases. In a study of the pore-structural and surface-chemical changes accompanying autoclaving at various temperatures,[146] it was found that the development of microporosity reached a maximum at autoclaving temperatures of 130—150 °C. The surface density of hydroxy-groups remained roughly constant, however (at 7—9 μmol m^{-2}), even though the specific surface varied twenty-fold. When ageing is carried out in organic liquids such as ethanol, dioxan, and glycerol, the reduction in specific surface is less than in the case of ageing in water; it is suggested that stabilization occurs by complex formation between the protonated hydrogen atom of the surface silanol groups and the solvent.[147] Ageing also occurs in acetone, with formation of elongated particles and decrease in the surface density of hydroxy-groups, although the specific surface does not change significantly.[148] High-temperature drying, or even calcination, of the silica gel has further effects on its structure and properties. Thus, Mikhail and Shebl[149] found that heating wide-pored gels to 400 °C rendered them markedly hydrophobic, although their specific surface, as measured by nitrogen adsorption, was scarcely changed. Exposure of the gels to saturated water vapour restored their hydrophilic character, the change being accompanied by some irreversible uptake of water. A microporous gel, on the other hand, remained hydrophilic throughout. Heating the gels to 1000 °C reduced both specific surface (markedly) and pore volume. In another study[150] it was concluded that changes in specific surface

[143] O. P. Stas, R. Yu. Sheinfain, and I. E. Neimark, *Kolloid. Zhur.*, 1970, **32**, 104 [*Colloid J. (U.S.S.R.)*, 1970, **32**, 81].
[144] E. Robinson and R. A. Ross, *Canad. J. Chem.*, 1970, **48**, 2210.
[145] G. M. Belotserkovskii, V. Kh. Dobruskin, and T. G. Plachenov, *Zhur. priklad. Khim.*, 1970, **43**, 1380.
[146] R. L. Gorelik, L. T. Zhuravlev, A. V. Kiselev, Yu. S. Nikitin, E. B. Oganesyan, and K. Ya. Shengeliya, *Kolloid. Zhur.*, 1971, **33**, 58 [*Colloid J. (U.S.S.R.)*, 1971, **33**, 42].
[147] R. Yu. Sheinfain, O. P. Stas, and I. E. Neimark, *Kolloid. Zhur.*, 1970, **32**, 451 [*Colloid J. (U.S.S.R.)*, 1970, **32**, 375].
[148] V. B. Aleskovskii and N. G. Roslyakova, *Kolloid. Zhur.*, 1971, **33**, 186 [*Colloid J. (U.S.S.R.)*, 1971, **33**, 151].
[149] R. S. Mikhail and F. A. Shebl, *J. Colloid Interface Sci.*, 1970, **34**, 65.
[150] G. M. Belotserkovskii, V.Kh. Dobruskin, G. E. Kireeva, and T. G. Plachenov, *Zhur. priklad. Khim.*, 1970, **43**, 445.

and density after heating to 750 °C depended mainly on the presence of residual alkalies remaining in the gel after manufacture. Kiselev *et al.*[151] studied gels which had been first autoclaved and then calcined at 900—1000 °C, and finally rehydroxylated. They found that all microporosity had been eliminated after 6—10 h calcination, but that longer calcination caused microporosity to reappear. Silica sols can also be prepared by treating sodium silicate solutions with ion-exchange resins in the hydrogen form.[152] Such sols can be stabilized at high concentration by adjusting to a slightly alkaline pH, and on drying they yield xerogels of average pore size 1.3—1.5 nm, and specific surface about $700 m^2 gm^{-1}$.[153] The effect of included salts in producing modified silica gels is well known. It has recently been found[154] that the gas-chromatographic adsorption behaviour of silica gel towards hydrocarbon mixtures is very substantially changed when controlled amounts of copper are deposited in the gel from $CuSO_4$ solutions. Another method of modifying the surface properties is by reaction with halogenated silanes. Phenyltrichlorosilane,[155] for example, reacts with adjacent pairs of surface hydroxy-groups, to the extent of about 75 % of the total number available, giving a surface consisting of phenylchlorosilyl groups. When reaction is complete, the product is strongly hydrophobic. Although in gels with pore sizes of 20 and 110 nm there was no change in pore size after treatment, a gel with pores originally of 1.4 nm was apparently without accessible pore space after reaction with phenyltrichlorosilane. When a silica gel is allowed to imbibe a concentrated silicic acid sol and is subsequently dried, a material described as 'water-resistant' is obtained, in which the macropore volume is reduced and the mechanical strength increased.[156] The relationship between the structure of silica gels and their suitability as support media in gel-permeation chromatography has been examined;[157] their chromatographic behaviour could not be simply correlated with their pore structures as revealed by mercury porosimetry. The use of silica gels as soil reinforcement agents is well known. A paper by Caron[158] describes recent developments in this technique.

Other Oxides.—Oxides of many other metals can be produced in a porous form. Sometimes these are prepared by methods similar to those used for SiO_2 and Al_2O_3, such as dehydration of hydroxides or of more indefinite hydrous oxides, which are themselves either precipitated from aqueous salt

[151] J. Jekabsons, A. V. Kiselev, B. V. Kuznetsov, and Yu. S. Nikitin, *Kolloid. Zhur.*, 1970, **32**, 41 [*Colloid J.* (*U.S.S.R.*), 1970, **32**, 32].

[152] V. A. Burylov, V. Kh. Dobruskin, G. M. Belotserkovskii, S. V. Drozhzhenikov, V. T. Zolotov, N. D. Kostina, B. A. Lipkind, A. G. Monetov, P. M. Pishchaev, and A. T. Slepneva, *Khim. Prom.*, 1969 (6), 19 (*Chem. Abs.*, 1970, **72**, 47 889).

[153] S. D. Kolosentsev, G. M. Belotserkovskii, and T. G. Plachenov, 'Ionnyi Obmen Ionity' ed. G. V. Samsonov, Leningrad, 1970, p. 112 (*Chem. Abs.*, 1971, **74**, 103 410).

[154] S. K. Ghosh and N. C. Saha, *Technology*, 1970, **7**, 138.

[155] K. Unger, K. Berg, and E. Gallei, *Kolloid.-Z.*, 1969, **234**, 1108.

[156] G. M. Belotserkovskii, V. N. Novgorodov, V. Kh. Dobruskin, and T. G. Plachenov, *Zhur. priklad. Khim.*, 1969, **42**, 2749.

[157] D. Berek, I. Novak, Z. Grubisic-Gallot, and H. Benoit, *J. Chromatog.*, 1970, **53**, 55.

[158] C. Caron, *Silicates Ind.*, 1970, **35**, 101.

solutions or obtained by hydrolysis of alkoxides or of covalent halides. In addition, many metal oxides can be prepared by thermal decomposition of carbonates, oxalates, and other compounds.

Thus, $Mg(OH)_2$ can be prepared[159] either by hydrolysis of magnesium methylate, or by adding alkali to $MgCl_2$ solutions. The latter form has much the lower specific surface, which is decreased still further by drying above room temperature. (This behaviour is similar to that observed during the 'ageing' of silica hydrogels, referred to earlier.[141–146]) Porous magnesium oxide is obtained by further dehydration which, if carried out slowly and not taken to completion, yields a microporous product. Subsequent heating tends to destroy the micropores, especially if the original $Mg(OH)_2$ was impure. (A parallel observation was recorded for the effect of impurities in SiO_2.[150]) The microporous MgO rehydrates on exposure to water vapour, with the reappearance of $Mg(OH)_2$ which, however, has a more crystalline (lamellar) structure, and a correspondingly lower specific surface, than the original hydroxide.[160]

Thermal decomposition at 745 °C to limestone or of dolomite (a mixed carbonate of magnesium and calcium) yields an oxide with relatively large diameter pores ($\sim 1\,\mu m$).[161] Calcination at higher temperatures produces oxides of much lower specific surface, whereas decomposition of pure magnesium carbonate (at 565 °C) yields an oxide with much smaller pores.

Highly surface-active ZnO can be prepared by thermal decomposition of the oxalate in an oxygen atmosphere.[162] Its enhanced adsorbent power is ascribed to a non-stoichiometric cation-deficient surface layer. Sintering at temperatures above 400 °C not only reduces the specific surface, but permits the formation of a more homogeneous and less active surface by diffusion of zinc ions from the interior to the surface.

Nickel oxide, formed by controlled thermal dehydration of the hydroxide, has a wide range of pore sizes, including micropores small enough to show molecular-sieve effects.[163] When dehydration is carried out *in vacuo*, an apparently oxygen-deficient surface is formed (compare ZnO above).

In a study of the pore structure developed on dehydration of a microcrystalline hydrated ceric oxide,[164] measurements of nitrogen adsorption were found to be consistent with cylindrical pores. There was some evidence for the existence of a small volume of micropores, but dehydration at high temperatures (up to 1000 °C) led to sintering, in which larger pores grew at the expense of smaller ones. In a parallel study of porous thoria,[165] the same authors found that water adsorption (at 35 °C) on a sample believed to be microcrystalline differed sharply from low-temperature nitrogen adsorption. There was clear evidence of a specific interaction between water and the

[159] M. Faure and B. Imelik, *Bull. Soc. chim. France*, 1969, 4263.
[160] M. Faure, *Bull. Soc. chim. France*, 1970, 69.
[161] R. K. Chan, K. S. Murthi, and D. Harrison, *Canad. J. Chem.*, 1970, **48**, 2972.
[162] D. Dollimore and P. Spooner, *Trans. Faraday Soc.*, 1971, **67**, 2750.
[163] R. B. Fahim and A. I. Abu-Shady, *J. Catalysis*, 1970, **17**, 10.
[164] R. S. Mikhail, R. M. Gabr, and R. B. Fahim, *J. Appl. Chem.*, 1970, **20**, 222.
[165] R. B. Fahim, R. M. Gabr, and R. S. Mikhail, *J. Appl. Chem.*, 1970, **20**, 216.

crystalline ThO_2, accompanying the physical adsorption process. Nevertheless, analysis of the desorption branch of the water isotherm in the capillary condensation region yielded a surface area value which agreed with that obtained by BET analysis of measurements of low-temperature adsorption of nitrogen, which showed no specific interaction. The water-adsorption results (as with those of nitrogen adsorption on CeO_2)[164] were analysed by means of Brunauer's 'corrected model-less' method,[82] and it would be interesting to apply the same method (of a comparison of the adsorption of nitrogen and of water) to swelling clays, with which specific interactions also occur.

In an interesting study of hydrous Cr_2O_3 gel[166] precipitated from solution at nearly neutral pH, it was shown that removal of water by gentle outgassing close to room temperature gave a gel with a high adsorbent capacity for water but virtually none for nitrogen. More vigorous heat treatment, however, caused the surface area and the pore volume accessible to nitrogen to approach the values for water. This was interpreted as showing the existence in the former case of micropores too small to admit nitrogen but large enough to accommodate water. It was suggested that the micropores are formed when that water originally present as hydration shells surrounding Cr^{3+} ions is removed without causing structural collapse.

Titania and zirconia may be prepared by controlled hydrolysis of the corresponding alcoholates in organic media. In an extensive study of these oxides,[167] a super-critical autoclaving technique was used to remove the organic phase after hydrolysis, without modifying the pore structure by the imposition of large capillary forces during drying. The final pore structure was found to depend not only on the medium in which hydrolysis was performed, but also on the quantity of water added, too little yielding an oxide of low porosity, whilst too much promoted recrystallization during autoclaving.

Hydrolysis of dilute $FeCl_3$ under suitable conditions of pH and temperature[168] yields a product described as β-ferric oxyhydroxide, β-FeO(OH). This appears to form hollow tubular microcrystals having a rather well-defined and uniform pore structure. From nitrogen-adsorption data, a mean pore diameter of 2.84 nm was deduced.[169] Such a material could perhaps be used as a model substrate for studies of capillary processes in a pore system of defined geometry.

Mixed Oxides.—Porous oxides of mixed chemical composition are of interest for a variety of reasons, and some are of major technological importance. Incorporation of other components in a porous oxide can not only modify the resultant pore structure, but can also provide new properties not found in either component alone. Even traces of an additional component can modify the high-temperature phase behaviour of certain porous oxides. In certain

[166] F. S. Baker, K. S. W. Sing, and L. J. Stryker, *Chem. and Ind.*, 1970, 718.
[167] M. A. Vicarini, G. A. Nicolaon, and S. J. Teichner, *Bull. Soc. chim. France*, 1970, 1651.
[168] K. J. Gallagher, *Chimia (Switz.)*, 1969, **23**, 465.
[169] K. J. Gallagher, *Nature*, 1970, **226**, 1225.

applications it is desirable to attain a very homogeneous and intimate admixture of two components, or uniform distribution of one component within another, and this can readily be achieved by suitable preparative techniques. Many catalysts are prepared in this way.

Several different methods of preparation are in use. An obvious modification of the alkoxide hydrolysis method mentioned in the preceding section is to decompose a mixture of alkoxides in a suitable medium. Thus, mixtures of magnesium methylate and zirconium isopropylate, dissolved in a mixture of benzene and methanol, and subjected to hydrolysis, yield, after removal of excess solvent with subsequent dehydration, an intimate and highly porous mixture of magnesia and zirconia.[170] Magnesia + alumina and magnesia + titania mixtures can be similarly prepared. Another possible route is by thermal decomposition of a mixture of hydroxides prepared by co-precipitation in an aqueous medium; this has been used for, among others, mixed oxides of $Al + Ti$,[171] $Si + Al$,[172-174] and $Si + Mg$.[175, 176] Thermal decomposition of co-precipitated carbonates has also been used, in the case of the alkaline-earth metals.[177] Certain clay minerals, notably kaolinite,[178,179] can be made to decompose to yield $SiO_2 + Al_2O_3$ mixtures of high porosity and specific surface.

An altogether different approach is to produce one of the oxides in gel form by one of the standard methods and impregnate it with a solution of an appropriate salt of another cation which can subsequently be converted into an oxide.[172,180] A related method has been used to form TiO_2 within the pores of SiO_2 gel by hydrolytic decomposition of $TiCl_4$ introduced from the vapour phase.[181] A similar technique has been employed to deposit TiO_2 within the micropores of active carbon.[182]

Several studies have been reported of the properties of such mixed oxides, both with regard to differences from the properties of the constituent oxides alone, and concerning methods of further modifying their properties. Thus, while samples of ZrO_2 and MgO, prepared by hydrolysis of the individual

[170] M. A. Vicarini, G. A. Nicolaon, and S. J. Teichner, *Bull. Soc. chim. France*, 1970, 3384.
[171] N. F. Ermolenko and G. G. Korunnaya, *Vestsi Akad. Navuk Belarusk. S.S.R., Ser. khim. Navuk*, 1970 (4), 14 (*Chem. Abs.*, 1970, **72**, 6447).
[172] Chien-Chon Shih, *Hua Hsueh*, 1966 (3), 85 (*Chem. Abs.*, 1970, **72**, 71 007).
[173] J. Parasiewicz-Kaczmarska, *Zeszyty Nauk. Uniw. Jagiel., Pr. Chem.*, 1969 (14), 237 (*Chem. Abs.*, 1970, **72**, 25 364).
[174] V. G. Gurevich, I. A. Grishkan, K. M. Arutyunova, and S. P. Dzhavadov, *Zhur. priklad. Khim.*, 1971, **44**, 260.
[175] S. K. Ghosh, D. S. Mathur, and N. C. Saha, *Technology*, 1969, **6**, 177.
[176] P. C. Banerjee, S. K. Ghosh, and N. C. Saha, *Technology*, 1970, **7**, 3.
[177] M. D. Judd and M. I. Pope, *J. Appl. Chem.*, 1970, **20**, 69.
[178] I. B. Dubnitskaya and V. S. Komarov, *Vestsi Akad. Navuk Belarusk. S.S.R., Ser. khim. Navuk*, 1971 (2), 123 (*Chem. Abs.*, 1971, **75**, 53 560).
[179] V. S. Komarov and I. B. Dubnitskaya, *Vestsi Akad. Navuk Belarusk. S.S.R., Ser. khim. Navuk*, 1969 (5), 29 (*Chem. Abs.*, 1970, **72**, 47 867).
[180] K. Nobe, M. Hamidy and Chieh Chu, *J. Chem. and Eng. Data*, 1971, **16**, 327.
[181] S. I. Kol'tsov, *Zhur. priklad. Khim.*, 1970, **43**, 1956.
[182] N. F. Ermolenko and M. I. Yatsevskaya, *Zhur. fiz. Khim.*, 1970, **44**, 189 (*Russ. J. Phys. Chem.*, 1970, **44**, 102).

alcoholates, had a specific surface of $400 m^2 g^{-1}$, areas as high as $800 m^2 g^{-1}$ could be obtained in mixed oxides formed by the decomposition of the alcoholates together.[170] The field of catalysis provides many examples of the modification of properties of mixed oxides to suit particular purposes. The dependence of catalytic activity on pore structure and related properties is explored in a recent study of $Fe_2O_3 + CrO_3$ catalysts.[183]

A recent investigation deals with the stability of co-precipitated $TiO_2 + Al_2O_3$ to hydrothermal[184] and thermal[171] treatments. Hydrothermal ageing (an effect which was mentioned in the section dealing with Al_2O_3[130–133]) is modified, in the mixed oxide, in a manner depending on composition, the Al_2O_3-rich oxides being less susceptible, and the TiO_2-rich oxides more susceptible, than pure Al_2O_3 and pure TiO_2 respectively. In mixed oxides having molar proportions of about $2\frac{1}{2}:1$, the pore structure was predominantly controlled by the major component in each case. On calcination, there is a tendency for Al_2O_3 to form corundum in the presence of TiO_2, with a consequent coarsening of the pore structure. This can be controlled by the incorporation of Li_2O, which stabilizes γ-Al_2O_3.[185] The effect of calcination on the pore structure of $Al_2O_3 + ZrO_2$[186] and $SiO_2 + Al_2O_3$[187] has also been investigated.

Preparation of a $SiO_2 + TiO_2$ mixed oxide by hydrolysing $TiCl_4$-vapour-impregnated silica gel[181] leads to a progressive change of pore structure. As increasing amounts of TiO_2 are deposited, the total pore volume decreases. The specific surface also decreases, but at a lesser rate, so that the 'mean hydraulic pore radius' (pore volume/specific surface) tends to decrease. This is supported by direct measurements of pore-size distribution which show an increase in the proportion of small pores, but it is not clear to what extent this is attributable to the formation of a secondary pore structure in the TiO_2, rather than to narrowing of the original pore structure in the SiO_2.

The pore structure of Al_2O_3 and $Al_2O_3 + SiO_2$ gels may also be modified by impregnating with various salt solutions before drying. The effect is believed to arise from the influence of the cation on the aggregation of primary particles during drying, and a systematic study of a series of metal chlorides[179] led to the conclusion that highly hydrated cations of low polarizability promoted the formation of the smallest pores. It has been suggested that the pore size of Al_2O_3, SiO_2, and $SiO_2 + MgO$ co-precipitate (Florisil) can be modified by addition of Na_2SO_4 and $CuSO_4$, producing materials suitable for use as substrates in gas–solid chromatography.[175,176] Organophilic porous oxides, of various compositions, were prepared by autoclaving in

[183] K. Hennig, W. Hildebrandt, J. Scheve, and H. Blume, *Z. anorg. Chem.*, 1970, **377**, 256.
[184] N. F. Ermolenko and G. G. Korunnaya, *Vestsi Akad. Navuk Belarusk. S.S.R., Ser. khim. Navuk*, 1969 (4), 16 (*Chem. Abs.*, 1971, **74**, 15 997).
[185] N. F. Ermolenko, A. V. Oboturov, M. D. Efros, F. G. Kramarenko, and M. I. Yatsevskaya, *Vestsi Akad. Navuk Belarusk. S.S.R., Ser. khim. Navuk*, 1969 (5), 5 (*Chem. Abs.*, 1970, **72**, 36 105).
[186] I. V. Nicolescu, N. Cioc, C. Pascu, and I. Sandulescu, *An. Univ. Bucuresti, Chim.*, 1969, **18**, 19 (*Chem. Abs.*, 1971, **74**, 68 166).
[187] D. Beruto, A. Peloso, and A. Carrea, *Ann. Chim.* (*Italy*), 1970, **60**, 679.

such liquids as ethanol, butan-1-ol, or carbon tetrachloride, venting to atmosphere when the critical temperature was exceeded.[188] Although this apparently left the pore structure unchanged, the surface acquired an organophilic coating which was stable up to 200—300 °C.

On the other hand, a Russian group,[189] beginning with the hypothesis that capillary forces during drying had a major effect on the structure of the resultant xerogel, investigated the effect of replacing intermicellar water by isoamyl alcohol in mixed $Al_2O_3 + SiO_2$ gels. The gels were then dried by normal heating, rather than supercritical autoclaving. The final product was found to have a substantially higher specific surface and porosity, which was interpreted as signifying a reduction in the co-ordination number of the primary particles, rather than any intrinsic change in their structure. The surface properties of the alcogel were not commented upon.

Differences in solubility between components in a mixed oxide raise the possibility of formation of a new pore structure by preferential leaching of one component; this is, of course, the method by which porous glasses of the Vycor and other types are prepared. A recent paper[190] reports the application of this technique to the production of $SiO_2 + Al_2O_3$ catalysts of more uniform, controlled pore-size.

Clays and Zeolites.—The clay minerals do not in general possess the random pore structure typical of the other adsorbents so far discussed; on the other hand, they include examples of the 'lattice' pore structure that is shown, in its most developed form, by the zeolites. In the case of the clay minerals, however, such intracrystalline spaces are generally accessible only to specific adsorbate molecules, which not only must be of a suitable size, but must also be able to cause separation of the laminar clay crystallites. In a recent review of the porosity of natural mineral adsorbents,[191] three types were recognized: laminar expanding structures, such as montmorillonite and vermiculite; layered ribbon structures (palygorskite and sepiolite); and non-expanding laminar minerals, including talc, pyrophyllite, hydromica, and kaolinite.*

Many clay minerals can, however, be partially decomposed by suitable thermal and/or chemical treatment, to yield more or less random pore structures. Thus, palygorskite, hydromica, and montmorillonite are all susceptible to attack ('activation') by 25% sulphuric acid, yielding materials of much

---

* Mention should here be made of a mercury porosimetric investigation of the structure of various clays and soils subjected to drying by several different techniques.[192] Although the pores concerned here are formed between, rather than within, particles, such structures are of great practical importance in soil physics and several other areas of powder technology.

[188] H. Utsugi, A. Watanabe, K. Ito, and S. Nishimura, *Nippon Kagaku Zasshi*, 1970, **91**, 431 (*Chem. Abs.*, 1970, **73**, 80 989).
[189] V. G. Gurevich, K. M. Arutyunova, V. I. Golikova, and S. P. Dzhavadov, *Kolloid. Zhur.*, 1970, **32**, 853 [*Colloid J. (U.S.S.R.)*, 1970, **32**, 717].
[190] G. Perriolat, D. Barthomeuf, and Y. Trambouze, *Bull. Soc. chim. France*, 1970, 2459.
[191] Yu. I. Tarasevich, *Ukrain. khim. Zhur.*, 1969, **35**, 1112 (*Chem. Abs.*, 1970, **72**, 25 157).
[192] S. Diamond, *Clays and Clay Min.*, 1970, **18**, 7.

increased adsorbent capacity.[193] Bentonite can be similarly activated[194]—in this case the reaction is predominantly that of removing soluble sesquioxides R_2O_3. The sorption activity of the product can be still further increased by heating to temperatures up to 500 °C.

It has been found that acid-leaching of iron from suitable diatomites, followed by high-temperature fusion with sodium carbonate or nitrate, yields a product having a uniform pore structure[195] which is considered to be suitable for use as a chromatographic support.[196]

Diatomites are also used as a starting material in the preparation of zeolitic materials by a hydrothermal process in an alkaline medium containing sodium aluminate.[197] More generally, any appropriate mixture of sodium silicate and aluminate may be used, and it is claimed that by choice of the reaction temperature and time, controlled particle size in the range 0.01—100 μm can be achieved.[198]

It has been shown that Type 4A zeolite loses its structure progressively on heating above 550 °C.[199] The particles at first acquire a dense impermeable coating, containing closed macropores, surrounding a core of untransformed zeolite. The adsorption capacity falls almost to zero at this stage. Further heating completes the structural collapse, and ultimately eliminates even the closed pores.

Acid 'activation' of zeolites produces changes in their chemical composition and physical structure. Thus, when clinoptilolite is extracted with progressively more concentrated hydrochloric acid, it first loses uni- and bi-valent cations, together with Fe_2O_3 and some Al_2O_3. At higher concentrations, more substantial loss of Al_2O_3 occurs.[200] Acid dealumination of mordenite has been used as a method of producing controlled changes in its pore size.[201]

Carbon.—Pores can be produced in carbon, and the resultant structure can be modified, by a variety of methods. Thermal degradation of certain organic polymers[202] can produce both micropores, whose shape depends on the molecular structure of the polymer, and larger pores which probably originate

[193] F. A. Belik, Yu. I. Tarasevich, and F. D. Ovcharenko, *Ukrain. khim. Zhur.*, 1971, **37**, 441 (*Chem. Abs.*, 1971, **75**, 53 672).
[194] I. K. Sataev, E. A. Aripov, and K. S. Akhmedov, 'Adsorptionnye Svoistva Nekot. Prir. Sin. Sorbentov', ed. E. A. Aripov, Izd. 'Fan' Uzbek. S.S.R., Tashkent, 1969, p. 161 (*Chem. Abs.*, 1970, **73**, 113 278).
[195] A. P. Atanasov, N. I. Bryzgalova, T. B. Gabrilova, and A. V. Kiselev, *Kolloid. Zhur.*, 1970, **32**, 807 (*Colloid J.* (*U.S.S.R.*), 1970, **32**, 677].
[196] A. V. Kiselev, T. B. Gabrilova, and N. I. Bryzgalova, *U.S.S.R.* Pat. 269 574 (*Chem. Abs.*, 1970, **73**, 48 842).
[197] P. Onu and V. Ababi, *Anal. Sti. Univ. 'Al. I. Cuza' Iasi, Sect. Ic*, 1970, **16**, 191 (*Chem. Abs.*, 1971, **74**, 116 293).
[198] C. V. McDaniel and P. K. Maher, Fr. P. 1 557 393 (*Chem. Abs.*, 1960, **72**, 25 389).
[199] J. L. Thomas, M. Mange, and C. Eyraud, *Adv. Chem. Series*, 1971, **101**, 443.
[200] E. A. Aripov, A. M. Mirsalimov, and K. S. Akhmedov, 'Opokovidnye Gliny Kermine', ed. K. S. Akhmedov, Izd. 'Fan' Uzbek. S.S.R., Tashkent, 1970, p. 92 (*Chem. Abs.*, 1971, **75**, 144 251).
[201] E. M. Flanigen, U.S.P. 3 597 155 (*Chem. Abs.*, 1971, **75**, 122 448)
[202] L. B. Adams, E. A. Boucher, and D. H. Everett, *Carbon*, 1970, **8**, 761.

from spaces between particles or aggregates of particles in the starting material. If the polymer melts before thermal degradation begins, the larger pores may be absent. If ion-exchange resins are used as starting materials, the resultant micropore structure seems to depend on the substituent cations present in the resin.[203] When carbon black impregnated with an organic binder is fired at 900 °C, it is found that rapid heating leads to the formation of macropores within the binder, with an adverse effect on the mechanical properties of the product.[204] An exceptionally high adsorption capacity is claimed[205] for carbons produced by the reaction with oxygen and water vapour at high temperature of selected hydrocarbon mixtures which have been pretreated in an autoclave.

Controlled oxidation of carbon (*activation* or *gasification*) can generate a relatively large volume of micropores, with a consequent increase in specific surface. In its initial stages, the reaction with CO_2 appears to be localized at the site of individual microcrystals of graphite.[206] A similar conclusion has been reached in the case of steam activation at 1000 °C.[207] These observations suggest that the crystalline structure of the carbon before activation is an important factor in determining the pore structure which is produced. As the reaction progresses, oxidation takes place mainly at the walls of the larger pores, which thus grow still larger;[206] the rate of this reaction is controlled by gaseous diffusion within pores at temperatures above 935 °C.[208] The reaction of poly(furfuryl alcohol) carbon with CO_2 at 850 °C[209] is catalysed at local sites in the surface produced by addition of cations of such metals as iron and nickel.[210] This leads to the formation of meso- and macro-pores before the metallic particles are themselves oxidized and become inactive as catalysts. Addition of CO to the gas stream can inhibit the surface reaction with CO_2 in larger particles, leading to a more uniform reaction and a higher micropore volume in the final product.[211] Similarly, it is found that $CO_2 + H_2O$ mixtures produce a more microporous carbon by activation at 850 °C than does CO_2 alone or $CO_2 + O_2$ or $H_2O + O_2$ mixtures.[212]

In an interesting comparative study of carbons activated with water vapour at 950 °C and with zinc chloride at 650 °C,[213] it was concluded that zinc chloride activation produced both micro- and macro-pores of pronounced saccate (or 'ink-bottle') shape. Evidence for this was drawn from comparison

[203] V. P. Musakina and T. G. Plachenov, *Zhur. priklad. Khim.*, 1969, **42**, 2756 (*Chem. Abs.*, 1970, **72**, 79 746).
[204] G. M. Butyrin, E. F. Chalykh, M. I. Rogailin, I. M. Rozenman, and L. N. Shein, *Khim. Tverd. Topl.*, 1970 (4), 118.
[205] M. Chigusa and S. Akai, Jap.P. 7 005 645 (*Chem. Abs.*, 1970, **72**, 125 408).
[206] W. M. Kalback, L. F. Brown, and R. E. West, *Carbon*, 1970, **8**, 117.
[207] L. P. Gilyazetdinov, V. K. Gus'kova, R. A. Virobyants, R. S. Orlova, and I. M. Yurkovskii, *Zhur. fiz. Khim.*, 1971, **45**, 1598 (*Russ. J. Phys. Chem.*, 1971, **45**, 908).
[208] N. V. Dovaston and B. McEnaney, *Vac. Microbal. Techn.*, 1970, **7**, 91.
[209] H. Marsh and B. Rand, *Carbon*, 1971, **9**, 47.
[210] H. Marsh and B. Rand, *Carbon*, 1971, **9**, 63.
[211] B. Rand and H. Marsh, *Carbon*, 1971, **9**, 79.
[212] N. V. Malin, *Zhur. priklad. Khim.*, 1970, **43**, 1039, 1538.
[213] O. Kadlec, A. Varhanikova, and A. Zukal, *Carbon*, 1970, **8**, 321.

of mercury porosimetry data with measurements of water vapour and benzene adsorption, and it was suggested that such pores were formed by partial entrapment of gaseous reaction products as bubbles within the plastic carbonaceous raw material during activation. Saccate pores were also reported in carbon formed by pyrolysis of cellulose fibres at 1100 °C.[214]

Some interesting observations were reported in a comparative study of several adsorbates of different size, at various temperatures, on a series of carbons, before and after partial oxidation.[215] This revealed several anomalous effects attributable to the existence of micropores having constrictions of molecular width. In particular, the isotherms showed very pronounced hysteresis at relative pressures down to zero, and the limiting uptake at saturation not only depended on the molecular size of the adsorbate, but also, for a given adsorbate, increased markedly with increasing temperature. Spencer and co-workers[216] similarly found that when HCN is removed from polyacrylonitrile fibres at successively higher temperatures, a large pore volume is at first opened up, which subsequently becomes inaccessible to large molecules as pore entrances gradually contract. As the temperature of heat treatment is increased, so smaller molecules are excluded by the molecular-sieve action of the micropores, which eventually are not penetrated even by helium at 25 °C. Thus, the distinction between 'open' and 'closed' porosity becomes dependent entirely on the molecular size of the immersion fluid in which density measurements are carried out. Since physical adsorption undoubtedly occurs in at least some of the systems studied, the interpretation of such measurements is by no means simple.

Although it is known that carbon fibres prepared from polyacrylonitrile are much stronger than those prepared from vinylidene chloride + vinyl chloride copolymer,[217] the relationship between mechanical properties and the geometry and volume of the pores is complicated by the effects of crosslinking and crystallite orientation.

Modification of the pore structure of carbon at high temperatures was studied by heating charcoals up to 3000 °C in an argon atmosphere.[218] Although it is certain that textural changes occur on heating, it seems that their nature may depend on the source of the original carbon or on the conditions of treatment, since different authors have variously reported the closure of pores, with decrease in net porosity,[208] a decrease in closed porosity with an increase in open porosity,[219] and a decrease in total porosity with increasingly hydrophilic character,[214] as a result of heat treatment.

[214] A. I. Baver, L. L. Demburg, A. V. Kiselev, N. V. Kovaleva, and Yu. S. Nikitin, *Khim. Tverd. Topl.*, 1971 (2), 149 (*Chem. Abs.*, 1971, **75**, 10 661).

[215] S. J. Gregg, F. M. Olds, and R. F. S. Tyson, *Third Conference on Industrial Carbons and Graphite*, 1970, Academic Press, London, 1971, p. 184.

[216] D. H. T. Spencer, M. A. Hooker, A. C. Thomas, and B. A. Napier, *Third Conference on Industrial Carbons and Graphite*, 1970, Academic Press, London, 1971, p. 467.

[217] E. A. Boucher, R. N. Cooper, and D. H. Everett, *Carbon*, 1970, **8**, 597.

[218] M. Francois, E. Bretey, Y. Grillet, and H. Guerin, *Compt. rend.*, 1970, **273**, C, 23.

[219] V. S. Ostrovskii, I. N. Krutova, T. D. Shashkova, and A. P. Fedoseev, *Konstr. Mater. Osn. Grafita*, 1967 (3), 204 (*Chem. Abs.*, 1970, **72**, 125 402).

A short review has appeared[220] describing methods of measuring pore-size distribution and their application to active carbons. As was pointed out in Chapter 1, the study of microporous solids presents special problems, and these have been discussed with reference to active carbons by Marsh and Campbell.[221] In particular, these authors have studied adsorption from solution, and vapour adsorption in the presence of presorbed non-volatile material, in addition to the more usual methods of micropore estimation. Differences in pore volume accessible to small molecules (He, MeOH) and larger molecules (C_6H_6, CCl_4) have also been used in studying the evolution of micropore structure during activation,[222] as have changes in the n.m.r. spectra of adsorbed molecules.[223]

Other Materials.—Many materials used in the construction industry are porous, and their pore properties play an essential role in determining their value in any given applicational environment.[224] In particular, much interest is focused on the generation of pore structures during the hardening of cements and plasters.

The most important of the several reactions occurring during the hardening of Portland cement pastes is the hydration of calcium silicates of various compositions. Mikhail *et al.*[225] prepared a series of such silicates by suspending quicklime and silica gel, in various proportions, in water (at a fixed ratio of solids to water), and ageing at 40 °C for 20 months, after which time a partial chemical combination had occurred. The products (known collectively as tobermorite gels, after the mineral form of calcium silicate) were dried and subjected to water adsorption. This revealed the existence of both micropores (of 0.4—0.5 nm radius) and somewhat wider pores (above 1 nm in radius). Extraction of the uncombined excess of lime and silica from the tobermorite gels exposed a considerable additional volume of micropores. On the other hand, nitrogen adsorption on fully hydrated cement pastes[226] fails to reveal any micropores, the only pore-space accessible to nitrogen being in 'ink-bottle' pores of hydraulic radius 1.5—4 nm. These presumably correspond to the larger group disclosed by water adsorption, and their size depends principally on the water-to-cement ratio in the original paste.

Calcium chloride, which finds wide use as a practical additive to accelerate the hardening of cement, has the effect of reducing the radius of pores in the larger group (at a given water/cement ratio), whilst leaving the smaller pores unaffected. Thus, whilst in a hydrated tricalcium silicate without additive the area accessible to nitrogen was only 0.055 times that accessible to water,

[220] O. Jantti and V. Peltonen, *Kem. Teollisuus*, 1970, **27**, 15 (*Chem. Abs.*, 1970, **73**, 123 875).
[221] H. Marsh and H. G. Campbell, *Carbon*, 1971, **9**, 489.
[222] T. G. Plachenov, V. P. Musakina, L. B. Sevryngov, and V. M. Fal'chuk, *Zhur. priklad. Khim.*, 1969, **42**, 2020.
[223] H. Estrada-Szwarckopf, J. Auvray, and A. P. Legrand, *J. Chim. phys.*, 1970, **67**, 1292.
[224] P. J. Sereda, *Canad. Building Digest* No. 127, N.R.C., Ottawa, 1970.
[225] R. Sh. Mikhail, A. M. Kamel, and S. A. Abo-El-Enien, *J. Appl. Chem.*, 1969, **19**, 324.
[226] E. E. Bodor, J. Skalny, S. Brunauer, J. Hagymassy, and M. Yudenfreund, *J. Colloid Interface Sci.*, 1970, **34**, 560.

this ratio rose to 0.22 for a sample hydrated in the presence of calcium chloride.[227] From water-adsorption measurements, it was deduced that the population of micropores of ~ 0.4 nm hydraulic radius was unchanged. Since, however, the radii of the 'larger' pores decrease as the degree of hydration increases, and the effect of calcium chloride is to promote hydration, there is a decrease in the average pore radius. Similar results were obtained from a parallel study of the hydration of β-dicalcium silicate.[228]

There are certain similarities in the process of the hardening by hydration of gypsum pastes.[229] Once again a bimodal pore-size distribution is found, although the radii are 1 and 7—8 nm, and all pores are accessible to nitrogen. The course of hydration depends on the water/plaster ratio in the paste, the specific surface of the product increasing as this ratio is decreased. Calcium chloride accelerates hydration, as it does with cement, and yields a product with a higher specific surface at a given water/plaster ratio. Hydration can be retarded by small additions of calcium lignosulphonate, without significantly changing the specific surface or the pore-size distribution.

The process of dehydration of calcium sulphate dihydrate has been studied by means of electron microscopy and gas adsorption.[230] In the first stage, the dihydrate goes to the hemihydrate, with considerable structural reorganization. This material is evidently not of a strongly stable structure, since it ages fairly rapidly, with disappearance of the finer pores and a decrease in specific surface. Further dehydration of the hemihydrate proceeds by a zeolitic mechanism.

Interest has recently been expressed in the alkaline-earth fluorides as porous media which might be expected to show similar properties to the oxides, but with superior thermal stability. Thus, a gel of magnesium fluoride, precipitated from magnesium sulphate solution by the action of hydrofluoric acid and then dried, was found to have a fairly uniform pore structure (as indicated by steep-sided hysteresis loops for methanol adsorption).[231] Just as with silica gels,[147] if the intermicellar water was replaced by alcohols before drying, the pore structure of the product was modified in a reproducible way. These authors also remark that the pore structure of their magnesium fluoride gel can be controlled by hydrothermal treatment, although no details are given.

[227] J. Skalny, I. Odler, and J. Hagymassy, *J. Colloid Interface Sci.*, 1971, **35**, 434.
[228] I. Odler and J. Skalny, *J. Colloid Interface Sci.*, 1971, **36**, 293.
[229] R. Sh. Mikhail and R. I. A. Malek, *J. Appl. Chem. Biotechnol.*, 1971, **21**, 277.
[230] B. Molony, J. Beretka, and M. J. Ridge, *Austral. J. Chem.*, 1971, **24**, 449.
[231] R. Yu. Sheinfain, O. P. Stas, and I. E. Neimark, *Kolloid. Zhur.*, 1969, **31**, 922 [*Colloid J. (U.S.S.R.)*, 1969, **31**, 745].

5
Particulate Dispersions

BY R. H. OTTEWILL

1 Introduction

Research into the behaviour of disperse systems requires both the detailed description of observed phenomena and the interpretation of the observations in terms of a theoretical model. A considerable impetus to the understanding of the behaviour of lyophobic colloidal systems was given by the theoretical models of colloid stability put forward by Derjaguin and Landau[1] and Verwey and Overbeek.[2] The basic premise of their theory, the so-called DLVO theory, was that the potential energy of interaction between a pair of particles could be considered to consist of two components:

(i) that arising from the overlap of the electrical double layers and leading to repulsion, V_R;

(ii) that arising from electromagnetic effects and leading to van der Waals attraction, V_A.

These were considered to be additive so that the total potential energy of attraction, V, could be written as

$$V = V_R + V_A. \qquad (1)$$

The general features of the curve of potential energy of interaction against the distance of separation between the particle surfaces, H_0, are given in Figure 1. At short distances of separation, the combination of strong short-range repulsive forces and van der Waals attraction leads to a deep minimum termed the *primary minimum*; the position of the primary minimum determines the distance of closest approach, H_0^*. At high surface potentials and low ionic strengths and at intermediate distances the electrical repulsion term is the dominant one and hence a maximum occurs in the potential energy curve; this is normally termed the *primary maximum* and has a magnitude V_m. At larger distances, the energy of electrical repulsion falls off more rapidly with increasing distance of separation than the van der Waals attraction and a second minimum appears in the curve of depth V_{SM}; this is termed the *secondary minimum*.

[1] B. V. Derjaguin and L. Landau, *Acta Physiochim. U.R.S.S.*, 1941, **14**, 633.

[2] E. J. W. Verwey and J. Th. G. Overbeek, 'Theory of the Stability of Lyophobic Colloids', Elsevier, Amsterdam, 1948.

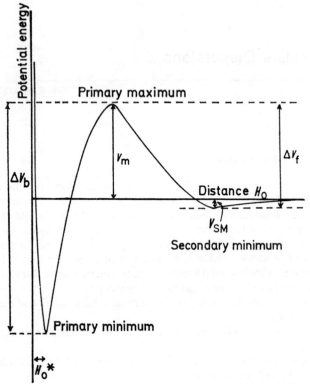

Figure 1 *Form of the curve of potential energy* $(V = V_R + V_A)$ *against distance of surface separation,* H_0, *for the interaction between two particles (schematic)*

Much of the current research in colloid science is centred on confirming and extending the basic ideas propounded in the DLVO theory. Thus there is considerable theoretical interest in obtaining more detailed models to evaluate both V_A and V_R. In parallel with this theoretical work considerable success has been achieved in devising experimental techniques which can measure, essentially independently, either the van der Waals attractive forces or the electrical repulsive forces, *i.e.* the magnitude of the gradient of the curve of interaction potential energy against distance in the direction of the particles. These aspects are reviewed in Sections 2 and 3.

The kinetic aspects of colloid stability, which are of prime importance, also continue to receive considerable attention. The kinetic energy barrier to particle association in the primary minimum is represented by ΔV_f in Figure 1 and the energy barrier to redispersion from a primary minimum is represented as ΔV_b. The terms *coagulation* and *flocculation* still tend to be used indiscriminately in the literature and general agreement on their usage has not yet been reached. For the purposes of this Report, the term *coagulation* will be used when it is clear that the particle association processes are occurring under primary minimum conditions, *i.e.* as in the particle aggregation processes which occur with lyophobic sols on the addition of simple electrolytes such as sodium chloride. The term *flocculation* will be used to describe secondary minimum association and the joining together of particles by 'bridging' with polymer molecules. Thus flocculation is a particle association process in which in the associated state the average distance between the particles is considerably greater than the order of atomic dimensions and hence the aggregate formed in this way has an open structure. Coagulation, however, leads to the formation of a compact aggregate structure in which the average distance of separation between the particles can be of the order of atomic dimensions. The term *peptization* will be used to describe the redispersion of particles from a primary minimum situation. The preparation of disperse systems and various aspects of coagulation and flocculation are reviewed in Section 4.

Although the fact that the adsorption of certain molecules could confer on particles stability, at electrolyte concentrations where coagulation would normally have occurred, was discussed by many early authors, it is only in recent years that a more detailed understanding of the mechanism of this process has been achieved. Although this understanding is far from complete it is clear that an additional term has to be included in the potential energy of interaction equation for what is called, somewhat loosely, *steric stabilization*. Thus, giving this term the symbol V_S, one can write,

$$V = V_R + V_A + V_S. \tag{2}$$

V_S is essentially a short-range term which can be expected to rise rapidly at a distance of separation between the native particle surfaces of the order of twice the dimensions of the adsorbed molecule. The effect of V_S is shown schematically in Figure 2 for the situations which occur in the presence and absence of an electrostatic barrier. It is clear from this qualitative picture that entry into a deep primary minimum is made impossible by the presence of an adsorbed layer. Work on this topic pertinent to aqueous dispersions is reviewed in Section 5; non-aqueous systems are discussed by Vincent.[3]

An important parameter in the theory of colloid stability as applied to aqueous systems is the nature of the surface potential which governs the stability and the relation of this to the normally accepted models of the electrical double layer. Experimentally, electrokinetic effects are frequently used to obtain information on the electrical potential of dispersed systems. Recent work in this field is reviewed in Section 6.

This Report covers mainly work on, or relevant to, the properties of disperse systems reported during the period 1970—1971 but some earlier work has been mentioned where relevant. Earlier reviews in this field had covered work up until the end of 1969.[4, 5]

[3] B. Vincent, this volume, Chap. 7.
[4] G. D. Parfitt, *Ann. Reports (A)*, 1967, **64**, 125.
[5] R. H. Ottewill, *Ann. Reports (A)*, 1969, **66**, 183.

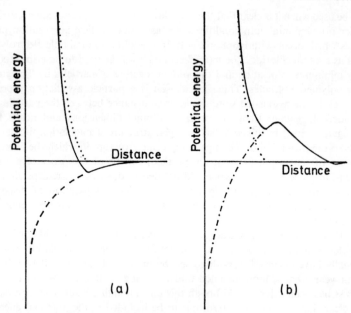

Figure 2 *The influence of the steric stabilization term, V_S, on the form of the potential energy against distance curve for the interaction between two particles (schematic)*
(a) *in the absence of electrostatic repulsion:* $---$, V_A; $\ldots\ldots$, V_S; $\longrightarrow$, $V = V_A + V_S$
(b) *in the presence of electrostatic repulsion:* $-\cdot-\cdot-$, $V_R + V_A$; $\ldots\ldots$, V_S; $\longrightarrow$, $V = V_R + V_A + V_S$

This Report is not exhaustive and with the extensive literature in this field it would be difficult to make it so. Emphasis has been given to reporting new theories, new experimental work amenable to interpretation, and novel ideas, even if qualitative, which might ultimately lead to a better understanding of the subject.

2 van der Waals Interactions

The theoretical treatment of van der Waals interactions between particles can be undertaken from two basic viewpoints:
(i) the *microscopic* approach of de Boer[6] and Hamaker,[7] based on the summation of the pairwise interactions between the atoms of separated particles;
(ii) the *macroscopic* approach of Lifshitz[8] which treats the interacting phases as being continuous and the interaction as occurring through a continuous medium.

[6] J. H. de Boer, *Trans. Faraday Soc.*, 1936, **32**, 10.
[7] H. C. Hamaker, *Physica*, 1937, **4**, 1058.
[8] E. M. Lifshitz, *Zhur. eksp. teor. Fiz.*, 1955, **29**, 94.

For most of the situations which arise, both approaches can be summarized by writing the potential energy of the van der Waals attraction, V_A, in the form

$$V_A = \text{geometrical factor} \times \text{interaction parameter.}$$

Thus for the interaction between two infinitely thick parallel plates, in the non-retarded region, with their surfaces separated by a distance H_0, on the basis of the de Boer–Hamaker theory, the potential energy of attraction per unit area is given by

$$V_A = -\frac{A_H}{12\pi H_0^2}, \tag{3}$$

and the force of attraction per unit area by

$$\Pi_A = \frac{A_H}{6\pi H_0^3}. \tag{4}$$

When the plates interact in a vacuum the Hamaker constant A_H is that for the plate material and can be given by[2]

$$A_H = \tfrac{3}{4} h\nu_0 \alpha_0^2 \, \pi^2 q^2, \tag{5}$$

where h = Planck's constant, α_0 = the static polarizability, and ν_0 = the dispersion frequency taken as a single frequency occurring in the ultraviolet; q = the number of atoms per unit volume of the plate material.

In the case of plates of different materials (1 and 3) interacting through a liquid medium 2, the presence of the liquid medium is taken into account by rewriting the Hamaker constant in the form

$$A_H = A_{13} + A_{22} - A_{12} - A_{23}, \tag{6}$$

or, when the plates are composed of the same material, *i.e.* $A_{11} = A_{33}$

$$A_H = A_{11} + A_{22} - 2A_{12}; \tag{7}$$

or, assuming a geometric mean relationship,

$$A_{12} = (A_{11}A_{22})^{1/2}, \tag{8}$$

$$A_H = (A_{11}^{1/2} - A_{22}^{1/2})^2, \tag{9}$$

where A_{11}, A_{22}, A_{33} are the constants calculated on the basis of interactions between pairs of atoms of those materials, *i.e.* the situation for interaction between atoms in a dilute gas. The various methods of calculating Hamaker constants from optical dispersion data have been reviewed by Gregory.[9]

As the distance separating the surfaces of the plates increases to a value of the order of λ_0 ($\lambda_0 = c/\nu_0$ where c = velocity of light) a finite time is required for the propagation of the electromagnetic radiation between the particles and the attractive energy is reduced. Under these conditions, the force of

[9] J. Gregory, *Adv. Colloid Interface Sci.*, 1969, **2**, 396.

attraction is said to be retarded and is given by

$$\Pi_A = \frac{B_H}{H_0^4}, \tag{10}$$

where B_H is termed the retarded Hamaker constant; its relation to A_H is given by $B_H = 1.24 \times 10^{-2} A_H \lambda_0$. Hence under retarded conditions the force of attraction between two plates is inversely proportional to the fourth power of the separation distance; this is to be compared with the third power in the non-retarded region. In the period under review Clayfield *et al.*[10] have used the microscopic approach to obtain expressions for the retarded dispersion force between spherical particles and between a spherical particle and a thick plate.

Using the macroscopic approach it was found by Lifshitz[8] that for small distances of separation between the surfaces of flat plates, $H_0 \ll \lambda_0/2\pi$, the force of non-retarded attraction was given by

$$\Pi_A = \frac{h\overline{\omega}}{16\pi^3 H_0^3}. \tag{11}$$

The formal similarity between equations (11) and (4) is immediately apparent. In this equation, if the plates are of material 1 interacting in a vacuum then,

$$\omega = \int_0^\infty \left(\frac{\varepsilon_1(i\xi) - 1}{\varepsilon_1(i\xi) + 1} \right)^2 d\xi, \tag{12}$$

and for plates of materials 1 and 3 interacting in a medium 2,

$$\overline{\omega} = \int_0^\infty \left(\frac{\varepsilon_1(i\xi) - \varepsilon_2(i\xi)}{\varepsilon_1(i\xi) + \varepsilon_2(i\xi)} \right) \left(\frac{\varepsilon_3(i\xi) - \varepsilon_2(i\xi)}{\varepsilon_3(i\xi) + \varepsilon_2(i\xi)} \right) d\xi. \tag{13}$$

These expressions contain the relative permittivity $\varepsilon(i\xi)$ as a function of the frequency of the field expressed in rad s^{-1}. In order to use equations (12) and (13) a convenient representation is required. The frequency-dependent relative permittivity $\varepsilon(\omega)$ can therefore be expressed as a complex function, $\varepsilon'(\omega) + i\varepsilon''(\omega)$ of a complex frequency,

$$\omega = \omega_{real} + i\xi.$$

The dispersion energy depends only on the values of the relative permittivity on the imaginary frequency axis (see ref. 13), *i.e.* on $\varepsilon(i\xi)$, and thus the integration has to be carried out in the complex angular frequency range (ω) along the imaginary frequency axis, $i\xi$. The Krönig–Kramers relationship gives $\varepsilon(i\xi)$ in terms of the imaginary part $\varepsilon''(\omega)$ of the relative permittivity, as a function of the

[10] E. J. Clayfield, E. C. Lumb, and P. H. Mackay, *J. Colloid Interface Sci.*, 1971, **37**, 382.

real angular frequency, ω, in the form

$$\varepsilon(i\xi) - 1 = \frac{2}{\pi} \int\limits_0^\infty \frac{\omega \varepsilon''(\omega)}{\omega^2 + \xi^2} \, d\omega, \tag{14}$$

where $\varepsilon''(\omega)$ can be obtained, for example, from spectral measurements (see ref. 18).

It is immediately clear that the computation of an interaction constant using Lifshitz theory requires considerably more basic data than those needed to calculate a Hamaker constant. Indeed, a notable attraction of the London-de Boer–Hamaker theory has always been that it has allowed ready computation of a value for A_H. Since the theory is based on electromagnetic fluctuations occurring over a narrow band of frequencies in the u.v., both ν_0 and α_0 are calculable from refractive index data obtained over a range of visible wavelengths.

In the past few years, a new emphasis has been given to the importance of Lifshitz theory by the work of Ninham and Parsegian and their co-workers[11, 12] It becomes clear from the work of these authors that the use of the London-de Boer–Hamaker theory is restricted by the assumptions:

(i) of pairwise additivity of individual interatomic interactions in condensed media;

(ii) the approximation that the contributions centred around a single dominant frequency of the electromagnetic spectrum are the only important ones; and

(iii) that the insertion of material between the plates, such as a liquid, can be dealt with by the insertion of an arbitrary 'dielectric constant' at single frequency.

Ninham and Parsegian[13] consider, as emphasized by Dzyaloshinskii *et al.*,[14] that the calculations of van der Waals forces based on these assumptions, although they are valid for dilute gases, are intrinsically unsound for condensed systems. In fact, they consider that, 'this method obscures almost totally the most interesting qualitative features of van der Waals forces'. The theory of Lifshitz,[8, 15] on the other hand, 'includes all many-body forces through a continuum picture, retains contributions from all interaction frequencies, and deals correctly with the effects of intermediate substances'. They[16] conclude that:

(i) because of the highly polar nature of liquid water much of the van der Waals force in this medium comes from polarizations at i.r. and microwave frequencies rather than the u.v.;

(ii) it is incorrect to think of the van der Waals force between liquid layers as being the sum of individual interactions between unit segments of the

[11] V. A. Parsegian and B. W. Ninham, *Nature*, 1969, **224**, 1197.
[12] B. W. Ninham and V. A. Parsegian, *J. Chem. Phys.*, 1970, **52**, 4578.
[13] B. W. Ninham and V. A. Parsegian, *Biophys. J.*, 1970, **10**, 646.
[14] I. E. Dzyaloshinskii, E. M. Lifshitz, and L. P. Pitaevskii, *Adv. Phys.*, 1961, **10**, 165.
[15] E. M. Lifshitz, *Soviet Phys. J.E.T.P.*, 1956, **2**, 73.
[16] V. A. Parsegian and B. W. Ninham, *J. Colloid Interface Sci.*, 1971, **37**, 332.

constituent materials. The idea of a characteristic 'Hamaker Constant' is therefore quite misleading for this interaction;

(iii) by virtue of the low-frequency contribution, the van der Waals force also contains a temperature-dependent component;

(iv) dielectric data are well enough known through the range of frequencies to draw these conclusions and to make quantitative numerical estimates with little ambiguity.

These authors argue[17] that forces which arise from electromagnetic fluctuations should be called by the general name of *van der Waals interactions*. In addition, they suggest that the temperature-independent part should be called a *dispersion force* and the remainder or temperature-dependent part a *molecular force*. Thus, for example, there is a large long-range van der Waals contribution from the molecular interactions of the permanent dipoles in liquid water. This occurs essentially at zero frequency and increases almost linearly with temperature.

From Lifshitz theory the general formula for the attractive force per unit area between two semi-infinite flat plates of material 1, separated by a medium 2, can be put in the form of an integral in terms of the parameter p:

$$\Pi_A = \frac{kT}{\pi c^3} \sum_{n=0}^{\infty}{}' \varepsilon_2^{3/2} \xi_n^3 \int_1^{\infty} p^2 \left\{ \left[\Delta^{-2} \exp \left(\frac{2p\xi_n H_0 \varepsilon_2^{1/2}}{c} \right) - 1 \right]^{-1} \right.$$
$$\left. + \left[\bar{\Delta}^{-2} \exp \left(\frac{2p\xi_n H_0 \varepsilon_2^{1/2}}{c} \right) - 1 \right]^{-1} \right\} dp, \quad (15)$$

where ε_1 and ε_2 are the relative permittivities evaluated on the imaginary frequency axis at $\omega = i\xi_n$, with

$$\xi_n = 4\pi^2 k T n/h, \quad (16)$$

$$s = (p^2 - 1 + \varepsilon_1/\varepsilon_2)^{1/2}, \quad (17)$$

$$\Delta = \frac{s-p}{s+p}, \quad (18)$$

and

$$\bar{\Delta} = \frac{s\varepsilon_2 - p\varepsilon_1}{s\varepsilon_2 + p\varepsilon_1}. \quad (19)$$

The sum is taken over the integers n and the prime on the summation sign indicates that the $n=0$ term must be multiplied by $\frac{1}{2}$. The corresponding potential energy of interaction is obtained as

$$V_A = \frac{kT}{8\pi H_0^2} \sum_{n=0}^{\infty}{}' I(\xi_n, H_0), \quad (20)$$

where

$$I(\xi_n, H_0) = \left(\frac{2\xi_n H_0 \varepsilon_2^{1/2}}{c} \right)^2 \int_1^{\infty} p \left\{ \ln \left[1 - \bar{\Delta}^2 \exp \left(\frac{-2p\xi_n H_0 \varepsilon_2^{1/2}}{c} \right) \right] \right.$$
$$\left. + \ln \left[1 - \Delta^2 \exp \left(\frac{-2p\xi_n H_0 \varepsilon_2^{1/2}}{c} \right) \right] \right\} dp. \quad (21)$$

[17] V. A. Parsegian and B. W. Ninham, *Biophys. J.*, 1970, **10**, 664.

This gives

$$V_A = -\frac{A(H_0, T)}{12\pi H_0^2} = -\frac{A_L}{12\pi H_0^2}, \qquad (22)$$

which can be compared with equation (3). It is important to note, however, that the interaction parameter A_L is now also a function of the distance of surface separation and temperature. From equations (20), (21), and (22) it follows that

$$A_L = 1.5 \, kT \sum_{n=0}^{\infty} {}' I(\xi_n, H_0), \qquad (23)$$

so that the primary problem is the evaluation of the summation term, and in order to use these formulae a convenient means of expressing the relative permittivity as a function of frequency is required. Ninham and Parsegian[17] suggested that the most direct representation was that of a damped linear oscillator, giving the equation

$$\varepsilon(\omega) = 1 + \frac{C_{mw}}{1 - i\omega/\omega_{mw}} + \sum_j \frac{C_j}{1 - (\omega/\omega_j)^2 + i\gamma_j\omega}, \qquad (24)$$

where the first sum refers to a Debye-type microwave relaxation of resonance frequency ω_{mw} and the second to a resonating oscillator with resonance frequency ω_j and a damping term $\gamma_j\omega$.

On the imaginary-frequency axis the expression becomes

$$\varepsilon(i\xi) = 1 + \frac{C_{mw}}{1 + \xi/\omega_{mw}} + \sum_j \frac{C_j}{1 + (\xi/\omega_j)^2 - \gamma_j\xi}. \qquad (25)$$

It is found that the damping term can often be neglected.
At very high frequencies,

$$\varepsilon(\omega) = 1 - (4\pi Ne^2/m\omega^2) \qquad (26)$$

and

$$\varepsilon(i\xi) = 1 + (4\pi Ne^2/m\xi^2) \qquad (27)$$

where e and m are the electron charge and mass and N is the number of electrons per cm^3.

For water the expression used for $\varepsilon(i\xi)$ was found to be

$$\varepsilon(i\xi) = 1 + \frac{C_{mw}}{1 + \xi/\omega_{mw}} + \frac{C_{ir}}{1 + (\xi/\omega_{ir})^2} + \frac{C_{uv}}{1 + (\xi/\omega_{uv})^2}, \qquad (28)$$

which on insertion of the numerical constants C_{mw}, C_{ir}, and C_{uv} gave

$$\varepsilon(i\xi) = 1 + \frac{75.2}{1 + (\xi/1.06 \times 10^{11})} + \frac{3.42}{1 + (\xi/5.66 \times 10^{14})^2} + \frac{0.78}{1 + (\xi/1.906 \times 10^{16})^2} \qquad (29)$$

at 20 °C. In the case of normal hydrocarbons, a much simpler expression was obtained owing to the small dielectric relaxation shown by these materials between audio and visible frequencies, namely

$$\varepsilon(i\xi) = 1 + \frac{n_{hc}^2 - 1}{1 + (\xi/\omega_{uv})^2}, \qquad (30)$$

where n_{hc} = refractive index of the liquid hydrocarbon. This basic procedure of interpolation to obtain values of $\varepsilon(i\xi)$ appears to be simpler than the method used by Krupp of obtaining $\varepsilon(i\xi)$ from measured values of the imaginary part of the complex relative permittivity.[18]

Calculation of the values of A_L from these data for a layer of liquid decane sandwiched between two layers of water gave the values reported in the Table. Similar values were obtained by Gingell and Parsegian.[19] The Hamaker constant A_H, using the London procedure,[20] for this situation gave a value of 1.05×10^{-21} J.

Table A_L *for decane in water*

$H_0/\text{Å}$	A_L/J
0	6.36×10^{-21}
10	6.33×10^{-21}
100	5.78×10^{-21}
1000	4.43×10^{-21}

As would be anticipated, it is possible to obtain the de Boer–Hamaker-type equations from Lifshitz theory as a limiting case.

Several other topics have been discussed in the light of Lifshitz theory, *viz.* a macroscopic theory of temperature dependence,[21] the extension of the theory to magnetic media,[22] calculations of the forces across films of liquid helium,[23] and van der Waals forces between two spheres.[24] The case of a triple-layer film, together with calculations specifically referring to soap films in air, was also considered in detail by Ninham and Parsegian.[25] This paper includes a derivation of the dispersion energy of a film using the method of van Kampen *et al.*[26] which, for the non-retardated situation, is probably more easily followed than the Lifshitz derivation.

A number of fundamental and important papers using continuum theory have also been contributed by Langbein,[27–31] particularly in terms of the interaction between spherical particles.

It is quite clear that during the past few years a significant shift of viewpoint has occurred in terms of the way in which van der Waals interactions between particles should be considered. There is no doubt that considerable

[18] H. Krupp, *Adv. Colloid Interface Sci.*, 1967, **1**, 111.
[19] D. Gingell and V. A. Parsegian, *J. Theor. Biol.*, 1972, **36**, 41.
[20] F. Hauxwell and R. H. Ottewill, *J. Colloid Interface Sci.*, 1970, **34**, 473.
[21] B. W. Ninham, V. A. Parsegian, and G. H. Weiss, *J. Stat. Phys.*, 1970, **2**, 323.
[22] B. W. Ninham and P. Richmond, *J. Phys. (C)*, 1971, **4**, 1988.
[23] B. W. Ninham and P. Richmond, *J. Low Temp. Phys.*, 1971, **5**, 177.
[24] D. J. Mitchell and B. W. Ninham, *J. Chem. Phys.*, 1972, **56**, 1117.
[25] B. W. Ninham and V. A. Parsegian, *J. Chem. Phys.*, 1970, **52**, 4578.
[26] N. G. van Kampen, B. R. A. Nijboer, and K. Schram, *Phys. Letters (A)*, 1968, **26**, 307.
[27] D. Langbein, *J. Adhesion*, 1969, **1**, 237.
[28] D. Langbein, *Phys. Rev. (B)*, 1970, 3371.
[29] D. Langbein, *J. Chem. and Phys. Solids*, 1971, **32**, 1657.
[30] D. Langbein, *J. Phys. (A)*, 1971, **4**, 471.
[31] D. Langbein, *J. Phys. and Chem. Solids*, 1971, **32**, 133.

expansion will occur in this area in the next few years to deal with effects of geometry, *i.e.* forces between spheres, cylinders, plates of limited thickness *etc.*, and with the effects produced by adsorbed layers *etc.* Hopefully a unified theory will be produced which considers the electrostatic interactions simultaneously with the electromagnetic. It is to be hoped also that some effort will be devoted to obtaining the necessary experimental data on relative permittivity as a function of frequency to enable more extensive computations to be made.

Direct Measurements of Attractive Forces.—Simultaneously with advances in understanding the theory of attractive forces between particles and macroscopic bodies, considerable advances have been made in techniques for the direct measurement of these forces.

The force of attraction between two molecularly smooth surfaces of muscovite mica has been measured by Tabor and Winterton.[32-34] The mica sheets were silvered on the back surfaces and then glued on to hemicylindrical glass formers. For the measurements the hemicylinders were arranged with their axes mutually at right angles. Under these conditions the force of attraction is equivalent to that between a sphere of radius a and a flat surface and the unretarded force is given by

$$\Pi_A = Aa/6H_0^2. \qquad (31)$$

In the retarded region the attractive force is given by

$$\Pi_A = 2\pi Ba/3H_0^3. \qquad (32)$$

One of the hemicylindrical surfaces was supported on a rigid mount and the other was attached to a light cantilever beam. During the measurement one surface was moved towards the other using a piezoelectric transducer. At a critical distance of separation the surfaces flicked together and since this distance depended directly on the stiffness of the cantilever, a direct measure of the attractive force was obtained. The actual distance of separation was measured by multiple-beam interferometry. Using this technique it was possible to measure the attractive force with a distance of only 5 nm separating the two surfaces. The measured force was found to agree with the unretarded force equation at distances of less than 10 nm and with the retarded force equation at distances greater than 20 nm. The results for the unretarded force measurements were in agreement with an A value of 10^{-19} J; the B constant was found to be 0.81×10^{-28} J m, in good agreement with calculations from the Lifshitz formula.[8] The results show clearly that the retardation effect, in air, becomes important at separation distances of the order of 10 nm, and that the transition from the unretarded to the fully retarded force occurs over a distance of *ca.* 10 nm.

[32] D. Tabor and R. H. S. Winterton, *Nature*, 1968, **219**, 1120.
[33] D. Tabor and R. H. S. Winterton, *Proc. Roy. Soc.*, 1969, **A312**, 435.
[34] D. Tabor, *J. Colloid Interface Sci.*, 1969, **31**, 364.

The work of Tabor and Winterton was extended by Israelachvili and Tabor,[35] who measured the force between the surfaces down to a separation distance of 1.4 nm with an increased degree of accuracy. For separations in the range 10—130 nm a new method was adopted which involved feeding the piezoelectric transducer with an a.c. signal so that the lower surface could be set vibrating at very small amplitudes over a convenient range of frequencies. The upper surface was also supported by a piezoelectric bimorph whose natural frequency depended both on stiffness and on the van der Waals force exerted upon it by the lower surface. The idea of using a dynamic method was to determine the natural frequency as a function of separation and hence deduce the force law. In the region where results by the dynamic method overlapped with those obtained by Tabor and Winterton,[33] good agreement between the two sets of measurements was obtained. In the range of separation, 2—12 nm, the attractive force was non-retarded with an A value of $1.35 \pm 0.15 \times 10^{-19}$ J. At separation distances greater than 12 nm the exponent of H_0 increased above 2.0 and by $H_0 = 50$ nm had become 2.9. At distances greater than 50 nm the forces were retarded with a B value of $0.97 \pm 0.06 \times 10^{-28}$ J m. Some experiments were also carried out in which a monolayer of stearic acid was deposited on to each of the mica surfaces (layer thickness 2.5 nm). The results suggested that for separation distances greater than 5 nm between the coated surfaces the forces were the same as for bulk mica. At a separation distance of less than 3 nm, however, the measured forces were smaller than in the absence of the monolayer and appeared to be dominated by the properties of the monolayers themselves.

Direct measurements of the force of attraction between a fused silica flat plate and a fused silica plano-convex lens, with the convex side towards the flat plate, were reported by Rouweler and Overbeek.[36] The measurements were made using an apparatus similar to that described by van Silfhout.[37] Before mounting the plates in the apparatus, the plates were carefully examined in a phase-contrast microscope and only those with surfaces free from pits and scratches were selected. In some cases the surface topography of the plates was examined using an optical set-up following Tolansky;[38] no surface irregularities were found. The curve of force against distance gave an exponent of 3.0 for H_0 at distances greater than 50 nm, indicating that the force was fully retarded. A value of $B = 1.05 \pm 0.04 \times 10^{-28}$ J m was obtained compared with a Lifshitz value of 0.57×10^{-28} J m. At shorter distances a transition towards a $1/H_0^2$ dependence was detected. The previous experimental value of B of 0.66×10^{-28} J m found by van Silfhout[37] was found to be incorrect and on correction gave 1.32×10^{-28} J m.

From these experimental observations it appears that there is excellent agreement between different sets of equipment and observers as to the

[35] J. N. Israelachvili and D. Tabor, *Nature Phys. Sci.*, 1972, **236**, 106.
[36] G. C. J. Rouweler and J. Th. G. Overbeek, *Trans. Faraday Soc.*, 1971, **67**, 2117.
[37] A. van Silfhout, *Proc., k. ned. Akad. Wetenschap.*, 1967, **B69**, 501.
[38] S. Tolansky, 'Surface Microtopography', Longmans, London, 1960, p. 30.

distance exponents in the retarded and unretarded force laws, and that these agree with those predicted by theory. The agreement of the B values obtained experimentally with those calculated from the Lifshitz theory must also be regarded as highly encouraging.

The deformation which is produced by attractive forces in the region of contact between two elastic bodies of small elastic modulus, *e.g.* rubber or gelatine, has been analysed by Johnson, Kendall, and Roberts.[39] The theory predicts a finite contact area between surfaces under zero load and gives equations for the external force required to separate the bodies. Since the theory depends on the geometry of the bodies and on surface energy, it indicates a possible method for determining the interfacial surface energy of a solid if one of the bodies is deformable. An alternative approach to the determination of surface energy is described by Bailey *et al.*,[40, 41] who determined the force required to cleave mica sheets coated with unimolecular layers of stearic acid in humid air and in polar and non-polar liquids. The results obtained were compared with those predicted from Young's equation. The latter relation was shown to hold for systems in which the phases remained homogeneous right up to the interface. For non-polar liquids, the predicted and measured values were the same, whereas in the case of water, an additional term of 7.5 mN m^{-1} ($\equiv 7.5 \text{ dyn cm}^{-1}$) had to be introduced into Young's relation to obtain equality. This additional energy was associated with hydrophobic interactions arising as a consequence of the formation of an ordered layer of water molecules at the interface with the molecules reoriented in such a manner as to enable them to form hydrogen bonds with water molecules in the bulk phase. On the assumption that only the molecules in immediate contact with hydrocarbon were involved in the ordering process, a value of 895 J mol^{-1} was obtained for the free energy of formation of this layer.

3 Forces of Electrostatic Origin

Direct Measurement of Normal Surface Forces Due to Double-layer Repulsion.— The repulsive force exerted by diffuse electrical double layers on two solid surfaces as a function of their distance of separation in the direction of a normal to the surface in an aqueous medium has been measured by Roberts and Tabor.[42] An optically smooth, hemispherical surface was produced[43] at the end of a polyisoprene rubber cylinder with an elastic modulus of *ca.* $2 \times 10^5 \text{ N m}^{-2}$ ($2 \times 10^6 \text{ dyn cm}^{-2}$); this surface was pressed on to an optically flat glass surface through an aqueous solution. In this way, a thin liquid film was squeezed between the glass and rubber surfaces. An advantage of using

[39] K. L. Johnson, K. Kendall, and A. D. Roberts, *Proc. Roy. Soc.*, 1971, **A324**, 301.
[40] A. I. Bailey and A. G. Price, *J. Chem. Phys.*, 1970, **53**, 3421.
[41] A. I. Bailey, A. G. Price, and S. M. Kay, *Special Discuss. Faraday Soc.*, 1970, No. **1**, 118.
[42] A. D. Roberts and D. Tabor, *Proc. Roy. Soc.*, 1971, **A325**, 323.
[43] A. D. Roberts, *Eng. Mat. Design*, 1968, **11**, 579.

rubber was that it easily deformed over dust particles or local protrusions and provided essentially parallel surfaces over the compression region. The distance between the surfaces was measured[42, 44] by means of interference fringes for separations greater than 100 nm and by measuring reflected light intensities at shorter distances. By the application of a load to the rubber hemisphere, it was possible to measure the pressure exerted by the rubber as a function of the distance of separation between the charged surfaces on the rubber and the glass. Hence an experimental curve of force against distance was obtained giving the magnitude of the repulsive forces arising from the overlap of electrical double layers; for this situation, the attractive forces involved were very small. With a 10^{-2} mol dm^{-3} solution of sodium dodecyl sulphate between the surfaces at a pressure of 6×10^4 N m^{-2} (6×10^5 dyn cm^{-2}) the total film thickness (water + surface-active agent) was 12 nm. At the same concentration of sodium dodecyl sulphate but with the addition of 4×10^{-2} mol dm^{-3} sodium chloride the film thickness was found to be *ca.* 7 nm, indicating the compression of the double layer with increasing ionic strength.

Further studies using the same equipment were made to investigate the mechanical properties of very thin films[45] and the effectiveness of the film as a lubricating layer. Since the double-layer repulsive force acted against the normal load, it helped to protect the surface from abrasion. Further protection was afforded by the adsorbed layer of surface-active agent on the surfaces at the points of contact where the separating fluid had been locally penetrated. In the absence of a monolayer, the friction increased rapidly with decreasing film thickness towards the dry value.[46]

The thinning of a liquid film with time can be used to determine the viscosity of the liquid by means of the Reynolds equation,[47, 48]

$$t = \frac{3\pi}{64} \frac{\eta D^4}{W} \left(\frac{1}{h_1^2} - \frac{1}{h_2^2} \right), \tag{33}$$

where t = time taken for a film to thin a given distance $(h_2 - h_1)$, with $h_2 > h_1$, under a constant normal load W with D equal to the diameter of the approaching circular plates. Measurements made by Roberts[47] on aqueous films of 10^{-2} mol dm^{-3} sodium dodecyl sulphate between glass and rubber, which thinned from 200 nm down to 20 nm, showed that their effective viscosity over this whole range was constant and nearly the same as the bulk viscosity of the solution. This suggested that for this system no structure was induced into the aqueous film by the presence of the solid surfaces. An apparatus for measuring the viscosity of a thin liquid film between two silica plates has also been described by Peschel and Aldfinger.[49] It was found that the

[44] A. D. Roberts and D. Tabor, *Wear*, 1968, **11**, 163.
[45] A. D. Roberts and D. Tabor, *Special Discuss. Faraday Soc.*, 1970, No. **1**, 243.
[46] A. D. Roberts, *J. Phys. (D)*, 1971, **4**, 433.
[47] A. D. Roberts, *J. Phys. (D)*, 1971, **4**, 423.
[48] O. Reynolds, *Trans. Roy. Soc.*, 1886, **177**, 157.
[49] G. Peschel and K. H. Aldfinger, *Ber. Bunsengesellschaft phys. Chem.*, 1970, **74**, 347, 351.

viscosity of the liquid in the thin layer, η_G, depended on the bulk viscosity, η and the distance between the plates, h, according to the expression

$$\eta_G = \eta[1 + B \exp(-mh)], \qquad (34)$$

where B and m were parameters obtained by experiment. At a film thickness of 30 nm the ratio η_G/η was found to be 5.4 for water at 20 °C and 4.0 for benzene at 20 °C.

Roberts[50] has also reported some interesting experiments, using the glass–rubber technique, on films of synovial fluids. Two samples were taken, one from the joint of a normal person, and the other from the joint of an arthritic patient. The film thickness of the fluid from the normal patient was found to be approximately twice that of the arthritic patient. Stable thin films were also formed from saliva. These experiments suggested that double-layer repulsion could be a feature common to many biological lubricants.

An estimate of the force of interaction between styrene–butadiene latex particles has been made by El-Aasser and Robertson[51] using an ultracentrifuge technique. The latex was centrifuged in a sucrose solution whereupon the latex rose to the top of the dispersion in the tube and separated into two layers, one forming a complete film and the other forming a fraction underneath this which was redispersible. The film was recovered, washed free of redispersible material, and dried at 105 °C. Thus the mass of uncoalesced latex, m, was obtained from the difference between the original weight of latex put in the tube and the weight of the film. The force of film formation was calculated from

$$F = m(1 - \rho_1/\rho_2)a, \qquad (35)$$

where $\rho_1 = $ density of the medium, $\rho_2 = $ density of the polymer and $a = $ acceleration of the centrifugal field. The styrene–butadiene latex was found to have its maximum stability between pH 8.8 and 10.7. Outside of these limits, the stability decreased and the latex was found to be unstable below pH 2 and above pH 12. The pressure required for coalescence of the particles was found to be of the order of 10^4 N m^{-2} (10^5 dyn cm^{-2}). A similar type of technique was used by Melville, Willis, and Smith.[52] These authors examined a silver iodide sol at pI 4 in an ultracentrifuge cell and, on the basis that the centrifugal force at the bottom of the cell was sufficient to induce coagulation, which then proceeded upwards through the cell, an analysis of the critical forces required for coagulation was obtained. The measured forces were found to be lower than those calculated theoretically on the basis of nearest-neighbour interactions by a factor of up to 7.

Barclay and Ottewill[53] measured directly the pressure of an assembly of sodium montmorillonite plates as a function of the distance of separation between the plates down to separation distances of the order of 2 nm. The

[50] A. D. Roberts, *Nature*, 1971, **231**, 434.
[51] M. S. El-Aasser and A. A. Robertson, *J. Colloid Interface Sci.*, 1971, **36**, 86.
[52] J. B. Melville, E. Willis, and A. L. Smith, *J.C.S. Faraday I*, 1972, **68**, 450.
[53] L. Barclay and R. H. Ottewill, *Special Discuss. Faraday Soc.*, 1970, No. **1**, 138.

results were compared directly with those expected on the basis of theory for the interaction between flat plates at constant surface potential. For the results obtained in 10^{-4} mol dm^{-3} sodium chloride solution the measured forces were less than those predicted theoretically at distances greater than 15 nm. At shorter distances the forces were greater than those predicted theoretically on this model and it was suggested that at very short distances a force additional to double-layer repulsion and van der Waals attraction occurred and that this arose from solvation. It is of interest in this connection to note that Derjaguin *et al.*[54] found that the dielectric constant of the first one or two layers of water adsorbed by sodium montmorillonite was about half the normal value, suggesting that considerable orientation occurred at the surface. Some confirmation of solvation forces was obtained by carrying out experiments with adsorbed layers of dodecylhexaoxyethylene glycol monoether adsorbed on the plates.[53] These conditions led to an enhancement of the force at short distances.

Similar compression measurements[55] were carried out using polystyrene latex particles of different diameters. With particles having a diameter of 185 nm very little interaction was detected between the particles until the distance of separation was about 30 nm; at this distance, the particles appeared to assemble in a hexagonally close-packed array. In the case of the larger particles, diameter 1.10 μm, the particles appeared to pack randomly under pressure and the effect of the repulsive forces became significant at much larger distances, *ca.* 80 nm.

The force of adhesion of carbon black particles from an aqueous dispersion on to a cellophane film was measured by Visser.[56] A direct method was used which involved depositing carbon black particles on a cellophane film, determining the number per unit area and then rotating the film in a concentric cylinder apparatus, and finally counting the number of particles which remained attached to the film. The extent of adhesion was found to depend on the pH of the solution. Maximum adhesion occurred at pH 3.3 and it was found possible to interpret the results obtained at this pH by means of the theory of dispersion forces for interaction between a sphere and a plate, *i.e.* on the grounds that at this pH the adhesive force was determined solely by van der Waals' forces.[57]

Theoretical Studies.—From the theoretical viewpoint, the basic problem is to derive expressions for the repulsive force between charged surfaces, or particles, based on a model of the electrical double layer. For comparison with experimental measurements of the force against distance curves for montmorillonite plates, Barclay and Ottewill[53] used the expressions derived earlier by Verwey and Overbeek[2] for the overlap between two diffuse double layers for

[54] B. V. Derjaguin, N. A. Krylov, and V. F. Novik, *Doklady Akad. Nauk S.S.S.R.*, 1970, **193**, 126 (*Doklady Phys. Chem.*, 1970, **193**, 501).
[55] L. Barclay and R. H. Ottewill, *Special Discuss. Faraday Soc.*, 1970, No. 1, 164.
[56] J. Visser, *J. Colloid Interface Sci.*, 1970, **34**, 26.
[57] J. Visser, *Reports Progr. Appl. Chem.*, 1968, **53**, 714.

a situation in which the surface potential of two approaching plates was maintained constant. An alternative method of approach is to consider the interaction whilst maintaining the surface charge constant. Both the constant-potential and constant-charge situations have been examined by Honig and Mul[58] for interactions between two parallel flat plates and between two spheres. The results obtained were presented in tabular form, which allows ready computation of the energy of interaction. Limiting equations were given for four different sets of conditions, *viz.* small separations, large separations, moderate potentials, and moderate charge densities. The ratios of the flocculation concentrations for univalent, bivalent, and tervalent ions were also computed for both constant potential and constant charge. Bell and Levine[59] pointed out that in the Verwey–Overbeek expressions[2] used by Barclay and Ottewill[53] the effects of the ionic environment were not considered, *i.e.* effects such as those due to discreteness of charge, which depend on the difference between the real potential situation in the vicinity of an ion, were neglected and an average value for the electric field at the surface was assumed. They also suggested that to produce a complete theory, ionic volume and polarization effects would have to be taken into account. Moreover, in interactions between very thin plates such as those obtained with sodium montmorillonite, the effect of the thinness of the plate on the image potentials would have to be included. They also commented, 'both the constant potential and the constant surface charge assumptions are artificial'. The influence of the molar volume and the polarization of the electrolyte on the repulsion potential for two spherical particles has been considered by Sanfeld *et al.*[60] and the energies obtained have been used to evaluate a second virial coefficient.[61, 62] It was found that at constant surface charge, the coefficients calculated for different electrolytes were slightly higher than those calculated by Vrij[63] using the Verwey–Overbeek theory. The observed sequence for the magnitude of the coefficient was found to be

$$HCl > LiCl > NaCl > Verwey–Overbeek.$$

On the basis of a constant-potential model, bigger differences were found and the sequence was inverted. The deviations from the Verwey–Overbeek value became greater as the particle size decreased and the electrolyte concentration and surface potential increased. Application of a correction for non-ideality of the solution also gave important differences from the classical theory.

Approximate methods for calculating the energy of interaction between two spherical particles were proposed by Bell *et al.*[64] Utilizing the theoretical work

[58] E. P. Honig and P. M. Mul, *J. Colloid Interface Sci.*, 1971, **36**, 258.
[59] G. M. Bell and S. Levine, *Special Discuss. Faraday Soc.*, 1970, No. **1**, 166.
[60] A. Sanfeld, C. Devillez, and M. Lubelski, *Electrochim. Acta*, 1968, **13**, 1937.
[61] A Sanfeld, C. Devillez, and P. Terlinck, *J. Colloid Interface Sci.*, 1970, **32**, 33.
[62] A. Sanfeld, C. Devillez, and P. Terlinck, 'Proceedings of the International Congress on Surface Activity', Barcelona, 1968, p. 805.
[63] A. Vrij, Doctoral thesis, University of Utrecht, 1959.
[64] G. M. Bell, S. Levine, and L. N. McCartney, *J. Colloid Interface Sci.*, 1970, **33**, 335.

of Gronwall, La Mer, and Sandved[65] and the numerical results of Dutch workers[66] for the potential distribution around single spheres, they were able to obtain results outside the linear Debye–Hückel range. A simple expression for the energy of interaction at large distances suitable for all radii and potentials was obtained.

The form of the curve of electrostatic potential against the distance between two flat plates was also discussed by Sigal and Alekseenko,[67] who suggested numerical and analytical solutions for the non-linear Poisson–Boltzmann equations involved. Graphical solutions for the interaction between diffuse double layers have also been given by Kemper and Quirk.[68]

Structure of Liquids near Surfaces.—The question of the ordering of liquids near a solid interface, and the possible role that such an ordered structure might play in the stabilization of a colloidal dispersion, is an interesting problem which has not yet been completely answered. Nevertheless, evidence does seem to be accumulating in favour of a short-range ordering of molecules near a surface.

As mentioned earlier, in their determination of the viscosity of thin films of liquids, Peschel and Aldfinger[49] found that the viscosity of benzene in a thin film was four times that of normal benzene and, in measurements of the disjoining pressure of some dipolar organic liquids[69] between two fused silica surfaces as a function of temperature, some evidence was found for an oriented surface layer in which a rotation of the molecules was restricted. In a study of the boundary viscosity of a number of liquid polydimethylsiloxanes, Derjaguin *et al.*[70] found that on glass and steel substrates the bulk viscosity was retained down to thicknesses of the order of 15—20 nm and then the viscosity increased slightly. However, at a distance of 1.0–1.5 nm from the substrate, the viscosity became anomalously low: this was attributed to orientation of polydimethyl-siloxane molecules. For long-chain liquid alkanes Ash and Findenegg[71] found evidence for increased orientation of the molecules near a Graphon interface, although this did not occur with smaller molecules.

In the case of water, Peschel and Aldfinger[49] found that the viscosity in a thin layer was 5.4 times that of normal water and Churayev, Sobolev, and Zorin,[72] using microcapillaries of quartz, also obtained evidence for an enhanced viscosity. These data appear to be in disagreement with the results of Roberts[47] who found that in films *ca.* 20 nm thick between solid surfaces,

[65] T. H. Gronwall, V. K. La Mer, and K. Sandved, *Phys. Z.*, 1928, **29**, 358.
[66] A. L. Loeb, J. Th. G. Overbeek, and P. H. Wiersema, 'The Electrical Double Layer around a Spherical Colloid Particle', M.I.T. Press, Cambridge, Mass., 1961.
[67] V. L. Sigal and A. M. Alekseenko, *Kolloid Zhur.*, 1971, **33**, 737 [*Colloid J.* (*U.S.S.R.*), 1971, **33**, 614].
[68] W. D. Kemper and J. P. Quirk, *Proc. Soil Sci. Soc. Amer.*, 1970, **34**, 347.
[69] K. H. Aldfinger and G. Peschel, *Special Discuss. Faraday Soc.*, 1970, No. **1**, 89.
[70] B. V. Derjaguin, V. V. Karasev, I. A. Lavygin, I. I. Skorokhodov, and E. N. Khromova, *Special Discuss. Faraday Soc.*, 1970, No. **1**, 98.
[71] S. G. Ash and G. H. Findenegg, *Special Discuss. Faraday Soc.*, 1970, No. **1**, 105.
[72] N. V. Churayev, V. D. Sobolev, and Z. M. Zorin, *Special Discuss. Faraday Soc.*, 1970, No. **1**, 213.

water had essentially normal bulk properties. In experiments on the interaction between montmorillonite plates at distances of the same order of magnitude Barclay and Ottewill[53] also found no evidence for ordered layers. However, at distances of the order of 2 nm the experimental results favoured the formation of oriented water layers. Olejnik, Stirling, and White,[73] using neutron scattering to study the dynamics of interlamellar water molecules in layered silicate clay minerals, found significant changes in the water structure for water thicknesses of about one to two molecular layers. In studies on vermiculite, but using different experimental techiques, Hougardy *et al.*[74] found evidence for organization of the water in the two hydration layers near the clay surface. Using a sodium montmorillonite, Thomas and Cremers[75] also obtained evidence for an increased viscosity of water near the clay surface and Derjaguin, Krylov, and Novik[54, 76] found that the dielectric constant of the first two layers of water adsorbed by sodium montmorillonite was about half the normal value.

The interaction between particles covered with solvated layers has been considered theoretically in a series of papers by Totomanov.[77] It was found that stabilization against coagulation could occur by interaction between dipole layers alone, provided that the number per unit area and the magnitudes of the dipoles were sufficiently large. It was further considered that interaction between oriented dipole layers could form the basis of a long-range solvation interaction force provided that solvent molecules were adsorbed so as to give a uniformly directed layer of dipoles. Totomanov also derived expressions for the energy of interaction between two coaxially oriented dipole layers[78] and for the interaction between two planar layers of dipoles in the presence of electrical double layers[79, 80] Moreover, conditions were shown to exist in which dipole–dipole interactions could lead to more effective protection against coagulation than diffuse double layer interactions.

4 Studies on Disperse Systems

Monodisperse Systems.—Polymer latices, which can be prepared as monodisperse spheres under conditions in which the radius of the ultimate particles can be varied from *ca.* 50 nm to several microns, provide a very suitable model colloidal system for a wide range of studies. In fact, it has been suggested that these systems should now be called 'polymer colloids'.[81] In the period under

[73] S. Olejnik, G. C. Stirling, and J. W. White, *Special Discuss. Faraday Soc.*, 1970, No. **1**, 194.
[74] J. Hougardy, J. M. Serratosa, W. Stone, and H. van Olphen, *Special Discuss. Faraday Soc.*, 1970, No. **1**, 187.
[75] H. C. Thomas and A. Cremers, *J. Phys. Chem.*, 1970, **74**, 1072.
[76] B. V. Derjaguin, N. A. Krylov, and V. F. Novik, *Doklady Akad. Nauk S.S.S.R.*, 1970, **193**, 126.
[77] D. S. Totomanov, *Izvest. Otdel. Khim. Nauki, bulg. Akad. Nauk*, 1970, **3**, 385.
[78] D. S. Totomanov, *Izvest. Otdel. Khim. Nauki, bulg. Akad. Nauk*, 1970, **3**, 391.
[79] D. S. Totomanov, *Izvest. Otdel. Khim. Nauki, bulg. Akad. Nauk*, 1970, **3**, 397.
[80] D. S. Totomanov, *Izvest. Otdel. Khim. Nauki, bulg. Akad. Nauk*, 1970, **3** 403.
[81] 'Polymer Colloids', ed. R. M. Fitch, Plenum Press, New York, 1971.

review, detailed instructions for the preparation of polystyrene latices in a highly monodisperse form have been given by Deželić *et al.*[82] by methods using both direct and seeded emulsion polymerization. The best results were obtained using Aerosol MA as the emulsifier. A major conference covering many aspects of emulsion polymerization also gave a number of methods for preparing latices.[83] A considerable innovation came from Kotera, Furusawa, and Takeda,[84] who described methods for the formation of highly monodisperse latices by the aqueous-phase polymerization of styrene, in the absence of an emulsifying agent, using potassium persulphate as the initiator. The method appears to be very versatile since it was found that the diameter of the particle radius could be varied within the limits 350—1400 nm according to the conditions used. Preparation of latices in this way avoids a major problem in that the latices do not have to be 'cleaned' to remove stabilizing surface-active agent from the surface. A method which appears to be successful for the latter process is to treat the latex with well-washed ion-exchange resins, and details of the procedure have been given by Vanderhoff and his co-workers.[85, 86] These workers have also described a conductometric method for determining the number of charged groups on the latex surface.[86, 87] The analysis of surface groupings by potentiometric titration[88] and dye-partition[89] has also been reported, together with an electrophoretic method of surface characterization.[90]

The possibility that the use of ion-exchange resins to remove emulsifier from latices might affect the form of the size distribution was investigated by McCann *et al.*[91] Synthetic bimodal particle-size distributions were formed, which after ion-exchange treatment were examined by electron microscopy to check if the distributions had been modified; the individual latices were also examined. Electron microscopy showed that the number of small particles present decreased significantly with time. It was also found that if the resins had not been adequately purified, the stability of the latices containing small particles appeared to increase as a consequence of the adsorption of polyelectrolytes leached from the ion-exchange resin.

The DLVO theory of colloid stability was invoked by Dunn and Chong[92] to discuss the mechanism of particle formation in aqueous vinyl acetate systems. The maximum rate of polymerization was found to be directly proportional to the concentration of persulphate initiator and to that of the

[82] N. Deželić, J. J. Petres, and G. Deżċeli, *Kolloid-Z.*, 1970, **242**, 1142.
[83] 'Emulsion Polymers', *Brit. Polymer J.*, 1970, **2**, pp. 1–205.
[84] A. Kotera, K. Furusawa, and Y. Takeda, *Kolloid-Z.*, 1970, **269**, 677.
[85] H. J. van den Hul and J. W. Vanderhoff, *J. Colloid Interface Sci.*, 1968, **28**, 336.
[86] J. W. Vanderhoff, H. J. van den Hul, R. J. M. Tausk, and J. Th. G. Overbeek, in 'Clean Surfaces', ed. G. Goldfinger, Marcel Dekker, New York, 1970, p. 15.
[87] H. J. van den Hul and J. W. Vanderhoff, *Brit. Polymer J.*, 1970, **2**, 116.
[88] G. D. McCann, E. B. Bradford, H. J. van den Hul, and J. W. Vanderhoff, ref. 81, p. 29.
[89] G. Roy, B. M. Mandel and S. R. Palit, ref. 81, p. 49.
[90] J. Hearn, R. H. Ottewill, and J. N. Shaw, *Brit. Polymer J.*, 1970, **2**, 116.
[91] G. D. McCann, E. B. Bradford, H. J. van den Hul, and J. W. Vanderhoff, *J. Colloid Interface Sci.*, 1971, **36**, 159.
[92] A. S. Dunn and L. C.-H. Chong, *Brit. Polymer J.*, 1970, **2**, 49.

monomer in the aqueous phase, as expected from Smith–Ewart theory. The surface of the polymer particles was found to be the main locus of polymerization. The rate of particle coalescence depended primarily on the stability of the particles in the colloidal sense and was only indirectly related to the rate of polymerization. In keeping with the observations of Kotera *et al.*,[84] latices prepared in the absence of surface-active agent were found to be monodisperse, whereas those prepared, even with emulsifier at high ionic strength, were polydisperse. The theory of latex particle formation has also received a considerable new stimulus from the work of Fitch and Tsai.[93]

The formation of ordered structures with polymer latices has been commented on by Vanderhoff *et al.*[86] and investigated in detail by Krieger *et al.* in both aqueous[94] and non-aqueous systems.[95]

The stability of poly(vinyl chloride) latices at different surface concentrations of dodecyl sulphate has been investigated by Bibeau and Matijević.[96] The critical coagulation concentration of the dispersion for sodium and magnesium ions was found to increase as the amount of dodecyl sulphate adsorbed on the surface increased. The surface concentration as obtained from electrokinetic measurements, however, remained almost constant as the actual surface concentration was increased and the authors concluded that, for this system, the electrophoretic measurements were not sufficiently sensitive to be of use in indicating the inherent stability of the sol. The adsorption of hafnium on to poly(vinyl chloride) latices was measured by Stryker and Matijević.[97] They suggested that at pH values greater than 4.5 the hafnium existed in solution as neutral $Hf(OH)_4$ species. The plateau on the adsorption isotherm gave a cross-sectional area of 0.23 nm^2 per $Hf(OH)_4$ and the authors suggested that this could be utilized for the determination of latex surface areas. The effects of the chelating agents 2,2'-bipyridyl (bipy) and 1,10-phenanthroline (phen) and the cobalt complexes tris(bipyridyl)cobalt(III) perchlorate, $Co(bipy)^{3+}$, and tris(phenanthroline)cobalt(III) perchlorate, $Co(phen)^{3+}$, on the stability of poly(vinyl chloride) latices was examined by Lauzon and Matijević[98] over a broad pH range; in the chelated form, coagulation and charge-reversal studies were possible under conditions where hydrolysis occurs with many tervalent ions. It was found that bipy did not coagulate the latex above pH 3.9, even at the highest concentrations used, and that phen was not effective above pH 5. With the chelate ions, coagulation and then restabilization occurred over a broad pH range; the concentration at which charge reversal of the latex particles occurred, however, did not coincide with the maximum extent of coagulation. The amount of $Co(bipy)^{3+}$ required varied, with $3.6 \times 10^{-3} \text{ g cm}^{-3}$ of latex, from *ca.* $2 \times 10^{-5} \text{ mol dm}^{-3}$ at a dodecyl sulphate

[93] R. M. Fitch and C. H. Tsai, ref. 81, pp. 103.
[94] I. M. Krieger and P. A. Hiltner, ref. 81, p. 63.
[95] P. A. Hiltner, Y. S. Papir, and I. M. Krieger, *J. Phys. Chem.*, 1971, **75**, 1881.
[96] A. A. Bibeau and E. Matijević, ref. 81, p. 127.
[97] L. J. Stryker and E. Matijević, *J. Colloid Interface Sci.*, 1969, **31**, 39.
[98] R. V. Lauzon and E. Matijević, *J. Colloid Interface Sci.*, 1971, **37**, 296.

concentration of 4×10^{-5} mol dm^{-3} to 2×10^{-3} mol dm^{-3} at *ca.* 10^{-2} mol dm^{-3} surface-active agent.

Cationic surface-active agents also make excellent coagulating agents for latices. The adsorption of a series of alkyltrimethylammonium compounds has been reported in detail by Connor and Ottewill and the results obtained correlated with electrophoretic measurements.[99] The coagulation of latices stabilized by a non-ionic surface-active agent both by adding electrolyte and by temperature changes has been reported by Neiman and Taranovskaya.[100] Surface-active agents can also act as collectors for latex particles and the flotation of latices using lauric acid in the presence of aluminium sulphate and nitrate has been observed.[101]

The *in situ* polymerization of surface-active agents on to the surface of stytrene–butadiene latices has been carried out by Greene and Saunders[102] using sodium acrylamidostearate; an ingenious procedure which allowed a variation in the number concentration of integrally bound surface groups to be obtained. The stability of these latices in the presence of sodium, barium, and lanthanum chlorides was investigated and from these studies a value of 3.2×10^{-21} J was obtained for the Hamaker constant of styrene–butadiene particles.

The possibility of producing inorganic materials, with a narrow size distribution, as spherical particles offers many possibilities for experimental work on model systems, particularly since these materials usually have larger Hamaker constants than polymeric materials. In this respect, the work of Matijević and Demchak[103] in producing monodisperse chromium hydroxide sols, which do contain spherical particles, by heating solutions of chromium salts at elevated temperatures ($75\,^{\circ}$C) is particularly interesting. They found that the high degree of monodispersity and the well defined higher-order Tyndall spectra (HOTS) were only obtained in the presence of sulphate ions and that the method did not work with chromium(III) nitrate or perchlorate. Extension of this work[104] showed that the best conditions for sol formation were obtained by heating chrome alum solutions, of concentration between 4×10^{-4} and 8×10^{-4} mol dm^{-3}, for 18 h at $75\,^{\circ}$C at a pH of between 3.1 and 3.3. It was suggested that during this ageing process $Cr_2(OH)_2^{4+}$ species were formed and that positions occupied by OH could then be taken up by SO_4^{2-} to give configurations such as

[99] P. Connor and R. H. Ottewill, *J. Colloid Interface Sci.*, 1971, **37**, 642.

[100] R. E. Neiman and S. I. Taranovskaya, *Kolloid Zhur.*, 1969, **31**, 436 [*Colloid J. (U.S.S.R.)*, 1969, **31**, 347].

[101] E. A. Cassel, E. Matijević, F. J. Mangravite, T. M. Buzzel, and S. B. Blabac, *Amer. Inst. Chem. Engineers J.*, 1971, **17**, 1486.

[102] B. W. Greene and F. L. Saunders, *J. Colloid Interface Sci.*, 1970, **33**, 393.

[103] R. Demchak and E. Matijević, *J. Colloid Interface Sci.*, 1969, **31**, 257.

[104] E. Matijević, A. D. Lindsay, S. Kratohvil, M. E. Jones, R. I. Larson, and N. W. Cayey, *J. Colloid Interface Sci.*, 1971, **36**, 273.

Once the concentration of basic chromium sulphate species reached the critical limiting supersaturation the sudden formation of nuclei relieved the supersaturation. From this point, the basic chromium sulphate species formed by ageing diffused to the existing nuclei, which then grew uniformly to produce hydrosols of narrow size distribution. Electrophoretic measurements showed these particles to have an isoelectric point in the pH range 7.5—7.8. The critical coagulation concentrations for a number of electrolytes were also determined. During this work, a simple closed wet cell for direct transmission studies in the electron microscope was developed and used to observe the chromium hydroxide particles.[105]

One of the principal difficulties in the control of precipitation processes is often the lack of reproducibility in mixing the reactants. To cope with this problem, Burgarić, Füredi-Milhofer, and Adamski[106] have designed an apparatus which they claim enables extremely slow and controlled mixing of reactants to be achieved without the formation of regions of supersaturation. The rates of nucleation and crystal growth achieved appeared to be comparable with those found in precipitation from homogeneous solutions.

The precipitation of calcium and barium oleate investigated by Nemeth and Matijević[107] was found to lead to the formation of particles which had a disc-like appearance in the electron microscope. The authors suggested that these might have existed in the dispersed form as ellipsoids with their interiors filled with solution but had flattened on dehydration for electron-microscope examination.

The stability of monodisperse silica sols (Ludox HS and Ludox AM) has been examined in some detail by Allen and Matijević.[108] Studies were carried out as a function of pH in the presence of lithium, sodium, potassium, caesium, and calcium chlorides, sodium bromide, sodium iodide, and calcium nitrate. The stability of the sol towards added electrolyte was found to decrease markedly with increasing pH. Potassium and caesium chlorides, however, appeared to behave differently and showed a minimum in stability at intermediate pH values (7—11). The electrophoretic mobilities determined in a moving-boundary apparatus did not show any correlation with the coagulation concentration, that is the mobilities increased with pH in the region of coagulation and the primary reason for destabilization appeared to be associated with ion exchange at the silica surface. Essentially similar experimental results were obtained by Depasse and Watillon[109] using amorphous silica hydrosols, prepared by the polymerization-dehydration of silicic acid, which resembled the Ludox preparations used by Allen and Matijević. The excess electrolyte present in the sol was removed by percolation through a mixed-bed ion-exchange resin. They also found that all the alkali-metal cations were able to coagulate the silica between pH 7 and 11, but that above

[105] E. F. Fullam, *Rev. Sci. Instr.*, 1972, **43**, 245.
[106] B. Burgarić, H. Füredi-Milhofer, and T. Adamski, *Croat. Chem. Acta*, 1971, **43**, 127.
[107] R. Nemeth and E. Matijević, *Kolloid-Z.*, 1971, **245**, 497.
[108] L. H. Allen and E. Matijević, *J. Colloid Interface Sci.*, 1969, **31**, 287.
[109] J. Depasse and A. Watillon, *J. Colloid Interface Sci.*, 1970, **33**, 430.

pH 11 they could be classified into two groups: Li^+ and Na^+ caused coagulation, whereas K^+, Rb^+, and Cs^+ remained ineffective. Again the negative mobility of the particles increased, so that no correlation with coagulation was found. They found an isoelectric point for their silica particles of pH 4.4, compared with the values of pH 1.6 and 1.2 found by Allan and Matijević for Ludox HS and Ludox AM respectively. Harding,[110] who used an aqueous pyrogenic silica dispersion, has reported similar coagulation effects using alkali metals. Methylation of the surface with hexamethyldisilazane, although it did not affect the electrophoretic mobility, gave a material which showed the expected dependence of stability on pH. The anomalous behaviour, *i.e.* the decrease of stability with increase of mobility, appeared to be confined to particles with diameters of less than 50 nm. Later work of Allen and Matijević[111] was directed towards determining the critical coagulation concentration against pH diagrams for Ludox HS and AM in order to delineate the domains of stability and coagulation for univalent, bivalent, and tervalent cations. The cation exchange was also measured as a function of pH and electrolyte concentration. At low pH the cation-exchange process obeyed the law of mass action, but at pH values greater than 6 the observed exchange was smaller than that predicted for the exchange of simple metal ions by H_3O^+ ions; a mathematical analysis of the exchange process was therefore undertaken.[112] At a given pH[111] the minimum amount of cation exchange which occurred at the surface and which was necessary to produce coagulation was found to be the same for all the alkali-metal cations regardless of the coagulation concentration. Up to a pH of 8.0, the same relationship was found for Ca^{2+} and La^{3+} ions as for univalent cations. Depasse and Watillon,[109] however, offer a different explanation and suggest that coagulation with univalent ions occurs as a consequence of different types of bonding, namely, the formation of dissociated silanol–silanol bonds, which progressively turn into siloxane bonds, *i.e.*

$$-Si-OH\cdots O^- -Si- \rightarrow -Si-O-Si- + OH^-$$

together with salt bonding by Li^+ or Na^+. Stumm, Huang, and Jenkins[113] have also discussed specific chemical interactions with oxide surfaces and their effect on the stability of dispersions (see also ref. 114).

The effects of hydrolysable cations were also studied by Allen and Matijević[115] and stability–coagulation domains were mapped-out, as a function of pH, for aluminium nitrate and sulphate and for lanthanum nitrate. The ion-exchange process, leading to instability, still appeared to occur with these ions. Some differences were observed between the behaviour of aluminium nitrate and aluminium sulphate and these were quantitatively explained by the

[110] R. D. Harding, *J. Colloid Interface Sci.*, 1971, **35**, 172.
[111] L. H. Allen and E. Matijević, *J. Colloid Interface Sci.*, 1970, **33**, 420.
[112] L. H. Allen, E. Matijević, and L. Meites, *J. Inorg. Nuclear Chem.*, 1971, **33**, 1293.
[113] W. Stumm, C. P. Huang, and S. R. Jenkins, *Croat. Chem. Acta*, 1970, **42**, 223.
[114] W. Stumm and J. J. Morgan, 'Aquatic Chemistry', Wiley–Interscience, 1970.
[115] L. H. Allen and E. Matijević, *J. Colloid Interface Sci.*, 1971, **35**, 66.

formation of an $AlSO_4^+$ species. Mobility determinations in the coagulation region showed little correlation with the electrolyte concentrations required for coagulation. Ludox AM was investigated in somewhat more detail. Regions of coagulation and stabilization were again mapped out and the reversal-of-charge concentrations for aluminium salts determined.[116]

The usefulness of the coagulation process for studying ion–surface interactions is also stressed by Dumont and Watillon.[117] Using ferric oxide hydrosols, which they considered to be composed of a haematite core surrounded by an hydrated oxide shell, they found that at low surface potentials added ions behaved as indifferent electrolytes, but that at higher surface potentials they were specifically adsorbed. The competitive adsorption of ions at oxide surfaces clearly plays a very important part in many processes which occur with these materials. Lyklema[118] has also discussed the function of the counterion in the coagulation of hydrophobic sols, and suggests that since in studies of α-haematite a reversal of the lyotropic series occurs, possible steric factors have to be considered. Hingston, Posner, and Quirk[119] measured the competitive adsorption of phosphate and arsenate, and phosphate and selenite, on gibbsite and goethite. They carried out a fundamental analysis of the problem and found that the competition was described by a Langmuir-type exchange equation. The correlation between the pK_a values of the conjugate surface groups and the pH values at which the extent of specific adsorption showed a marked change was investigated by the same group of authors[120] and it was found that specific adsorption depended on the ease of dissociation of the conjugate acids at the mineral surface.

The flocculation of monodisperse silica sols by a cationic polyelectrolyte, poly(methacryloyloxyethyldiethylammonium methyl sulphate), has been investigated by Iler.[121] For particles with a diameter of 40 nm or less, the amount of polyelectrolyte required to flocculate unit weight of silica varied inversely with particle diameter and it was considered that only a segment of a chain (length *ca.* 150 nm) was required to form a bridge. With larger particles the amount required varied with the square of the diameter—possibly because one or more chains were required to form the bridge at a contact point between a pair of particles. A useful spot-test was also described for detecting either polymer or silica in the supernatant liquid. The flocculation by cationic polyelectrolytes of polystyrene latex particles with surface carboxy-groups has been examined by Gregory.[122] Two polyelectrolytes were used, one of low molecular weight and one of high molecular weight. The results, in conjunction with electrophoretic measurements, indicated that polymer-bridging played a significant role in flocculation by high-molecular-weight materials, whereas

[116] E. Matijevi, F. J. Maćngravite, and E. A. Cassell, *J. Colloid Interface Sci.*, 1971, **35**, 560.
[117] F. Dumont and A. Watillon, *Discuss. Faraday Soc.*, 1971, No. 52, 352.
[118] J. Lyklema, *Croat. Chem. Acta*, 1970, **42**, 151.
[119] F. J. Hingston, A. M. Posner, and J. P. Quirk, *Discuss. Faraday Soc.*, 1971, No. 52, 334.
[120] F. J. Hingston, A. M. Posner, and J. P. Quirk, *Search*, 1970, **1**, 324.
[121] R. K. Iler, *J. Colloid Interface Sci.*, 1971, **37**, 364.
[122] J. Gregory, *Trans. Faraday Soc.*, 1969, **65**, 2260.

with the low-molecular-weight materials coagulation occurred as a consequence of charge reversal.

The use of polymeric flocculants for selective flocculation of minerals has become a possibility and the general principles governing the process have been discussed by Kitchener *et al.*[123, 124] Experiments[123] were carried out to test the principles derived using quartz, calcite, and galena by themselves and in mixed suspensions.

Silver Iodide.—An inorganic material frequently used for the preparation of colloidal dispersions is silver iodide, and a method for the preparation of silver iodide sols which allows a variation of particle size has been reported by Volkova *et al.*[125] An interesting phenomena, described by Ozaki and Hachisu,[126] which can be observed using silver iodide sols, is the motion and agglomeration of plate-like particles under the influence of strong illumination, *i.e.* photophoresis, the crystals redispersing into single particles when the intensity of the illumination is reduced. The influence of butanol adsorption on the surface charge and stability of silver iodide sols has been examined by Vincent *et al.*[127] They found that there was no simple relation between surface charge and stability, *e.g.* with simple 1:1 electrolytes the surface charge at a given pI decreased continuously upon addition of butanol, but the critical coagulation concentration went through a maximum. Moreover, the lyotropic sequence, in terms of its effect on the surface charge, was not affected by adsorption of butanol, whereas the sequence for coagulation values was inverted; similar trends were found with simple 2:1 electrolytes. A possible explanation for the observed trends was proposed in terms of the displacement of counter-ions from the Stern layer by butanol molecules. It was considered that the butanol molecules were adsorbed with their alkyl chains towards the surface.

The stability of silver iodide and silver bromide sols in solutions of hafnium chloride, as a function of pH, has been investigated by Stryker and Matijević.[128] Over the pH range 2—4, the charge on the particles was reversed and restabilization occurred when the particles became positively charged owing to the adsorption of hydrolysed hafnium ions. The effect of cation hydrolysis was also examined[129] on silver iodide sols stabilized by a non-ionic surface-active agent [$C_{12}H_{25}(CH_2CH_2O)_6OH$] where adsorption occurred with the alkyl chain towards the surface. Considerable enhancement of the stability of the sol towards the addition of unhydrolysed ions was obtained in the presence of the surface-active agent. The protective action,

[123] B. Yarrar and J. A. Kitchener, *Trans. Inst. Mining Met.* (*C*), 1970, **79**, 23.
[124] R. J. Pugh and J. A. Kitchener, *J. Colloid Interface Sci.*, 1971, **35**, 656.
[125] S. S. Volkova, O. N. Ovchinnikova, and T. A. Burlak, *Izvest. V.U.Z. Khim. i khim. Tekhnol.*, 1970, **13**, 772 (*Chem. Abs.*, 1970, **73**, 113 205).
[126] M. Ozaki and S. Hachisu, *Sci. of Light*, 1970, **19**, 59.
[127] B. Vincent, B. H. Bijsterbosch, and J. Lyklema, *J. Colloid Interface Sci.*, 1971, **37**, 171.
[128] L. J. Stryker and E. Matijević, *Kolloid-Z.*, 1969, **233**, 912.
[129] K. G. Mathai and R. H. Ottewill, *Kolloid-Z.*, 1970, **236**, 147.

however, was considerably diminished in the presence of the hydrolysed species of lanthanum, aluminium, and thorium ions. It was concluded that the hydrolysed species were either more strongly adsorbed than the surface-active agent, and hence displaced the latter from the surface, or that they formed hydrolysed structures which were large enough to cause coagulation by interparticle bridging; Napper and Netschey,[130] however, have provided more recent evidence that the former mechanism appears the more likely.

The flocculation of positively charged silver iodide sols by poly(acrylic acids) of different molecular weights in the range 1.7×10^4—2.4×10^6 was also studied.[131] The flocculation concentration, c_f, was related to the viscosity average molecular weight, M_V, by an equation of the form

$$c_f = a M_V^{-b}, \tag{36}$$

where a and b were constants which were evaluated experimentally.

Clays.—Aggregation in hydrogen-montmorillonite suspensions has been studied by Gilbert and Laudelout[132] using the rate of decomposition of ethyl diazoethanoate in order to demonstrate the presence of elementary platelets; the decomposition of the diazoethanoate was catalysed by hydrogen ions adsorbed on the montmorillonite and hence gave an estimate of the face sites available. A parallel determination was made of the size distribution of the hydrogen-clay particles. The presence of elementary platelets in the suspensions could not be demonstrated and it was suggested that sufficient face-to-face aggregation in a parallel array had occurred to prevent all the exchange sites being available. Although the term *tactoid* is often used for parallel arrays of plate-like crystals, Aylmore and Quirk[133] suggest that the term *quasi-crystalline* would be a better term to describe the regions of parallel alignment of individual aluminosilicate lamellae in montmorillonite. The term *domain* was suggested to describe the regions of parallel alignment of crystals for illites and other fixed-lattice clays.

The idea of almost exclusive face-to-face association for calcium bentonite was proposed by Bahin and Lahav[134] to explain their results, whereas Fitzsimmons *et al.*[135] showed by electron-microscope studies that calcium montmorillonite originally dispersed into separate platelets and then flocculated with time to form large, flat, sheet-like particles which exhibited pronounced flow birefringence. For rapid flocculation in the presence of salt, face-to-face condensation was deemed unlikely to occur, whereas after long periods in salt-free systems it was shown that a small amount of perpendicular stacking occurred, giving aggregates four layers thick,

[130] D. H. Napper and A. Netschey, *J. Colloid Interface Sci.*, 1971, **37**, 528.
[131] D. J. A. Williams and R. H. Ottewill, *Kolloid-Z.*, 1971, **243**, 141.
[132] M. Gilbert and H. Laudelout, *J. Colloid Interface Sci.*, 1971, **35**, 486.
[133] L. A. G. Aylmore and J. P. Quirk, *Proc. Soil Sci. Soc. Amer.*, 1971, **35**, 652.
[134] A. Bahin and N. Lahav, *J. Colloid Interface Sci.*, 1970, **32**, 178.
[135] R. F. Fitzsimmons, A. M. Posner, and J. P. Quirk, *Israel J. Chem.*, 1970, **8**, 301.

A new technique, that of examining the effect of an a.c. electric field on the light scattering of a suspension, has been applied by Schweitzer and Jennings[136] to montmorillonite suspensions. Measurements as a function of field strength up to 600 V cm^{-1} and frequencies up to 5 kHz showed the clay particles to be rigid plates with the predominant polarity along the transverse axis. In fresh sols the plates on average were found to have a length some six times the breadth and a thickness that of a single layer. In aged sols, aggregates of at least nine platelets were found to exist. The aggregate structure favoured was a linear stacked arrangement in which the rectangular plates were alternately associated with the edges jointed perpendicularly to the centres of the faces. The optical phenomena which can be induced in colloidal dispersions by the application of an electric field have been generally reviewed by Stoylov.[137] In addition, the optical dispersion of the electric birefringence of bentonite dispersions has been examined by Tibu.[138]

An alternative means of obtaining information on flocculated clay systems is to use rheological measurements and the effectiveness of this method has been demonstrated in studies on the synthetic clay material laponite[139] using a variety of electrolytes; a relationship was found between the effectiveness of the cations in increasing sol stability and their size and charge. Nicol and Hunter[140] report a rheological study of kaolinite dispersions. At neutral pH values, a high yield value was obtained, compatible with the formation of a flocculated structure composed of extended chains of clay particles; at high pH values a good dispersion was formed and the yield value became small. The effect of cationic surface-active agents was also examined. As the concentration was increased, these caused charge reversal leading to positively charged particles which gave a good dispersion with a very small yield value. An attempt was also made to calculate a yield value using the DLVO theory of colloid stability. A very interesting series of studies on the adsorption of cationic surface-active agents on clay systems of the montmorillonite type has been reported by Lagaly and Weiss;[141, 142] in these studies, X-ray techniques were used to obtain the interlamellar spacing and thus deduce information about the orientation of the chains and packing on the surface.[143, 144] The formation of mixed adsorbed layers of quaternary ammonium compounds and n-alkanols was also examined.[145-147]

Many of these studies reported on clay systems indicate the variability of

[136] J. Schweitzer and B. R. Jennings, *J. Colloid Interface Sci.*, 1971, **37**, 443.
[137] S. P. Stoylov, *Adv. Colloid Interface Sci.* 1971, **3**, 45.
[138] M. Tibu, *Compt. rend.*, 1971, **272**, C, 569.
[139] B. S. Neumann and K. G. Sansom, *Clay Minerals*, 1970, **8**, 389.
[140] R. F. Hunter and S. K. Nicol, *J. Colloid Interface Sci.*, 1968, **28**, 250.
[141] G. Lagaly and A. Weiss, *Kolloid-Z.*, 1970, **237**, 266.
[142] G. Lagaly and A. Weiss, *Kolloid-Z.*, 1970, **237**, 364.
[143] G. Lagaly and A. Weiss, *Kolloid-Z.*, 1970, **238**, 485.
[144] G. Lagaly and A. Weiss, *Kolloid-Z.*, 1971, **243**, 48.
[145] G. Lagaly and A. Weiss, *Kolloid-Z.*, 1971, **248**, 968.
[146] G. Lagaly and A. Weiss, *Kolloid-Z.*, 1971, **248**, 979.
[147] G. Lagaly and A. Weiss, 'Reunion Hispano—Belga de Minerales de la Arcilla', Madrid, 1970, p. 179.

flocculated structures and emphasize the need for work, not simply on the process of flocculation *per se*, but on the structure of the floccules formed. Little work of a systematic nature appears to have been reported, although in this context Medalia and Heckman[148] have suggested an electron-microscope method for the determination of aggregate morphology.

The adsorption of aluminium compounds on to clay surfaces has received attention from Colombera *et al.*,[149] who measured the adsorption of hydroxy-aluminium species on to Fithian illite from sodium perchlorate solutions. A Langmuirian isotherm was obtained which was explained on the basis that the hydroxy-species present in solution were adsorbed in constant relative proportions and that the relative proportions of hydroxy-aluminium species remained unchanged on dilution of the hydroxy-aluminium solutions at constant ionic strength. They suggested that hydrogen bonds were formed between hydrolysed aluminium species and the surface.

Secondary Minimum Phenomena.—The importance of the secondary minimum in the curve of potential energy of interaction against the distance of surface separation has been emphasized in the Introduction (Figure 1). This plays a part in the formation of loose aggregates and more experimental evidence is becoming available as to the validity of the secondary minimum concept and the part it plays in flocculation processes. Kotera *et al.*[150] have carried out a study of the effects of salt concentration on the stability of a series of polystyrene latices covering a range of particle sizes. They determined the electrolyte concentration required to destabilize the sol and the zeta-potential at this concentration using sodium, barium, and lanthanum chlorides. The flocculation concentrations obtained varied with particle size and reached a maximum at a particle diameter of 750 nm. For particle sizes larger than this the aggregation process was found to be completely reversible, suggesting that it was a secondary minimum process. The effect of particle size on stability has also been considered theoretically by Wiese and Healy.[151] A rational explanation was obtained for the experimental observation that as the particle size was increased[152] the stability of latices increased, as judged by coagulation concentrations, reached a maximum, and then decreased. The adhesion of carbon black particles to a glass substrate and their subsequent detachment by addition of an ionic surface-active agent, which was observed experimentally by Clayfield and Smith,[153] was found to be explicable on the basis that the initial adhesion occurred because the carbon black particles became located in secondary minima. The additional electrical double-layer repulsion brought about by adsorption of an ionic surface-active agent was

[148] A. I. Medalia and F. A. Heckman, *J. Colloid Interface Sci.*, 1971, **36**, 173.
[149] P. M. Colombera, A. M. Posner, and J. P. Quirk, *J. Soil Science*, 1971, **22**, 118.
[150] A. Kotera, K. Furusawa and K. Kudo, *Kolloid-Z.*, 1970, **240**, 837.
[151] G. R. Wiese and T. W. Healy, *Trans. Faraday Soc.*, 1970, **66**, 490.
[152] R. H. Ottewill and J. N. Shaw, *Discuss. Faraday Soc.*, 1966, No. 42, 154.
[153] E. J. Clayfield and A. L. Smith, *Environ. Sci. Technol.*, 1970, **4**, 413.

H

sufficient to decrease the depth of the minimum and facilitate particle removal. The reversible agglomeration of plate-like gold particles and the formation of tactoids, studied optically by Okamoto and Hachisu,[154] was also explicable in terms of association in a secondary minimum.

Using the DLVO theory for spherical particles, Thomas and McCorkle[155] have demonstrated that either isotropic or anisotropic floccules can be formed, depending on the particle radius, zeta-potential, and electrolyte concentration. They argued that when two spherical particles associate to form a doublet, the situation with regard to the potential energy of interaction loses its symmetry, and a third sphere approaching along the axis through the centre of the two spheres 'sees' a single sphere, whereas a sphere approaching normal to the axis would 'see' both spheres of the doublet. Calculation of the energy barriers indicated that the primary maximum of the interaction along the axis was about half of that in the normal direction. This indicated that as the electrolyte concentration in the system was increased a point would be reached at which the growth of floccules would occur by the preferential attachment of new particles to the ends of the doublet, thus leading to aniso-tropic flocculation, *i.e.* the formation of 'strings of beads'; the interaction normal to the axis would remain repulsive. Addition of enough electrolyte to lower both the primary maxima, so that the repulsive barriers could be overcome easily from any direction, would lead, however, to nearly isotropic floccule formation.

Many gels and other thixotropic structures are thought to exist as a con-sequence of the secondary minimum and a reasonable explanation of their behaviour can often be given in these terms. Lukashenko *et al.*[156] have carried out work which attempted to establish laws governing the effects of electrolytes on the secondary minimum and on gel formation. For association at a distance, the relationship between the ion concentration n at which association occurred and the valency v was found to be

$$nv^f = \text{constant}, \tag{37}$$

where $f = $ a numerical constant having a value between 2.5 and 3.5. Klimentova *et al.*[157] found that the maximum gelation time for silicic acid hydrosols occurred close to the isoelectric point at pH 1.7.

Kinetics of Coagulation.—The basic approach to slow coagulation is to treat the process as that of diffusion occurring in a force field.[158] This gives for the flux, J, of spherical particles of radius a, towards a central spherical particle,

[154] S. Okamoto and S. Hachisu, *Science of Light*, 1970, **19**, 49.
[155] I. L. Thomas and K. H. McCorkle, *J. Colloid Interface Sci.*, 1971, **36**, 110.
[156] G. M. Lukashenko, M. V. Serebrovskaya, O. G. Us'yarov, and I. F. Efremov, *Kolloid Zhur.*, 1971, **33**, 136 [*Colloid J. (U.S.S.R.)*, 1971, **33**, 88].
[157] Y. P. Klimentova, L. F. Kirichenko, and Z. Z. Vysotskii, *Ukrain. khim. Zhur.*, 1971, **31**, 433.
[158] N. Fuchs, *Z. Phys.*, 1934, **89**, 736.

the expression

$$J = 8\pi D_0 N_0 a \bigg/ \int\limits_0^\infty \frac{\exp{(V/kT)}}{(u+2)^2}\, du, \qquad (38)$$

where $u = H_0/a$, N_0 = number of particles per cm^3, and D_0 = diffusion coefficient of one particle in an infinitely dilute dispersion. The assumption is made in deriving this equation that as the particles approach each other the frictional coefficient given by the expression $f = kT/D_0$ remains constant. Although this assumption might be a good approximation for aerosols and for conditions where the electrostatic forces exist at distances large in comparison with the particle dimensions, *i.e.* as in some non-aqueous systems, Derjaguin and Muller[159] point out that in aqueous solutions of electrolytes interparticle interactions occur at relatively small distances and it is therefore not possible to neglect the dependence of the viscous drag on the distance between particles. Under these conditions, a frictional coefficient $f(u)$ has to be introduced which depends on the distance of separation and is given by

$$f(u) = f \cdot \beta(u),$$

where $\beta(u)$ is a dimensionless function which becomes equal to unity at large separations and tends to $a/4H_0$ for distances of approach much smaller than the radius of the particle. Hence equation (38) must be rewritten as

$$J = 8\pi D_0 N_0 a \bigg/ \int\limits_0^\infty \frac{\beta(u)}{(u+2)^2} \exp{(V/kT)}\, du, \qquad (39)$$

giving for the stability ratio, W, normally defined as

$$W = 2 \int\limits_0^\infty \frac{\exp{(V/kT)}}{(u+2)^2}\, du,$$

the approximate expression,

$$W \approx 1 + (1/8L)(3\pi kT u_m/A)^{1/2} \exp{(V_m/kT)}, \qquad (40)$$

where V_m = height of the energy barrier, u_m = the distance parameter at the energy maximum, and

$$L = \int\limits_0^\infty \frac{\beta(u) \exp{(V_A/kT)}}{(u+2)^2}\, du.$$

Since V_m is directly proportional to the particle radius, W also depends on a.

Spielman[160] also considered the problem of viscous interactions in coagulation and demonstrated theoretically that viscous effects could produce a retardation of the coagulation rate by as much as a factor of ten for thin

[159] B. V. Derjaguin and V. M. Muller, *Doklady Phys. Chem.*, 1967, **176**, 738.
[160] L. A. Spielman, *J. Colloid Interface Sci.*, 1970, **33**, 562.

electrical double layers. Thus he concluded that Hamaker constants estimated from measured coagulation rates could be subject to considerable errors if viscous interactions were ignored.

The treatment of Derjaguin and Muller[159] was extended by Honig *et al.*,[161] who derived a number of expressions for $\beta(u)$. A good approximation was found to be

$$\beta(u) = \frac{6u^2 + 13u + 2}{6u^2 + 4u}. \tag{41}$$

These authors also carried out extensive computations of the absolute rate of rapid coagulation. In agreement with Spielman[160] they found that hydrodynamic interaction diminished the rate by factors of about 0.4—0.6, depending on the magnitude of the Hamaker constant. The stability ratios W were also computed by numerical integration using the full Hamaker expressions for the van der Waals energy of interaction. The effect of hydrodynamic interaction on the slope of the $\log W$ against $\log c_e$ curves, where c_e = concentration of added electrolyte, was, however, found to be small. They therefore concluded, in agreement with Derjaguin and Muller, that the existing lack of correlation between experimental values of $d \log W/d \log c_e$ and those calculated theoretically could not be due to a hydrodynamic effect. Some measurements of the absolute rate constant of coagulation by Lips *et al.*,[162] using a refined light-scattering method, gave a value of the rate constant $0.67 - 0.69$ times that predicted by the original Smoluchowski theory.[163]

However, it is interesting to note that Matthews and Rhodes,[164] who investigated the use of the Coulter counter for examining coagulating systems, found, over a range of polymer latex particle sizes from 0.719 to 3.49 μm, good agreement between the experimental rate constant for rapid flocculation and that calculated from Smoluchowski theory. Neither passage of the dispersion through the orifice nor the dilution of the system with electrolyte appeared to affect the results obtained.

Since flocculation controlled by Brownian motion (*perikinetic* flocculation) is normally too slow for economic technological usage in processes such as the treatment of waste waters, shear flocculation (*orthokinetic* flocculation) is often employed to obtain a rapid build-up in aggregate size. However, at high rates of shear, floccules in which the particles are held together by weak forces can also be broken apart. Hence there usually exists an optimum shear rate which produces the maximum effect. In this context, Goren[165] has computed the force along the lines of centres of two touching spheres in a uniform shear gradient, G. This was found to be

$$F = 6\pi\eta a^2 G F(a_1/a_2), \tag{42}$$

[161] E. P. Honig, G. J. Roebersen, and P. H. Wiersema, *J. Colloid Interface Sci.*, 1971, **36**, 97.
[162] A. Lips, C. Smart, and E. Willis, *Trans. Faraday Soc.*, 1971, **67**, 2979.
[163] M. von Smoluchowski, *Z. phys. Chem.*, (*Leipzig*), 1917, **92**, 129.
[164] B. A. Matthews and C. T. Rhodes, *J. Colloid Interface Sci.*, 1970, **31**, 332, 339.
[165] S. L. Goren, *J. Colloid Interface Sci.*, 1971, **36**, 94.

where a_1 and a_2 are the radii of the touching spheres and $F(a_1/a_2)$ is a tabulated function. The capture of small particles by van der Waals attraction to a larger collector particle has also been considered by Spielman and Goren.[166]

The reverse process to coagulation and/or flocculation is peptization, and Frens and Overbeek[167] have drawn attention to some aspects of the kinetics of peptization. They point out that it is well known that fresh precipitates are easier to disperse than old ones, which indicates that an aggregate of colloidal particles is not in equilibrium and that irreversible, temperature-dependent ageing processes occur in coagulation. It follows, therefore, that the interpretation of peptization phenomena with aggregated systems is not possible unless the data are obtained in experiments with a shorter time-scale than the ageing time of the aggregate. The authors demonstrated that it was possible to follow the kinetics of peptization by suddenly diluting the sol containing electrolyte after a short period of coagulation. From their experiments they concluded that peptization was a rapid, spontaneous process and that electrical double-layer repulsion probably provided the driving force.

A particularly interesting coagulation process, and one which appears not to have been very extensively studied, is that in which particles coagulate at a free interface. This was originally called mechanical coagulation by Freundlich and is now termed *surface coagulation* by Heller.[168] The latter author, in conjunction with several co-workers, has carried out an extensive investigation of this phenomenon. A theory of surface coagulation was developed based on the following assumptions:

(i) that the coagulation process proceeded exclusively in the liquid/air interface and was a 'bimolecular' reaction;

(ii) that the contribution of aggregates to the rate of coagulation (Smoluchowski rate increment) could be neglected. Moreover, since the aggregates returned to the bulk they did not significantly affect the surface area of the interface available for occupation by unreacted primary particles;

(iii) that sufficient convection occurred to exclude the diffusion of particles to and from the interface as a rate-determining factor;

(iv) that a definite steady state occurred for the distribution of primary particles between the bulk dispersion and the surface and that this distribution could be described by a Langmuir adsorption isotherm;

(v) that the rate of adsorption was large enough relative to the rate of formation of fresh surface for the adsorption equilibrium to be unaffected.

Thus, putting the Langmuir adsorption in the form

$$c/a = K_1 + K_2 c, \qquad (43)$$

where c = concentration of colloidal particles in the bulk phase, a = surface concentration, and K_1 and K_2 are constants, and taking the 'bimolecular'

[166] L. A. Spielman and S. L. Goren, *Environ. Sci. Technol.*, 1970, **4**, 135.
[167] G. Frens, and J. Th. G. Overbeek, *J. Colloid Interface Sci.*, 1971, **36**, 286.
[168] W. Heller and J. Peters, *J. Colloid Interface Sci.*, 1970, **32**, 592.

surface reaction as

$$-\mathrm{d}a/\mathrm{d}t = K_0 a^2,$$ (44)

they obtained

$$-\mathrm{d}a/\mathrm{d}t = K_0 c^2 / (K_1 + K_2 c)^2.$$ (45)

With S = surface area at a constant rate of surface renewal and V = volume of solution, the rate at which the bulk concentration was changed was given by

$$-\frac{\mathrm{d}c}{\mathrm{d}t} = \frac{K_0 S}{V} \frac{c^2}{(K_1 + K_2 c)^2}.$$ (46)

Experiments were carried out to test these equations and to examine the general phenomena of surface coagulation using sols of α-FeOOH, when it was found that:

(a) colloidal dispersions which required a relatively large amount of electrolyte in order to exhibit conventional bulk coagulation were not susceptible to surface coagulation;

(b) coagulation in the surface required a critical low sol stability although the stability could well be sufficient to exclude coagulation in the absence of renewal of the liquid/air surface;

(c) the rate data indicated a lack of participation of the secondary and tertiary aggregates in the surface reaction.

The aggregates formed by surface coagulation were found to differ in form from those formed in bulk coagulation processes. They were in fact essentially laminar aggregates, as indicated by the 'silkiness' exhibited by mildly agitated dispersions.

The effects of stirring and temperature were examined by Peters and Heller[169] using α-FeOOH sols. The promotion of surface coagulation by addition of a destabilizing electrolyte was examined using polymer latices[170] and the prevention of surface coagulation by the addition of surface-active agents using α-FeOOH sols.[171] The role of the solid/liquid interface and the effect of turbulence were examined by Delauder and Heller.[172]

Review Articles.—In addition to the references cited in the text in this section, a number of review articles and books dealing with the properties of disperse systems[173–177] have appeared in the period under review.

[169] J. Peters and W. Heller, *J. Colloid Interface Sci.*, 1970, **33**, 578.
[170] W. Heller and W. B. de Lauder, *J. Colloid Interface Sci.*, 1971, **35**, 60.
[171] W. Heller and J. Peters, *J. Colloid Interface Sci.*, 1971, **35**, 300.
[172] W. B. Delauder and W. Heller, *J. Colloid Interface Sci.*, 1971, **35**, 308.
[173] B. V. Derjaguin, *Res. Surface Forces*, 1971, **3**, 448.
[174] H. Sonntag and K. Strenge, 'Coagulation and Stability of Dispersed Systems', Deutsche Verlag Wissenschaft, Berlin, 1970, pp. 173.
[175] 'The Chemistry of Solid–Liquid Interfaces', *Croat. Chem. Acta*, 1970, **42**, 81–395.
[176] 158th A.C.S. Symposium: 'Colloid Stability', *J. Colloid Interface Sci.*, 1970, **33**, 335–444.
[177] S. Hachisu, *Zairyo*, 1970, **19**, 500.

5 Steric Stabilization

The term *steric stabilization* appears to have been used initially by Heller and Pugh[178] to denote the fact that uncharged particles could be prevented from flocculating by the adsorption of non-ionic polymer molecules. The term *protection* was used at an earlier date.[179] As pointed out by Napper,[180] however, the term 'steric' is used in this context with a broad thermodynamic connotation rather than with the restricted meaning its use has in organic chemistry. Knowledge of the forces which can lead to short-range repulsion is still somewhat uncertain and evaluation of the force and energy of steric repulsion as a function of the distance of separation between the surfaces has not yet reached the same quantitative level as that achieved in the evaluation of the electrostatic and van der Waals forces. It does appear to be certain, however, that steric stabilization forces, when they occur, are of shorter range than the electrostatic and van der Waals forces and that the force increases very rapidly with decrease of distance.[181] As a somewhat crude first approximation, the energy of interaction can be assumed to rise almost vertically at the distance at which steric overlap first occurs.

In writing the total energy of interaction as

$$V = V_R + V_A + V_S, \qquad (2)$$

where all steric and solvation terms are put together in the term V_S, the position for the origin of the energy terms on the distance axis is not always clear. For example, in the case of V_R and V_A either both interactions can be assumed to emanate from the particle surface or the origin of the van der Waals interaction can be taken as the surface and the origin of the electrical interaction as the Stern layer.[127] Once the particle has an adsorbed layer, however, this can also affect the position of the Stern layer and the thickness and nature of the adsorbed layer may well have to be taken into account in calculating the van der Waals interactions; interpenetration of the adsorbed layers may also occur. Fundamentally, division of the interaction energy into three separate terms is not a very satisfactory procedure. Despite these problems, however, the past few years have seen considerable progress in this area both in the theoretical and in the experimental approaches to the topic.

In a number of cases it has been found that the stability of lyophobic aqueous dispersions to coagulation by electrolytes can be enhanced by the addition of non-ionic surface-active agents[182-186] and macromolecules.[187,188]

[178] W. Heller and T. L. Pugh, *J. Chem. Phys.*, 1954, **22**, 1778.
[179] H. Freundlich, 'Colloid and Capillary Chemistry', Methuen, London, 1926, 589.
[180] D. H. Napper, *Ind. and Eng. Chem. (Product Res. and Development)*, 1970, **9**, 467.
[181] L. Barclay and R. H. Ottewill, *Special Discuss. Faraday Soc.*, 1970, No. **1**, 169.
[182] R. H. Ottewill and T. Walker, *Kolloid-Z.*, 1968, **227**, 108.
[183] Y. M. Glazman and G. M. Kabysh, *Kolloid. Zhur.*, 1969, **31**, 27 [*Colloid J. (U.S.S.R.)*, 1969, **31**, 21.
[184] K. L. Daluja and S. N. Srivastava, *Indian J. Chem.*, 1969, **7**, 790
[185] E. Moriyama, K. Hattori, and K. Shinoda, *J. Chem. Soc. Japan*, 1969, **90**, 35.
[186] B. A. Matthews and C. T. Rhodes, *J. Pharm. Sci.*, 1968, **57**, 569.
[187] J. E. Glass, R. D. Lundberg, and F. E. Bailey, *J. Colloid Interface Sci.*, 1970, **33**, 491.
[188] J. Bontoux, A. Dauplan, and R. Marignan, *J. Chim. phys.*, 1969, **66**, 1259.

The thickness of the adsorbed layer of dodecylhexaoxyethylene glycol mono-ether on polystyrene latex particles was determined and found to be *ca.* 50 nm, just greater than the length of the extended surface-active agent molecule;[182] from this it was estimated that the concentration of water in the adsorbed layer was 0.74 g cm^{-3}, a clear indication of the importance of solvation. Theoretically, it was shown[182] that the free energy of interaction ΔG_S arising from non-ionic interactions for two spheres of radius a depended on the thickness of the adsorbed layer, δ, the concentration of non-ionic stabilizer in the adsorbed layer, c, the distance H_0 between the basic particle surfaces, an enthalpic term, χ, which characterized the interaction of the stabilizing group with the solvent, and an entropy of mixing term, ψ, for the stabilizing layers. The expression obtained was

$$\Delta G_S = V_S = \frac{4kT\pi c^2}{3\bar{V}_1 \rho_2^2} (\psi - \chi)(\delta - H_0/2)^2 (3a + 2\delta + H_0/2), \qquad (47)$$

where $\bar{V}_1$ = molecular volume of the solvent molecules and ρ_2 = density of the adsorbed film. A positive value of ΔG_S is necessary for the stability and Napper[189,190] pointed out that if ΔG_S is written as

$$\Delta G_S = \Delta H_S - T\Delta S_S,$$

three ways of obtaining steric stabilization can occur:

(i) if $T\Delta S_S$ is negative and exceeds the $-\Delta H_S$ value, ΔG_S will be positive. Since entropic effects therefore oppose the flocculation which enthalpic effects are trying to promote it was suggested this should be called *entropic stabilization*. It was also pointed out that, since $-T\Delta S_S$ usually decreases in magnitude with decreasing temperature, on reducing the temperature to the θ-temperature the enthalpy and entropy would become equal and flocculation would occur. Thus entropically stabilized dispersions would be characterized by a flocculation process which would occur on cooling;

(ii) ΔG_S can also be positive for positive values of ΔH_S and ΔS_S provided that $\Delta H_S > T\Delta S_S$, *i.e.* enthalpy aids stabilization and entropy aids flocculation. The term *enthalpic stabilization* was suggested for this mechanism. On heating, $T\Delta S_S$ should normally increase more rapidly than ΔH_S and hence flocculation should occur;

(iii) for the situation with a positive ΔH_S and a negative $T\Delta S_S$, ΔG_S would always be positive. For this situation flocculation would not occur at an accessible temperature since the θ-temperature would be negative.

In principle, therefore, both entropic and enthalpic effects play their part in steric stabilization. From the somewhat sparse evidence available at the present time, it appears that enthalpic effects are probably more important in the stabilization of aqueous dispersions and entropic contributions more impor-tant in the stabilization of non-aqueous systems.[5,191]

[189] D. H. Napper, *Kolloid-Z.*, 1969, **234**, 1149.
[190] D. H. Napper, *Proc. Royal Austral. Chem. Inst.*, 1971, 327.
[191] R. Evans, J. B. Davison, and D. H. Napper, *J. Polymer Sci., Part B, Polymer Letters*, 1972, **10**, 449.

Napper[192] investigated the stability of aqueous polymer dispersions stabilized by block copolymers of poly(ethylene oxide) and a vinyl or acrylic monomer using particles which were shown by electrophoresis to be completely uncharged at an ionic strength of 10^{-2}. Flocculation occurred when the solvency of the dispersion medium for the stabilizing chains was decreased as, for example, by raising the temperature: for these systems, a critical flocculation temperature was obtained which was found to be insensitive, over a limited range, to the molecular weight of the stabilizing poly(ethylene oxide) material. The influence of a number of other factors was later investigated,[193] including the nature of the anchoring groups of the stabilizing polymers, the nature of the disperse phase, the particle size, the surface coverage, and the molecular weight of the stabilizing polymer. It was found that for stability to occur it was necessary for the solvent properties of the dispersion medium to be better than those of a θ-solvent for the stabilizing polymer on the surface. This implied that the second virial coefficient of the stabilizing polymer needed to be positive so that the segmental excluded volume was also positive. Under these conditions, the configurational entropy of the molecules in the overlap volume would become less than that of molecules in the bulk dispersion medium and an excess osmotic pressure would occur in the overlap zone (see earlier). As a consequence of this, molecules of the dispersion medium could diffuse into the overlap region, forcing the stabilizing groups and the particles apart. The spontaneous redispersion of the sterically stabilized particles after centrifugation or flocculation supported this view and emphasized the essential difference between the flocculation of sterically stabilized dispersions and the coagulation of lyophobic sols.

In further work the flocculation of poly(vinyl acetate) dispersions, stabilized by poly(ethylene oxide) chains, was examined using a series of electrolytes.[194] The order of decreasing flocculation efficiency for the anions examined paralleled that of the Hofmeister series. The order found for the cations,

$$Rb^+ = K^+ = Na^+ = Cs^+ > NH_4^+ = Sr^{2+} > Li^+ = Ca^{2+} = Ba^{2+} = Mg^{2+},$$

was exactly the opposite to that expected on the basis of the postulate that the more hydrated ions should be the more effective flocculants. It was considered that the ability of an ion to flocculate this type of dispersion was determined by its ability to convert water into a θ-solvent for the stabilizing chains.

In an extension of this work[130] the combined effects of steric stabilization and electrostatic repulsion were studied; once the electrical double layer had been compressed, the particle remained sterically stabilized and the flocculation concentrations obtained were those of the sterically stabilized dispersions. However, in the case of weakly anchored stabilizing molecules, flocculation occurred in the presence of hydrolysed cations owing to the displacement of the stabilizer by the hydrolysed species; this was in agreement with the work

[192] D. H. Napper, *J. Colloid Interface Sci.*, 1969, **29**, 168.
[193] D. H. Napper, *J. Colloid Interface Sci.*, 1970, **32**, 106.
[194] D. H. Napper, *J. Colloid Interface Sci.*, 1970, **33**, 384.

of Mathai and Ottewill,[129] who came to similar conclusions after studying the flocculation by hydrolysed cations of silver iodide sols stabilized by non-ionic surface-active agents. It was also shown by Napper and Netschey[130] that polyoxyethylene chains acted as an enthalpic stabilizer in water and as an entropic stabilizer in methanol; the aqueous dispersions flocculated on heating and the methanolic dispersions flocculated on cooling to *ca.* 255 K. The importance of a combination of electrostatic factors and steric stabilization on the stability of dispersions of pharmaceutical interest was investigated by Matthews and Rhodes,[195] using as surface-active agents salts of dioxyethylated dodecyl sulphate in conjunction with particles of sulphamerazine, hydrocortisone, and griseofulvin. Their results were interpreted, including the effect of particle size, using equation (47) to calculate the energy of interaction due to steric effects; qualitatively, the experimental results were found to be in agreement with those expected theoretically.

Important papers on the stabilization of dispersed particles by macromolecules have also been contributed by Hesselink,[196, 197] following the work of Clayfield and Lumb[198] and Meier.[199] The thermodynamic properties of a polymer adsorbed at a solid/liquid interface, including those in a θ-solvent, have also been computed by Hoffman and Forsman,[200] using an extension of the theory of Forsman and Hughes.[201] Hesselink[196, 197] has examined theoretically the density distribution of segments of an adsorbed macromolecule as a function of distance from the adsorbent surface. The density distribution for a copolymer attached to the surface by anchor groups was found to pass through a maximum with distance from the surface whereas with homopolymers an exponential decrease with distance occurred. It was considered that on the approach of a second particle in a Brownian collision there was insufficient time for loop rearrangement to occur and hence a free energy change occurred, *i.e.* ΔG_S. When the free energy increased the consequence was repulsion but when the free energy went through a minimum with distance this was consistent with polymer bridge formation and consequential flocculation. Hesselink, Vrij, and Overbeek[202] extended these ideas and predicted that enhancement of stability would occur with long adsorbed chains and an extreme size distribution, a high amount of polymer adsorption, a good solvent, a small Hamaker constant, and a small particle size. The calculation of the interaction energy between two parallel adsorbing planes immersed in a solution from which polymers were adsorbed has also been considered by Ash and Findenegg.[203]

[195] B. A. Matthews and C. T. Rhodes, *J. Pharm. Sci.*, 1970, **59**, 521.
[196] F. Th. Hesselink, *J. Phys. Chem.*, 1969, **73**, 3488.
[197] F. Th. Hesselink, *J. Phys. Chem.*, 1971, **75**, 65.
[198] E. J. Clayfield and E. C. Lumb, *Macromolecules*, 1968, **1**, 133.
[199] D. J. Meier, *J. Phys. Chem.*, 1967, **71**, 1861.
[200] R. F. Hoffman and W. C. Forsman, *J. Polymer Sci., Part A-2, Polymer Phys.*, 1970, **8**, 1847.
[201] W. C. Forsman and R. Hughes, *J. Chem. Phys.*, 1963, **38**, 2130.
[202] F. Th. Hesselink, A. Vrij, and J. Th. G. Overbeek, *J. Phys. Chem.*, 1971, **75**, 2094.
[203] S. G. Ash and G. H. Findenegg, *Trans. Faraday Soc.*, 1971, **67**, 2122.

It is perhaps unfortunate from the point of view of a quantitative understanding of steric stabilization that this last group of papers do not explicitly deal with the effect of the solvent since, from an experimental point of view, it has been so clearly shown to be important. Hopefully, the next few years will see considerably further development of this subject.

6 Electrokinetic Studies

Electrokinetic measurements, partly because of the ease with which they can generally be carried out, still provide a very convenient means of obtaining information on the electrical properties of surfaces. A number of basic problems, however, continue to remain unclear in this field, such as the exact relationship of measured electrokinetic potentials, or, as they are more usually called, zeta-potentials, to the parameters used in describing the electrical double layer under static conditions, namely, the surface potential, ψ_0, the Stern potential, ψ_δ, surface charge, *etc.*

Moreover, some of the terms used in the literature are rather loose and some clarification should be attempted. Attention is drawn to the report of the IUPAC Commission on Colloid and Surface Chemistry which contains some proposals on this point.[204] Although universal acceptance has not been achieved in the use of nomenclature, care should be exercised with three terms: point of zero charge (PZC), isoelectric point (IEP), and reversal-of-charge concentration (RCC). The following tentative definitions are suggested.[205] The *point of zero charge* can be defined as the condition for a zero net charge on the particle. This term should therefore be retained for values which are obtained for direct measurements of surface charge density, for example, those determined by potentiometric measurements.[206,207] The *isoelectric point* can be defined as the negative logarithm (base 10) of the concentration of *potential-determining ions* when the electrokinetic potential becomes equal to zero, *e.g.* for electrophoresis the position where the electrophoretic mobility becomes zero. Many of the values quoted in the literature from electrophoretic measurements as being points of zero charge are strictly isoelectric points and may well be different from the true point of zero charge if specific adsorption of ions has occurred. *The reversal-of-charge concentration* can be defined as the concentration of added material at the point where the electrokinetic potential is reduced to zero at a constant concentration of potential-determining ions. In the case of silver iodide dispersions, for example, at constant pI the mobility of the particles may be reduced to zero by the addition of either strongly adsorbed organic ions or hydrolysed metal ions to the dispersion.

Other areas of current activity include the relationship between zeta-potential and surface ionization, potential-determining-ion adsorption, the

[204] D. H. Everett, *J. Pure Appl. Chem.*, 1972, **31**, 579.
[205] B. H. Bijsterbosch, J. Lyklema, R. H. Ottewill, R. Parsons, and A. Watillon, 5th Report, International Critical Tables, 1970.
[206] L. Blok and P. L. de Bruyn, *J. Colloid Interface Sci.*, 1970, **32**, 518, 527, 533.
[207] B. Vincent and J. Lyklema, *Special Discuss. Faraday Soc.*, 1970, No. 1, 148.

influence of electrolyte concentration and specific-ion adsorption,[208] and the influence of particle shape.[209-212] The calculation of zeta-potentials from electrophoretic measurements still needs careful consideration in order to allow correctly for the retardation and relaxation of electrophoretic motion.[213, 214]

Many authors are directing their attention to the correlation between zeta-potentials and other physical properties or phenomena, for example, adsorption of surface-active agents[99, 215, 216] and chelates,[217] contact angles,[218] flotation,[218] flocculation and coagulation,[219, 220] sedimentation volume,[221] rheology,[222] water purification, and electrochemical separations.[223, 224]

Electrophoresis can be defined as the motion due to the translational response of charged objects to a uniform field. However, spatially non-uniform a.c. fields exert a unidirectional force upon neutral polarizable bodies in a liquid dielectric medium and the motion resulting from this force is called *dielectrophoresis*. Thus the latter term may be defined as motion of a particle due to the translational response of a neutral object in a non-uniform field.

The subjects of dielectrophoretic and electrophoretic deposition have been discussed in a collection of papers edited by Pohl and Pickard.[225]

The dielectrophoretic force on a spherical particle of volume V is given by[226]

$$F_e = \frac{3}{2} V \varepsilon_1 \left(\frac{\varepsilon_2 - \varepsilon_1}{\varepsilon_2 + 2\varepsilon_1} \right) \nabla(E_e^2), \qquad (48)$$

where ε_1 and ε_2 are the absolute electric permittivities given by $\varepsilon_1 = \varepsilon_0 \varepsilon_{r_1}$ and $\varepsilon_2 = \varepsilon_0 \varepsilon_{r_2}$ of the particle and medium respectively with $\varepsilon_0 =$ permittivity of free space and ε_{r_1} and ε_{r_2} the relative permittivities. Thus the dielectrophoretic force on a particle is dependent on the gradient of the square of the field strength, the volume of the particle, and the nature of the internal polarization.

[208] R. J. Hunter and H. L. J. Wright, *J. Colloid Interface Sci.*, 1971, **37**, 564.
[209] F. A. Morrison, *J. Colloid Interface Sci.*, 1971, **36**, 139.
[210] L. B. Harris, *J. Colloid Interface Sci.*, 1970, **34**, 323.
[211] I. Gallily, *J. Colloid Interface Sci.*, 1971, **36**, 325.
[212] R. A. Mills, *Biopolymers*, 1970, **9**, 1511.
[213] S. S. Jana, D. N. Biswas, and M. Sengupta, *J. Indian Chem. Soc.*, 1970, **47**, 527.
[214] M. Sengupta and A. K. Bose, *J. Electroanalyt. Chem. Interfacial Electrochem.*, 1968, **18**, 21.
[215] G. D. Parfitt, *Tenside*, 1971, **8**, 136.
[216] A. N. Zhokov, N. A. Kibirova, M. P. Sidorova, and D. A. Fridriksberg, *Doklady Akad. Nauk S.S.S.R.*, 1970, **194**, 130 (*Doklady Phys. Chem.*, 1970, **194**, 666).
[217] E. Matijević, N. Kolak, and D. L. Catone, *J. Phys. Chem.*, 1969, **73**, 3556.
[218] F. Z. Saleeb and H. S. Hanna, *J. Chem. U.A.R.*, 1969, **12**, 229.
[219] M. Beccari and R. Passino, *Ingegnere*, 1970, **44**, 1003.
[220] A. Yabe and M. Sugiura, *Kogyo Kugaku Zasshi*, 1969, **72**, 2356 (*Chem. Abs.*, 1970, **73**, 4302).
[221] F. Tokiwa and T. Imamura, *J. Amer. Oil Chemists' Soc.*, 1969, **46**, 571.
[222] S. K. Nicol and R. J. Hunter, *Austral. J. Chem.*, 1970, **23**, 2177.
[223] H. Yukawa, *Kagaku Kogyo*, 1971, **22**, 417 (*Chem. Abs.*, 1970, **73**, 93 724).
[224] V. G. Verdejo, *Ion*, 1969, **29**, 443.
[225] H. A. Pohl and W. F. Pickard, 'Dielectrophoretic and Electrophoretic Deposition', Electrochemical Society Inc., New York, 1969.
[226] H. A. Pohl, in ref. 225.

An interesting investigation of this equation is given by the studies of Kirko *et al.*,[227] who examined the dielectrophoretic motion of gas bubbles under conditions of weightlessness, *i.e.* in a satellite. Their observations made in a V-shaped condenser confirmed the validity of equation (48).

Chen and Pohl[228] studied the phenomenon of dielectrophoresis using aqueous suspensions of silver bromide particles and determined the rate of dielectrophoretic precipitation of the particles on to parallel wire electrodes as a function of the frequency of the applied field. Some measurements were also made in dioxan and water–dioxan mixtures. The effectiveness of precipitation showed a maximum at frequencies between 10^5 and 10^6 Hz and this in turn showed a good correlation with the frequency dependence of dielectric dispersion. The authors suggest that a major contribution to the observed polarization was the migration of point defects in the subsurface of the silver bromide particles. This view was supported by the fact that the maximum observed was dependent on pAg but not on the dielectric constant of the dispersion medium. The influence of silver vacancies in the lattice of silver bromide was also examined in detail by Honig[229, 230] and by Honig and Hengst,[231] who deliberately doped the lattice. It was shown that the point of zero charge of silver bromide shifted to higher pAg values when lead was incorporated in the lattice, *i.e.* by the creation of silver vacancies, but shifted to lower pAg values when sulphur was incorporated, *i.e.* interstitial silver ions were formed. The effect of the diffuse layer in the solid phase on the point of zero charge of silver iodide was also discussed by Levine *et al.*[232]

Conventionally, most studies of electrophoresis have been carried out using a d.c. electric field. The studies of Vorob'eva, Vlodavets, and Dukhin,[233] however, suggest that there might be some advantages in using low-frequency a.c. fields. For a d.c. field the net velocity of a particle, w, is given by

$$w = u + v, \tag{49}$$

where u = the velocity of electro-osmotic flow of the liquid and v the electrophoretic velocity of the particle. For a flat cell of rectangular cross-section,

$$u = \tfrac{1}{2} u_w (3z^2/a^2 - 1) \tag{50}$$

where u_w = the electro-osmotic velocity at the cell wall, $2a$ = the depth of the cell, and z = the distance from the mid-point axis measured in a positive or negative direction, Thus for a determination of the true electrophoretic velocity, measurements need to be carried out at a distance $z = a/3^{\frac{1}{2}}$ from the axis, *i.e.* the so-called stationary level, where $u = 0$.

[227] I. M. Kirko, T. V. Kuznetsova, V. D. Mikhailov, V. N. Novikov, G. G. Podobedov, V. P. Rassokha, L. A. Rachev, and P. F. Shul'gin, *Doklady Akad. Nauk S.S.S.R.*, 1971, **198**, 1055 (*Doklady Tech. Phys.*, 1971, **16**, 471).
[228] C. C. Chen and H. A. Pohl, *J. Colloid Interface Sci.*, 1971, **37**, 354.
[229] E. P. Honig, *Trans. Faraday Soc.*, 1969, **65**, 2248.
[230] E. P. Honig, *Nature*, 1970, **225**, 537.
[231] E. P. Honig and J. J. Th. Hengst, *J. Colloid Interface Sci.*, 1969, **31**, 545.
[232] P. L. Levine, S. Levine, and A. L. Smith, *J. Colloid Interface Sci.*, 1970, **34**, 549.
[233] T. A. Vorob'eva, I. N. Vlodavets, and S. S. Dukhin, *Kolloid. Zhur.*, 1970, **32**, 189 [*Colloid J. (U.S.S.R.)*, 1970, **32**, 152].

The solution of the Navier–Stokes equations for alternating field conditions leads to an expression for the electro-osmotic velocity and its dependence on the frequency, v, of the form

$$u = u_0 \exp(-i.2\pi vt) \frac{\sin \beta a(1+i) - \beta a(1+i) \cos \beta z((1+i)}{\sin \beta a(1+i) - \beta a(1+i) \cos \beta a(1+i)}, \qquad (51)$$

where $\beta = (\pi v \rho / \eta)^{1/2}$, with $\rho =$ density of the liquid and η its viscosity.

At the wall of the cell, where $z = \pm a$, this expression reduces to

$$u = u_0 \exp(-i.2\pi vt) \qquad (52)$$

and, for $v = 0$,

$$u = u_0 = u_w.$$

At low applied field frequencies (*ca.* 1 Hz) the calculated depth distribution of the electro-osmotic velocity amplitude in the cell does not differ significantly from the parabolic distribution given by equation (50), but at higher frequencies the amplitude distribution is markedly different. A flattening of the parabola occurs and at frequencies as low as 10 Hz the amplitude remains essentially invariant at distances up to $\pm a/2$ from the axis. This raises the possibility of determining electrophoretic velocities under alternating field conditions in a situation where the electro-osmotic transfer of liquid is constant across part of the cell. In these circumstances the accurate location of the stationary layer would no longer be required. The correctness of the theoretical predictions was confirmed by experimental studies on aqueous suspensions of spherical melamine–formaldehyde particles in a field with a frequency of 30 Hz.

Dukhin[234] has also considered in a theoretical paper the consequences of a stagnant liquid film on a surface in a region adjacent to the electrical double layer. After allowance for the effect of ion flow through the stagnant layer and for the polarization of the diffuse double layer, expressions were obtained for the electrophoretic mobility of particles. The calculation of the electrophoretic mobility of polystyrene particles using these expressions gave better agreement with experiment than those carried out without making allowance for the stagnant layer. In an extensive paper these ideas are elaborated by Dukhin and Semenikhin[235] and it is pointed out that by measuring the electrophoretic and diffusiophoretic mobilities of the same system, or by measuring the electrophoretic mobility of three fractions of spherical particles with electrically identical surfaces, information can be obtained about both the zeta-potential and the Stern potential. The theory of the low-frequency dispersion of permittivity of suspensions of spherical particles produced by polarization of the double layer has also been discussed and the validity of the equations confirmed from experimental data obtained on polystyrene dispersions in aqueous solu-

[234] S. S. Dukhin, *Special Discuss. Faraday Soc.*, 1970, No. 1, 158.
[235] S. S. Dukhin and N. M. Semenikhin, *Kolloid Zhur.*, 1970, **32**, 360 [*Colloid J. (U.S.S.R.)*, 1970, **32**, 298].

tions of potassium chloride.[236] A theory of polarization of the double layer and its effect on electro-optical and electrokinetic phenomena and on the dielectric constant of disperse systems was also put forward by Stoilov and Dukhin.[237]

The electrokinetic effects produced by the low-speed flow of particles as a consequence of convective processes have been examined theoretically by Hiler *et al.*,[238] and Hildreth[239] has considered electrokinetic flow in small capillaries in terms of the distributed conductance across the channel and the effects produced by differences in the ionic mobilities of the co-ions and counterions. It was suggested that large flow retardations should be attainable in ultrafine capillaries, thus leading to high efficiencies for electro-osmotic pumping and streaming potential generation.

Shakhov and Dushkin[240] have reported effects of magnetic fields on the zeta-potentials of sols of iron and aluminium hydroxide. A decrease of potential was observed on application of the field, which the authors attributed to a dehydration of the stabilizing potential-determining ions.

A method for the determination of the electrokinetic potential of a free surface has been suggested by Levashova and Krotov[241] by considering the laminar flow of a viscous liquid on the surface of a rotating plane disc. It was found that the ratio E/ω_v^2, where $E =$ streaming potential and $\omega_v =$ angular velocity, was constant for a given solid and electrolyte concentration and the zeta-potential could be calculated by analogy with the Helmholtz–Smoluchowski formula for capillary systems.

The electrical properties of the oxide/water interface, a subject which has always been important because of the technological significance of these materials and their natural abundance, now appears to be receiving considerable attention, and an important conference on the surface chemistry of oxides[242] was held in the period under review. This included papers on the acidic and basic properties of hydroxylated metal oxide surfaces,[243] the theory of the differential capacity of the oxide/aqueous electrolyte interface,[244] discreteness of charge and solvation effects in cation adsorption at the oxide/water interface,[245] the interfacial electrochemistry of haematite,[246] competitive

[236] V. N. Shilov and S. S. Dukhin, *Kolloid. Zhur.*, 1970, **32**, 293 [*Colloid J.* (*U.S.S.R.*) 1970, **32**, 243].
[237] S. Stoylov and S. S. Dukhin, *Kolloid. Zhur.*, 1970, **32**, 757 [*Colloid J.* (*U.S.S.R.*), 1970, **32**, 631].
[238] E. A. Hiler, R. D. Brazee, R. B. Curry, and M. Y. Hamdy, *J. Colloid Interface Sci.*, 1971, **35**, 544.
[239] D. Hildreth, *J. Phys. Chem.*, 1970, **74**, 2006.
[240] A. I. Shakhov and E. E. Dushkin, *Voprosy Tekhnol. Obvab. Vody Prom. Pit'erogo Vodosnabzh.*, 1969, **1**, 53 (*Chem. Abs.*, 1971, **75**, 80 611).
[241] L. G. Levashova and V. V. Krotov, *Vestnik Leningrad. Univ.* (*Fiz. Khim.*,) 1969, **3**, 139.
[242] Papers in *Discuss. Faraday Soc.*, 1971, No. 52.
[243] H. P. Boehm, ref. 242, p. 264.
[244] S. Levine and A. K. Smith, ref. 242, p. 290.
[245] G. R. Wiese, R. O. James, and T. W. Healy, ref. 242, p. 302.
[246] A. Breeuwsma and J. Lyklema, ref. 242, p. 324.

adsorption on oxide surfaces,[247] and studies on the thermodynamics and adsorption behaviour in the quartz–aqueous surfactant system.[248]

Detailed studies of the zinc oxide/aqueous solution interface have been reported by Blok and de Bruyn.[206] The point of zero charge as estimated by a potentiometric procedure was found to depend on the method of preparation of the zinc oxide and the nature of the supporting electrolyte and ranged from pH 8.6 to pH 10.0. The suggestion was made that the surface was composed of a random mixture of $Zn(OH)_2$ and $Zn(OH)_{1.6}X_{0.4}$, where X was an anionic impurity incorporated during oxide formation. The surface properties were thus dependent on the pH and pX of the solution, giving a possible explanation for the variability of the point of zero charge. The specificity of anion adsorption followed the order $ClO_4^- < NO_3^- < I^- < Br^- < Cl^-$. Of fundamental importance to the interpretation of studies of this type is the use of the Nernst equation. Levine and Smith[249] have investigated the conditions under which an oxide surface in an aqueous dispersion will obey the latter and derived a modified form of the equation. This was then combined with a model of the inner part of the electrical double layer, which involved adsorption of both anions and cations to give a correction for discreteness of charge in the adsorption isotherms. The adsorption of polar molecules on to a solid surface to form a monolayer of oriented dipoles at the point of zero charge had been considered in a previous paper[250] in which the effect on the χ-potential was also discussed.

Wiese *et al.*[245] have examined the electrokinetic behaviour of vitreous quartz as a function of pH in solutions of potassium nitrate and barium nitrate and find an isoelectric point at pH $= 2.5 \pm 0.2$. With lanthanum nitrate solutions the isoelectric point shifted to a higher pH value (*ca.* pH 3.8 in $10^{-3}M$ salt solution). With both lanthanum nitrate and barium nitrate solutions the curves exhibited a minimum in zeta-potential as the pH was increased; this was interpreted as a discreteness of charge effect. Mackenzie and O'Brien[251] examined the electrokinetic behaviour of quartz particles as a function of pH in the presence of nickel and cobalt salt solutions. Both materials showed similar behaviour in that the original negative charge on the particles was reversed and the zeta-potential became positive in the pH range 6.3—11.8. The reversal of charge was attributed to ion hydrolysis with the formation of $NiOH^+$ and $CoOH^+$, which then adsorbed in super-equivalent amounts to the quartz surface.

It is of interest in connection with these studies that Seimiya[252] measured the adsorption of cobaltous ions on to glass and found that the adsorption was small at acidic pH values but reached a maximum in the neighbourhood of pH 9. The results suggested that hydrolysis of the cobaltous ion occurred in the pH region 7—9. Detailed examination of the electrokinetic

[247] F. J. Hingston, A. M. Posner, and J. Quirk, ref. 242, p. 334.
[248] B. Ball and D. W. Fuerstenau, ref. 242, p. 361.
[249] S. Levine and A. L. Smith, ref. 242, p. 290.
[250] S. Levine, A. L. Smith, and E. Matijević, *J. Colloid Interface Sci.*, 1969, **31**, 409.
[251] J. M. W. Mackenzie and R. T. O'Brien, *Trans. A.I.M.E.*, 1969, **244**, 168.
[252] T. Seimiya, *Bull. Chem. Soc. Japan*, 1969, **42**, 2797.

properties of silica dispersions are also reported by Matijević and his co-workers,[108] including an extensive investigation of the effects of the hydrolysis of aluminium salts.[115,116] The electrokinetic effects produced on other substrates by the hydrolysis of scandium[253] and hafnium,[128] have also been reported by the same group of workers.

Bye and Simpkin[254] report that the point of zero charge of bayerite crystals, grown in water for a period of 28 days, occurs at pH 9.2 and that alumina fired at 1200 °C and then aged in water for 28 days has a point of zero charge of 9.4. The effects of treatments at different temperatures (> 300 °C) and different periods of immersion in water on the electrophoretic mobility of γ-alumina are also reported by Maroto and Griot.[255]

The importance of ignition temperature in influencing the point of zero charge of some oxides was demonstrated by further work of Bye and Simpkin[254] using chromia. Chromium trihydroxide, after ageing in water for 28 days, had a point of zero charge at pH 7.2; on firing at 850 °C, storing under argon, and ageing in water for 28 days, the value was 6.05; on firing at 850 °C and ageing for 28 days, it was 4.0; and on firing at 1200 °C and ageing in aqueous suspension it was 3.0. It appears, therefore, that the final value obtained for the point of zero charge depends on (i) the temperature of ignition, (ii) the oxidation which occurs during cooling, and (iii) additional oxidation in the aqueous dispersion. The preparation of chromium hydroxide sols with a narrow size distribution was described by Matijević *et al.*[104] The sol particles had an isoelectric point at pH = 7.65 ± 0.15.

Numerous other studies have been made on oxide surfaces. A determination of the isoelectric point of uranium dioxide was made by Maroto[256] using both microelectrophoresis and subsidence rate; he found a value of 4.5. Electrophoretic studies on thoria sols, in nitric acid and ammonium nitrate mixtures at 20 °C, were made by Smith and Krohn,[257] using the moving-boundary method. At low ionic strengths, two or more charged components were observed but at high ionic strength, only one. These results were interpreted by assigning one boundary to the thoria particles and the other to charged species present in the solution. Tamura *et al.*[258] examined the effects of aluminium sulphate solutions and silicic acid hydrosols on titanium dioxide suspensions. The magnitude of the zeta-potential increased when both reagents were added together, and electron-microscope observations indicated that a coating had been formed on the original titania particles.

The surface conductance of glass was studied by Watillon and de Backer,[259] who compared the electrokinetic potentials obtained from surface conductance

[253] E. Matijević, A. B. Levit, and G. E. Janauer, *J. Colloid Interface Sci.*, 1968, **28**, 10.
[254] G. C. Bye and G. T. Simpkin, *Chem. and Ind.*, 1970, 532.
[255] A. J. G. Maroto and O. Griot, *Acta Cient.*, 1970, **3**, 15.
[256] A. J. G. Maroto, *Anales Asoc. quím. argentina*, 1970, **58**, 187.
[257] F. J. Smith and N. A. Krohn, *J. Colloid. Interface Sci.*, 1971, **37**, 179.
[258] H. Tamura, Y. Matsuda, and T. Inoue, *Kogyo Kagaku Zasshi*, 1969, **72**, 2341 (*Chem. Abs.*, 1971, **75**, 80 682).
[259] A. Watillon and R. de Backer, *J. Electroanalyt. Chem. Interfacial Electrochem.*, 1970, **25**, 181.

with those obtained from streaming-potential measurements. The results they obtained were interpreted in terms of a superficial gel layer on the glass, the thickness of which was estimated from the experimental data. The thickness of the layer did not appear to depend on the valency of the counter-ion. Streaming-current and surface-conductance studies on quartz have also been reported by Zhukov *et al.*[260]

The exchange of calcium by sodium on montmorillonite has been examined, using electrophoresis, by Bar-On, Shainberg, and Michaeli.[261, 262] Salt-free gels of sodium and calcium montmorillonite were prepared, freeze-dried, and then made up to a concentration of 0.1 % w/v. The mobility of the sodium form was found to be -2.70×10^{-2} μm m V^{-1} s^{-1} ($\zeta = ca. -38$ mV) and that of the calcium form -0.93×10^{-2} μm m V^{-1} s^{-1} ($\zeta = ca. -13$ mV). Bionic mixtures were prepared from the two homoionic clays and the mobility of the particles was determined. The mobility of the calcium montmorillonite reached that of the sodium form at an exchangeable sodium percentage of 35%. The authors suggested that the calcium clay existed in the form of tactoids and that the introduction of *ca.* 20% of sodium ions caused these to commence to break down. The breakdown process was apparently completed when 50—60% of the adsorbed calcium ions were replaced by sodium ions. Touret[263] examined the electrokinetic potential of montmorillonite, initially in the hydrogen form, as a function of pH and ionic strength. A value of -35 mV was found for the hydrogen-clay in distilled water and this value did not change with pH. At constant pH an increase in sodium chloride concentration caused a decrease in the magnitude of the zeta-potential but not a change in sign. In the presence of calcium chloride solutions, above pH 4, a similar behaviour was observed but it was inferred that the adsorption of calcium ions was stronger than that of sodium ions.

The relationship which can occur between the rheological properties of a dispersion and the electrokinetic potential of the particles was emphasized in the work of Nicol and Hunter.[222] The electrophoretic mobility of kaolinite (from Mount Egerton, Victoria) was measured as a function of pH. The mobility increased steadily from *ca.* -3.5×10^{-2} μm m V^{-1} s^{-1} at pH 7.0 to *ca.* -5.8×10^{-2} μm m V^{-1} s^{-1} at pH 9.3 and decreased to -5.2×10^{-2} μm m V^{-1} s^{-1} at pH 11. The yield stress of the dispersions showed a maximum value just below pH 7.0 and decreased steadily as the pH and zeta-potential increased. Results were also reported for the influence of a series of cationic surface-active agents on the mobility of kaolinite. From the change in concentration of cationic surface-active agent which reversed the sign of the electrophoretic mobility as a function of alkyl chain length, it was found that the van der Waals energy of lateral interaction per CH_2 group was 1.2 kT.

[260] A. N. Zhukov, N. A. Kibirova, M. P. Sidorova, and D. A. Fridrikhsberg, *Doklady Akad. Nauk S.S.S.R.*, 1970, **194**, 130.

[261] P. Bar-On, I. Shainberg, and I. Michaeli, *J. Colloid Interface Sci.*, 1970, **33**, 471.

[262] I. Shainberg, 'Transactions of the International Congress on Soil Science', 1968, Vol. 1, p. 577.

[263] C. Touret, *Compt. rend.*, 1969, **269**, C, 1155, 1591.

The influence of surface-active agents on the mobility of kaolinite has also been examined by Gasanova *et al.*[264] Further data on the mobility of kaolinite as a function of pH in the presence of bivalent and univalent electrolytes has been reported by Vestier[265] and in the presence of aluminium sulphate by Sono *et al.*[266] Surface conductance and electrokinetic measurements on kaolinite beds were reported by Lorenz.[267]

Barium sulphate is a crystalline substance which has been used, and continues to be used, as a material suitable for electrokinetic measurements. The electrokinetic properties of barium sulphate suspensions have been investigated by Grebenshchikova *et al.*[268] The magnitude of the positive electrokinetic charge was found to be increased by increasing the barium concentration of solution in contact with the solid; similarly, increasing the sulphate concentration increased the negative charge. However, in the presence of excess sulphate, charge reversal occurred as the pH was decreased. In the presence of a 3×10^{-5} mol dm^{-3} solution of a europium salt, the charge was also reversed at pH 6.2; this was attributed to the formation and adsorption of hydrolysed species of europium. Spitsyn *et al.*[269,270] have also carried out electrokinetic measurements on barium sulphate particles. The zeta-potential obtained varied according to the method of preparation. Negative potentials were obtained when the preparation was carried out in a neutral medium and positive potentials when it was carried out in sulphuric acid. Introduction of radioactive sulphur into the crystals did not change the signs of the potential but did alter the magnitude according to the amount of radioactive material incorporated.

A neglected interface in the realm of electrokinetic measurements is that of the air/liquid interface and there are very few reported measurements of the zeta-potentials of bubbles in water. More measurements would indeed be welcome. Some interesting experiments are reported by Iribane *et al.*,[271] who have carried out some experiments to measure the electrification of droplets produced in a spray and interpreted their results in terms of a charge-separation process involving a shearing of the electrical double layer at the air/water interface.

The general topic of electrophoresis has been reviewed by Strickland[272] and in the book by Shaw.[273]

[264] S. B. Gasanova, L. B. Abduragimova, and A. K. Miskarli, 'Mater Konf. Molodykh Uch. Inst. neorg. fiz. Khim. Akad. Nauk, Aszerb. S.S.R.', 1968, p. 110.
[265] D. Vestier, *Sci. Terre*, 1969, **14**, 289.
[266] K. Sono, Y. Mitsukami, H. Ishikawa, *Kogyo Yosui*, 1970, **143**, 36.
[267] P. B. Lorenz, *Clays and Clay Minerals Bull.*, 1969, **17**, 223.
[268] V. I. Grebenshchikova, Y. P. Davydov, and A. Pershin, *Radiokhimiya*, 1971, **13**, 375.
[269] V. I. Spitsyn, E. A. Torchenkova, and I. N. Glaskova, *Monatsh.*, 1970, **101**, 1164.
[270] V. I. Spitsyn, E. A. Torchenkova, and I. N. Glaskova, *Doklady Akad. Nauk S.S.S.R.*, 1969, **187**, 1335 (*Doklady Phys. Chem.*, 1969, **187**, 568).
[271] J. V. Iribane, M. Klemes, and C. L. Yip, *J. Electroanalyt. Chem. Interfacial Electrochem.*, 1970, **24**, 11.
[272] R. D. Strickland, *Analyt. Chem.*, 1970, **42**, 32.
[273] D. J. Shaw, 'Electrophoresis', Academic Press, London, 1969.

6
Emulsions

BY B. VINCENT

1 Introduction

As with most review articles it has been necessary to be selective in reviewing the progress made in the field of emulsions during 1970 and 1971. Thus, attention has been focused on those papers which, in the opinion of the Reporter, have contributed something to the understanding of the mechanisms of emulsion formation and emulsion stability. In particular, progress seems to have been made in correlating emulsion formation and stability with equilibrium phase diagrams and the changes brought about by variation in temperature, or by the addition of a fourth component (*e.g.* a fatty alcohol), to the usual three-component oil–water–surfactant systems. The structure of microemulsions, for example, is much better understood on this basis. Advances have also been made in the understanding of how macromolecules function as emulsifiers and stabilizers, and in the theoretical description of heterocoalescence. On the other hand, relatively little advance seems to have been made in the devizing of experimental techniques, either for studying their properties or for forming emulsions. A number of specialized methods, geared to specific industrial requirements, continue to appear, of course (see, for example refs. 1—4), but these are all essentially modifications of existing methods. Several novel techniques for obtaining particle size distributions have been suggested. For example, Kubo *et al.*[5] have described a direct photographic method whereby a microcamera is inserted into an emulsion, and it has also been claimed[6] that electron microscopy, hitherto unsuitable for liquid systems, has been successfully employed for studying microemulsions, of droplet size 5—20 nm, using a freeze-drying technique. The production of multiple emulsions, *i.e.* dispersions of droplets within droplets, is of growing industrial importance. Kessler and York[7] have described an injection technique for producing double emulsions and a high-

[1] L. E. M. de Chazel and J. T. Ryan, *Amer. Inst. Chem. Engineers J.*, 1971, **17**, 1226.
[2] N. Mimura and H. Chikamasa, G.P. 2012686/1970 (*Chem. Abs.*, 1971, **74**, 57654).
[3] R. E. Guerin, G.P. 2017777/1971 (*Chem. Abs.*, 1971, **74**, 91629).
[4] K. J. Lissant, U.S.P. 3565817/1971 (*Chem. Abs.*, 1971, **74**, 91628).
[5] T. Kubo, T. Tsukiyama, A. Takamura, and I. Takashima, *Yakugaku Zasshi*, 1971, **91**, 518 (*Chem. Abs.*, 1971, **75**, 53509).
[6] L. V. Kolpakov, *Kolloid. Zhur.*, 1970, **32**, 229 [*Colloid J. (U.S.S.R.)*, 1970, **32**, 187].
[7] D. P. Kessler and J. L. York, *Amer. Inst. Chem. Engineers J.*, 1970, **16**, 369.

speed photographic technique for assessing emulsion type and droplet size distribution. Equations for statistical parameters describing droplet size distributions in emulsions have been derived[8,9] and tested.

In the following sections, emulsion formation with surfactant and polymers is considered first. After that, sections on the various aspects of stability appear. Throughout it is assumed that the reader has at least some familiarity with the field, and so not all terms and concepts are rigorously explained. Further information may, of course, be obtained from the many excellent texts dealing with emulsions; in particular, one would recommend those by Becher (1965),[10] and Sherman (1968).[11]

2 Emulsification with Surfactants; Microemulsions

The emulsion 'type', *i.e.* which liquid will disperse in which, depends largely on the nature of the emulsifier. This controls not only the properties of the interfacial region, but also the mutual solubilization of each phase in the bulk of the other. A much used concept, in this respect, has been the hydrophile–lipophile balance (HLB) value.[12] This is a measure of the relative contribution of hydrophilic and lipophilic groups to the properties of the emulsifier molecule; w/o emulsifiers have low HLB values, whereas o/w emulsifiers have high HLB values. Various empirical equations have been proposed to quantify this concept.[12] Kruglyakov and Koretskii[13] have related the HLB value for a surfactant to its distribution coefficient between water and an oil phase (heptane or benzene). Much work in this direction has also been carried out by Shinoda's school. For example, the phase diagram[14] for the system water–cyclohexane–poly(oxyethylene)$_x$ nonylphenyl ether ($x = 8.6$), in the weight ratio, 47.5 : 47.5 : 5, respectively, is shown in Figure 1. At room temperature the surfactant is largely present in micellar form in the aqueous phase. On raising the temperature, solubilization of oil into these aqueous micelles increases, particularly as the cloud point (55 °C) of the surfactant in the aqueous phase is reached. The volume of the oil phase diminishes concurrently. Between 55 and 58 °C there are essentially three phases present: (*a*) the remnants of the oil phase; (*b*) a 'surfactant' phase occupying about 80 % of the volume and which contains most of the surfactant plus solubilized oil and water; (*c*) an aqueous phase which contains less than 0.1 % surfactant. Above 58 °C the original oil phase disappears so that only phases (*b*) and (*c*) remain. Phase (*b*) now becomes a predominantly oil-based phase as the temperature

[8] L. Djakovic, P. Dokic, and P. Radivojevic, *Kolloid-Z.*, 1971, **24**, 324.
[9] N. Ya. Avdeev, *Kolloid. Zhur.*, 1970, **32**, 635 [*Colloid J. (U.S.S.R.)*, 1970, **32**, 513].
[10] P. Becher, 'Emulsions, Theory and Practice', Amer. Chem. Soc. Monograph, Reinhold, New York, 1965.
[11] P. Sherman, 'Emulsion Science', Academic Press, New York, 1968.
[12] See *e.g.* ref. 11, p. 140.
[13] P. M. Kruglyakov and A. F. Koretskii, *Doklady Akad. Nauk S.S.S.R.*, 1971, **197**, 1106.
H. Saito and K. Shinoda, *J. Colloid Interface Sci.*, 1970, **32**, 647.

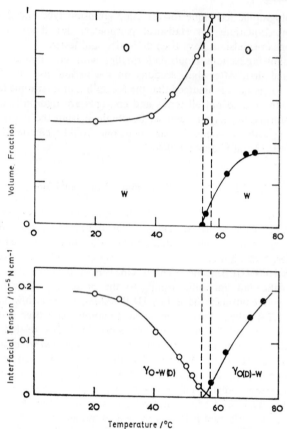

Figure 1 *The effect of temperature on the volume fractions of water, oil, and surfactant phases and the interfacial tension between these phases. The system is composed of 47.5 wt. % water, 47.5 wt. % cyclohexane, and 5 wt. % poly(oxyethylene)$_{8.6}$ nonylphenyl ether*
(Reproduced by permission from *J. Colloid Interface Sci.*, 1970, **32**, 647)

is raised (*i.e.* analogous to the aqueous phase below 55 °C). Water is rejected from the oil-based micelles, and so phase (*c*) grows in volume. In summary, the surfactant has been transferred from an aqueous to an oil environment on raising the temperature, owing largely to de-solvation of the hydrophilic poly(ethylene oxide) chains. 55 °C represents the phase separation temperature (cloud point) for the surfactant (at the stated concentration) in water, and 58 °C the same for the surfactant in the oil phase. Between 55 and 58 °C the three-phase region exists. Interfacial tension measurements show a minimum value, close to zero, near this three-phase region. Emulsion formation is thus favoured here, oil-in-water (o/w) emulsions being formed below

55 °C, and water-in-oil (w/o) emulsions above 58 °C. It is not clear what should happen in-between, where the interfaces are pseudo-continuous and poorly defined. Perhaps one should, therefore, talk about a phase-inversion region, rather than a distinct phase-inversion temperature (PIT). Shinoda's work[14] does, however, show clearly the relationship between cloud point and PIT and how these depend on x and, therefore, the HLB of the emulsifier. Nevertheless, it must be stressed that, although these factors correlate in terms of emulsion *formation* (that is, the initial droplet size is a minimum close to the PIT and increases steadily with temperature above the PIT for the w/o emulsion), the optimum *stability* of these emulsions is found to occur some 10 to 30 °C *above* the PIT. The reasons for this are discussed in Section 4.

Shinoda[14] makes use of a direct visual technique for determining PITs. Matsumoto and Sherman[15] have described a relatively simple differential thermal analysis (DTA) technique for obtaining PITs, which makes use of the fact that phase inversion is an endothermic process. They use the same oil for the thermal standard as that used in the emulsions, and claim that the DTA method is more reliable and less tedious than direct visual observation.

The primary condition for emulsion formation is, of course, that the interfacial tension should be minimal. In the system discussed above this can be achieved by varying the temperature. In the so-called 'microemulsions' discovered by Schulman[16] a state of zero interfacial tension is achieved when a long-chain alcohol is added to a hydrocarbon–water–surfactant system. This arises because of diffusion of alcohol molecules across the interface, and results in spontaneous emulsification. Gillberg, Lechtinen, and Friberg[17] have constructed triangular, three-component phase diagrams for alcohol–water–surfactant systems, and have determined the changes that occur on adding hydrocarbons to the system. N.m.r. chemical shift studies have also been made for the various types of proton in the system. It would seem that the presence of hydrocarbon in the intermicellar solution (hydrocarbon and alcohol) increases the solubilization of water in the micelles. This is the origin of the w/o microemulsions. The packing density of the (soap) hydrocarbon chains of the micelles is decreased and there is an increase in their alcohol content. In fact it would seem that the formation of these microemulsions (really micelles with solubilized water) is very sensitive to the distribution of the alcohol between the micelles and intermicellar solution. At high levels of added hydrocarbon the water solubilization is reduced owing to transfer of alcohol out of the micelles.

A clearer structural understanding of microemulsions has also resulted from the work of Shah and Hamlin.[18] They have carried out high-resolution (220 MHz) proton n.m.r. studies, coupled with electrical conductivity and birefringence studies. With increasing water content the system changes from an

[15] S. Matsumoto and P. Sherman, *J. Colloid Interface Sci.*, 1970, **33**, 294.
[16] J. H. Schulman, W. Stoeckenius, and C. Prince, *J. Phys. Chem.*, 1959, **63**, 1672.
[17] G. Gillberg, H. Lechtinen, and S. Friberg, *J. Colloid Interface Sci.*, 1970, **33**, 40.
[18] D. O. Shah and R. Hamlin, *Science*, 1971, **171**, 483.

optically clear w/o microemulsion through a turbid region to an optically clear o/w microemulsion. The turbid region exhibits birefringence, and it would seem that the water molecules exist in two distinct molecular environments on passing through this region. The electrical conductivity also changes. In one region the water molecules are less mobile than the hydrocarbon molecules and in the second the reverse is true. It is proposed that all these results can be accounted for in terms of a change in structure from spherical micelles containing solubilized water (*i.e.* the w/o optically clear microemulsions) to cylindrical structures and then to lamellar structures. These latter two structures represent the turbid region. At even high water contents the lamellar structures break down, and the system inverts to an aqueous solution containing solubilized hydrocarbon, *i.e.* the o/w microemulsion. The turbid region over which phase inversion occurs is thus analogous to the 'surfactant phase' described in Shinoda's work.[14] The difference in the two systems is that in the microemulsion system a fourth component (the alcohol) is the additional variable, whereas in Shinoda's system it is temperature. One other interesting fact that emerges from Shah and Hamlin's work[18] is that it would seem that the solubilized water in the spherical micelles (w/o microemulsion) has the *same* mobility as the water in the continuous phase. Most work with normal micelles, on the other hand, has indicated that solubilized water has a different mobility. It is still somewhat surprising that the alcohol which is also present in the spherical micelles has no effect on the mobility of the solubilized water.

Elbing and Parts[19] also claim to have made a type of microemulsion (droplet diameter: 40—100nm) from mixtures of vinyl stearate and an aqueous solution of the surfactant 'Renex 690'. The mechanism of formation is, however, closer to that of the Shinoda-type emulsions than the Schulman ones, since temperature rather than a fourth component is the additional variable. The mixture is heated above the cloud point for the aqueous solution. This presumably results in a surfactant phase containing solubilized vinyl stearate and water. On cooling, the soap-rich aqueous phase reforms, but excess vinyl stearate is carried across into the micelles (*i.e.* above the solubilization limit), so that kinetically, rather than thermodynamically, stable droplets result. Elbing and Parts[19] also refer to these microemulsions as 'giant micelles', although this seems a somewhat unsatisfactory term. The key to their formation would seem to be the high stirring rate required, and presumably also the rate of cooling, as well as a reasonable affinity between the soap and the oil even at room temperature. An alternative procedure in fact used by Elbing and Parts[19] was to dissolve the vinyl stearate in the pure soap, add boiling water and then cool. Dodecane, which is not very soluble in Renex 690, did not form a microemulsion. In terms of Figure 1, this would mean that, whereas vinyl stearate would be strongly solubilized into the aqueous phase near to the aqueous cloud point, little solubilization of dodecane would occur. Indeed, one might question whether Shinoda's model[14] of

[19] E. Elbing and A. G. Parts, *J. Colloid Interface Sci.*, 1971, **37**, 635.

phase inversion applies unless the soap does have a reasonable affinity for both phases over a wide temperature range. Otherwise the cloud points (for the surfactant in the water and oil, respectively) could be so far apart that the concept of a PIT really does break down. However, a surfactant that had a much greater affinity for one continuous phase than the other at any temperature would in any case tend to be a very poor emulsion stabilizer because of its weak adsorption at the interface. This argument is somewhat circular, therefore.

Nikitinova, Spiridonova, and Taubman[20] have indicated how mutual surfactant solubility is relevant in the emulsion polymerization of vinyl acetate. Quasi-spontaneous water–vinyl acetate emulsions would form at the bulk monomer–water interface with those surfactants which were more soluble in water and which were introduced into the oil phase. This is presumably due to mass transfer of the surfactant from the oil phase into the water through the interface. Polymerization subsequently occurred predominantly in these microemulsion droplets to give a latex essentially of the same particle size as the droplets.

3 Emulsification with Polymers

As stated above, the lower the interfacial tension the greater the ease of emulsification. However, emulsification is a hydrodynamic process and, as van den Tempel[21] has shown, it is not so much the absolute value of the interfacial tension that is important, but rather gradients in interfacial tension which arise during droplet formation from slicks. This is of obvious importance when macromolecules are used as emulsifiers, because of the relatively long times needed for segment rearrangement at an expanding interface. Lankveld[22] has investigated the role of hydrolysed poly(vinyl acetate) (PVA) in the emulsification of liquid paraffin in water. He has measured adsorption isotherms for PVA on the droplets and has also examined the way in which the specific surface area changes with PVA content. Samples with a 2% acetate content (PVA 98) behave as one might expect, that is, an increase in the PVA level over the range 200—1400 p.p.m. leads to a slight but steady increase in the specific surface area. This corresponds, more or less, to the plateau region in the corresponding adsorption isotherm. PVA samples with a 12% acetate content (PVA 88) give rise to a maximum in the specific surface area with increasing polymer concentration, even though the degree of adsorption increases continuously. Lankveld[22] in fact claims that this maximum in specific surface area corresponds to a discontinuity in the isotherm, although such a break is not always obvious and is certainly within

[20] S. A. Nikitinova, V. A. Spiridonova, A. B. Taubmann, *J. Polymer Sci., Part A-1, Polymer Chem.*, 1970, **8**, 3045.
[21] M. van den Tempel, Proceedings of the 3rd International Congress on Surface Activity, 1960, vol. 2, p. 573.
[22] J. M. G. Lankveld, Ph.D. Thesis, University of Wageningen, The Netherlands, 1970; J. M. G. Lankveld and J. Lyklema, *J. Colloid Interface Sci.*, 1972, **41**, 454, 460, 475.

experimental error for most of the isotherms. Nevertheless, it would seem that at concentrations above the maximum in specific surface area the total amount of PVA associated with the interface remains constant (*i.e.* the increase in adsorption per unit area is balanced by the decrease in total area). The concentration of PVA 88 corresponding to the maximum increases with molecular weight.

Lankveld[22] proposes the following mechanism to account for the above observations. During emulsification, elongated slicks are formed which break up into droplets. Dynamic equilibrium is reached when the rate of droplet production just balances the rate of droplet coalescence. The rate of coalescence in turn depends on (*a*) the time between collisions and (*b*) the time new surface has to 'adsorb' surfactant or polymer segments. If one imagines a slick breaking into two droplets, the freshly created interface between the drops will have a deficit of adsorbate and therefore have a higher interfacial tension than the neighbouring 'older' interface. This results in a large interfacial tension gradient which tends to suck liquid between the droplets thus forcing them apart. If polymer segments can adsorb at the new interface rapidly enough, however, the interfacial tension gradient will be less and this will favour droplet recoalescence rather than separation. Polymer segments can adsorb in two senses: (*i*) by the arrival of new polymer at the interface; but more important it would seem in this case is (*ii*) by unfolding of molecules already at the interface. Thus it is suggested that the weaker emulsifying power of PVA 98 at low concentrations, where mechanism (*ii*) dominates, is due to the fact that it is more flexible than PVA 88 and can unfold rapidly enough over the new interface to keep the interfacial tension gradients small.

Lankveld[22] also attempts to interpret the maximum occurring with PVA 88 in terms of interfacial tension gradients. However, this cannot be the *sole* mechanism operating. If it were, then as one increased the PVA 88 concentration one would expect occupation of the interface by mechanism (*i*) above to become more important, and for the rate of occupation of new interface to increase. This, on its own, would lead to a steadily *decreasing* specific surface area with increasing PVA level on the arguments given above. On the other hand, if mechanism (*i*) does become more important for PVA 88, this would also lead to thicker adsorbed layers at the interface and thus greater *steric* stabilization of the newly formed droplets. Specific surface area would then *increase* with increasing PVA 88 concentration. It could well be, therefore, that the observed maximum in the specific surface area is the resultant of both these effects.

4 Kinetic Studies of Emulsion Stability

It is necessary to distinguish several levels of emulsion stability. In particular, aggregation (or agglomeration), creaming, and coalescence. The first two have their counterparts in particulate dispersions (creaming being analogous to sedimentation). Coalescence, on the other hand, has more in common with the properties of thin liquid films. Direct studies of droplet coalescence are discussed in Section 5. In this section attention is directed towards kinetic studies of emulsion breakdown. Such studies may well involve two or even all three of the above mentioned processes.

Shinoda[14] has made a study of the various factors involved in emulsion stability. As discussed in Section 2, although emulsions are most readily formed in the region of the PIT, they are not very stable. This is presumably due to the fact that the anchoring (adsorption) of the surfactant in the 'interface' (poorly defined at the PIT) is weak. For example, the effect of temperature on the stability of a 50% water + 50% cyclohexane emulsion, stabilized by 3% poly(oxyethylene)$_x$ nonylphenyl ether ($x = 6.2$) is shown in Figure 2.[14]

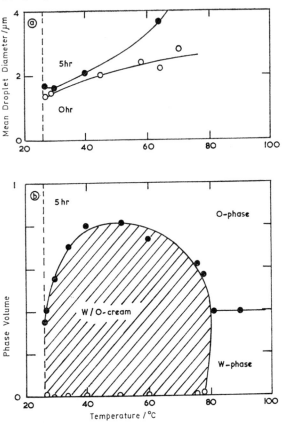

Figure 2 *The effect of emulsification temperature on* (a) *the mean initial droplet diameter of w/o type emulsions, and* (b) *the volume fractions of oil, cream, and water. The system is composed of 48.5 wt. % water, 48.5 wt. % cyclohexane, and 3 wt. % poly(oxyethylene)$_{6.2}$ nonylphenyl ether. PIT of the system was 26 °C*

(Reproduced by permission from *J. Colloid Interface Sci.*, 1970, **32**, 647)

As can be seen, there is a maximum in stability some 30 °C above the PIT. The rapid breakdown at high temperatures is due to increased thermal motion and a decrease in viscosity. Saito and Shinoda[14, 23] have also shown that the PIT increases with increase in x. Ideally, therefore, in choosing a surfactant stabilizer one should choose one with the appropriate HLB such that the PIT is around 0 °C in order to get a stable w/o emulsion at room temperature. Mixed-chain-length emulsifiers are often used as it is claimed that these can give better stability. As Shinoda points out,[23] one must compare mixtures which give the same PIT to make a meaningful comparison. The mixing of one long- and one short-chain-length (hydrophilic group) material generally appears to increase the stability, but the reason for this is by no means clear.

The effect of adding electrolyte to aqueous solutions of non-ionic surfactants is generally (but not always) to lower the cloud point. HCl and H_2SO_4 in fact *raise* the cloud point of aqueous poly(oxyethylene)$_x$ nonylphenyl ethers.[24] For this surfactant ($x=9.7$), Shinoda and Takeda[24] have shown that the effect of 6% (by wt.) of NaCl is to reduce the cloud point and the PIT by 14 °C. This is equivalent to a decrease in HLB of 1.2 units (*i.e.* towards a shorter hydrophilic chain length). Elworthy, Florence, and Rogers[25] have* followed the rate of breakdown of chlorobenzene–water emulsions stabilized by hexadecyl hexa(ethylene oxide) ($C_{16}E_6$) as a function of salt concentration. They find a critical electrolyte concentration, c_c, beyond which the rate of breakdown becomes rapid. c_c values are in the order $Na^+ > Ca^{2+} > Al^{3+}$. Even though the emulsions were charged in the absence of salt, the derived zeta potential (from electrophoresis measurements) fell rapidly with increasing electrolyte concentration well *below* the c_c values. Calculations of the electrical double layer repulsion term showed in fact that these forces could not be controlling the observed coalescence behaviour. Changes in the London attractive forces were also considered. The authors calculate that the adsorption of surfactant leads to an *increase* in attraction between the particles. However, this is because they use the Vold analysis[27] to calculate the interparticle attraction, keeping the centre-to-centre distance constant, rather than the outermost surface–surface distance constant.[28] The authors conclude that the explanation for their results must have its origin in the steric repulsion term. They suggest that changes in the thermodynamic mixing parameters of the adsorbed layer on adding salt lead to changes in the osmotic repulsion term. However, the c_c values obtained (*e.g.* 6.2×10^{-3} mol dm^{-3} for NaCl), are

---

* Florence and Rogers[26] have published a review of emulsion stabilization by non-ionic surfactants.

[23] K. Shinoda, H. Saito, and H. Arai, *J. Colloid Interface Sci.*, 1971, **75**, 624.
[24] K. Shinoda and H. Takeda, *J. Colloid Interface Sci.*, 1970, **32**, 642.
[25] P. H. Elworthy, A. T. Florence, and J. A. Rogers, *J. Colloid Interface Sci.*, 1971, **35**, 23.
[26] A. T. Florence and J. A. Rogers, *J. Pharm. Pharmacol.*, 1971, **23**, 153.
[27] M. J. Vold, *J. Colloid Interface Sci.*, 1961, **16**, 1.
[28] D. J. W. Osmond, B. Vincent, and F. A. Waite, *J. Colloid Interface Sci.*, 1973, **42**, 262.

much lower than the values Napper[29] quotes for incipient flocculation of aqueous polymer latices stabilized by poly(ethylene oxide) chains.

Attwood and Florence[30] have also concluded that for chlorobenzene–water emulsions stabilized either by $C_{16}E_{21}$ or by $C_{16}E_{21}SO_3^-$ the steric term is still the dominant one. The anionic form is a less efficient stabilizer than the non-ionic, even though it raises the zeta potential and the non-ionic lowers it. It is suggested that the anionic form is less hydrated and therefore acts as a weaker enthalpic stabilizer.[19] However, although the steric repulsion undoubtedly plays an important role in the *coalescence* of droplets, the longer-range electrical double layer and London terms must surely still dominate the interparticle interactions leading to aggregation. With respect to the electrical double layer term, Ho, Suzuki, and Higuchi[31] have considered the constant-charge–constant-potential question in the case of emulsions. They suggest that with freshly formed emulsion droplets, having low surface coverage, rapid adsorption–desorption can take place leading to the constant-potential model in which thermodynamic equilibrium is maintained. In aged emulsions where the surface coverage is higher, constant charge may be a better model. The logic of this argument is difficult to follow, and the authors present no experimental evidence to justify it. One would strongly suspect that the constant-charge model is more appropriate for all emulsions where the charge arises from surfactant or polyelectrolyte adsorption. The time of desorption is likely to be longer than the time of particle collision.

5 Droplet Coalescence

Volarovich and Avdeev[32] have tested the following equation for the kinetics of drop coalescence at a liquid/liquid interface,

$$N = N_0[1 - \exp\{-\alpha(\tau - \tau_0)\rho\}], \tag{1}$$

where N_0 is the total number of particles, and N the number coalescing in time $\tau > \tau_0$; α and ρ are constants for the system. Various statistical parameters describing the *time* distribution have been derived, *e.g.* the mean half-life $\tau_{1/2}$ (when $N = N_0/2$), the standard deviation, *etc.* A good fit for various emulsions was found.

Unfortunately, relatively few experiments have been performed to study the interaction between two approaching droplets and the factors leading to coalescence. Mateicek, Sivokova, and Eichler[33] have described an apparatus for measuring the time of coalescence of two droplets as a function of temperature, surfactant concentration, *etc.* In some excellent work, Sonntag *et al.*[34]

[29] D. H. Napper, *J. Colloid Interface Sci.*, 1970, **33**, 384.
[30] D. Attwood and A. T. Florence, *Kolloid-Z.*, 1971, **246**, 580.
[31] N. F. H. Ho, P. Suzuki, and W. I. Higuchi, *J. Pharm. Sci.*, 1970, **59**, 1125.
[32] N. Ya Avdeev and M. P. Volarovich, *Kolloid. Zhur.*, 1970, **32**, 28 [*Colloid J. (U.S.S.R.)*, 1970, **32**, 21].
[33] A. Mateicek, M. Sivokova, and J. Eichler, *Coll. Czech. Chem. Comm.*, 1971, **36**, 35.
[34] H. Sonntag, J. Netzel, and B. Unterberger, *Special Discuss. Faraday Soc.*, 1970, no. 1, p. 57.

have studied the thickness of the parallel oil–water–oil thin film formed between two octane droplets in water, using an interferometric technique. The stabilizer used was again a poly(oxyethylene)$_x$ nonylphenyl ether with $x = 20$. They found that the equilibrium film thickness decreased with increasing salt concentration at a given surface concentration and, more surprisingly, with increasing surfactant concentration (up to the c.m.c.) at a fixed salt concentration. The authors came to no definite conclusion as to the explanation of the latter effect, although it could lie in the change in the London attraction term.[28,35]

Sonntag[36] has also investigated the various factors contributing to the critical thickness, d_c, for film rupture. His results appear to contradict predictions based on the current fluctuation theory approach, in that, for example, he finds d_c is more or less independent of the interfacial tension. There also appears to be little effect due to ionic strength.

In two papers[37,38] Sonntag *et al.* consider the factors involved in the coalescence of a particle of material 1 with a particle of material 2 in a medium 0. Various conditions are written in terms of the various interfacial tensions and particle radii, to define the *thermodynamic* stability of various types of system. For example, for three condensed, immiscible phases (1, 2, and 0), the condition for thermodynamic stability is expressed as[37]

$$\sigma^{12} > \sigma^{10} + \sigma^{20} \quad \text{(equal particle radii)}, \qquad (2)$$

where σ is the interfacial tension.

One specific case that is considered is the system: polymer (1), oil (2), and water (0). Taking into account the possible effects of surfactant adsorption at the different interfaces and, therefore, on the various interfacial tensions, the authors conclude that thermodynamic stabilization is generally not possible for non-polar polymers since σ^{12} is small anyway. However, it may be possible for polar polymers where σ^{10} is small and if $\sigma^{12} > \sigma^{20}$. It is suggested that thermodynamic stabilization is additional to the electrical double layer and steric mechanisms. However, one is always dubious about mixing thermodynamic and kinetic analyses. A completely thermodynamic analysis of stability in terms of changes of interfacial tension (*including* the dispersion, electrical, and steric terms) ought to be possible if ion and stabilizer adsorption were completely reversible. This latter condition does not seem to be the case in general, and it may be better to stick to a totally kinetic argument based on the interparticle forces. One feels that the 'thermodynamic' term could equally well be expressed in terms of the interparticle van der Waals forces,* with proper account taken of the effect of the adsorbed layers and the non-dispersion forces.

---

* It was recognized by Hamaker[39] that the condition $A_1 > A_0 > A_2$ or $A_1 < A_0 < A_2$ would lead to interparticle repulsion.

[35] B. Vincent, *Special Discuss. Faraday Soc.*, 1970, no. 1, p. 78.
[36] H. Sonntag, *Kolloid. Zhur.*, 1971, **33**, 529 [*Colloid J. (U.S.S.R.)*, 1971, **33**, 440].
[37] H. Sonntag and N. Buski, *Kolloid-Z.*, 1971, **246**, 700.
[38] H. Sonntag, N. Buski, and B. Unterberger, *Kolloid-Z.*, 1971, **248**, 1016.
[39] H. C. Hamaker, *Physica*, 1937, **4**, 1058.

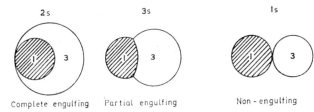

Figure 3 *Possible equilibrium configurations corresponding to the three sets of relations for S_i; '2s' indicate that there are two interfaces per (single) particle, and so on*
(Reproduced by permission from *J. Colloid Interface Sci.*, 1970, **33**, 67)

Torza and Mason[40,41] have also considered the equilibrium configuration of two droplets of different materials in a third material medium, in terms of the various interfacial tensions, or, alternatively, the various spreading coefficients. They assume that such factors as hydrodynamic effects and inter-particle forces are only important in determining the rate of attainment of equilibrium and not the actual equilibrium configuration. The three possible equilibrium configurations are illustrated in Figure 3. One may define three spreading coefficients as follows:

$$S_1 = \sigma^{20} - (\sigma^{10} + \sigma^{12}) \quad \text{(must be} < 0) \tag{3}$$

$$S_2 = \sigma^{12} - (\sigma^{20} + \sigma^{20}) \tag{4}$$

$$S_3 = \sigma^{10} - (\sigma^{12} + \sigma^{20}). \tag{5}$$

The equilibrium conditions for each of the three states shown in Figure 3 are:

2s state (complete engulfing): $S_1 < 0$, $S_2 < 0$, $S_3 > 0$,

3s state (partial engulfing): $S_1 < 0$, $S_2 < 0$, $S_3 < 0$,

1s state (non-engulfing): $S_1 < 0$, $S_2 > 0$, $S_3 < 0$.

(Note: the condition S_2 and S_3 both > 0 is algebraically impossible.) Some nineteen different systems consisting of various combinations of three mutually immiscible phases (together with a twentieth having two miscible phases) were prepared from various combinations of liquids. Examples which predicted all three states were chosen. The medium (0) was chosen to have a high viscosity and/or approximately the same density as phases 1 and 2, in order to eliminate sedimentation. Phase 1 was invariably water, together with Malachite Green to aid viewing and surfactant to lower σ^{12}. Droplet interaction was induced by application of shear (in a Couette apparatus) or an a.c. electric field. Still and ciné photography were used to observe the systems. In most of the systems the observed state corresponded to the predicted state. Discrepancies were fairly readily accounted for in terms of changes of sign of S_2 or S_3

[40] S. Torza and S. G. Mason, *J. Colloid Interface Sci.*, 1970, 33, 67.
[41] S. Torza and S. G. Mason, *Kolloid-Z.*, 1971, **246**, 593.

owing to spreading of a monolayer of the third phase over the interface of two approaching phases. Also the mechanism of engulfing when a 2s drop is formed was established. When two drops are brought together the film of medium between them drains and a hole suddenly forms in the same way as with two identical drops. In the three-phase system a new interface 1/2 is created at the expanding hole and engulfing occurs by a combination of two simultaneous processes: (a) penetration of phase 1 into phase 2, and (b) spreading of phase 2 over phase 1, as determined by the spreading coefficients and the geometry.

6 Ultracentrifuge Studies

The ultracentrifuge[42, 43] has a number of advantages in studying emulsion stability, the main one being that the breakdown of intrinsically stable systems can be studied. Garrett[42] has made measurements on stable, high phase volume o/w emulsions, and evaluated their flotation, creaming, and coalescence properties as a function of rotor speed. The flotation of toluene droplets at 50% phase volume and 50000 r.p.m. suggested at least two distinct phenomena. The first represented an initial rapid flotation of droplets (presumably the small, unhindered ones) which obeyed Svedburg's law. The second distinct stage was a relatively slow movement over short distances in the cell, and was interpreted as packing of the oil droplets with concurrent drainage of residual water from the cream. When the emulsion was examined at a low speed (13000 r.p.m.), intermediate transition steps occurred owing to the movement of droplet aggregates before packing. The creamed layer showed variations in opacity corresponding to different degrees of packing. After an initial lag period a separate toluene layer appeared at the top of the cell. The accumulation followed a first-order rate initially, and then occurred at a constant rate. Two coalescence mechanisms are possible: (a) coalescence at the toluene–emulsion interface, or (b) coalescence throughout the body of the creamed emulsion. The first mechanism would yield a constant rate, provided the cream was not stratified. If it were, then presumably the largest droplets would be nearest the interface and would coalesce more readily. This would lead to a pseudo-first-order rate dependence, *i.e.* coalescence would become more difficult with time. If coalescence occurs throughout the body of the cream, this would also lead to apparent first-order kinetics. The larger particles in an (unstratified) cream may coalesce preferentially and be able to drain through the cream to the top of the cell. The rate would then decrease with time. The initial lag period might also be due to the time necessary for the transport of coalesced droplets to the interface. It therefore depends on whether the cream is stratified or not as to which mechanism, if either,

[42] E. R. Garrett, *J. Soc. Cosmetic Chemists*, 1970, **21**, 393.
[43] K. L. Mittal, *Diss. Abs.* (*B*), 1970, **31**, 2611.

dominates. It is still necessary to explain the constant terminal rate, satisfactorily. It may be due to the fact that the remaining emulsion becomes not only smaller in droplet size but also more homodisperse with time, and that interfacial coalescence predominates.

Mittal[43] has made similar studies on various o/w emulsions, and found a similar pattern for the changes in the rate of coalescence with time. The gradient in the region where coalescence is occurring linearly as a function of time was used as a comparative index of stability. Effects of temperature, storage time, and long-chain alcohol addition were made.

7 Interfacial Structure and Stability

A further factor which contributes to stability against coalescence is the rigidity of the interfacial layer. For example, Elworthy, Florence, and Rogers[44] found that on adding hexadecanol to the emulsion system chlorobenzene–water–$C_{16}E_6$, a synergistic stabilizing effect of the $C_{16}E_6$ and hexadecyl alcohol resulted (provided the concentration of the alcohol was not too high). There was little change in zeta potential, but a marked increase in interfacial rigidity as determined by interfacial viscosity measurements. Electron microscopy has been used[45] to study the gradual development of gel-like structures at several o/w emulsion interfaces, *e.g.* with pentaerythritol monostearate plus pentaerythritol lanolin poly(ethylene oxide) stabilizer. Sanders,[46] and also Friberg's school,[47-53] have investigated further the suggestion that liquid crystalline surfactant phases may be present at the interface. When systems such as *p*-xylene + water + poly(ethylene oxide)$_x$ nonylphenyl ether,[47, 48] or tricontane + water + egg lechithin[51] were examined in a polarizing microscope, birefringence was observed around the droplets. A liquid crystalline phase was also observable as a separate layer after ultracentrifugal separation of the droplets. In each case the stability has been correlated with the three-component phase diagram and the formation of liquid crystalline phases. The formation of liquid crystalline network structures with surfactant + fatty alcohol mixtures was also apparent from rheological measurements[47] (see also Section 8).

There appear to be optimum pH ranges over which maximum interfacial structuring and maximum stability correlate in the case of emulsions stabilized

[44] P. H. Elworthy, A. T. Florence, and J. A. Rogers, *J. Colloid Interface Sci.*, 1971, **35**, 34.
[45] F. S. Gstirner and D. Kottenburg, *Arch. Pharm.* (*Weinheim*), 1971, **304**, 201.
[46] P. A. Sanders, *J. Soc. Cosmetic Chemists*, 1970, **21**, 377.
[47] S. Friberg and P. Solyon, *Kolloid-Z.*, 1970, **236**, 173.
[48] S. Friberg and I. Wilton, *Amer. Perfumer*, 1970, **85**, 27.
[49] S. Friberg and L. Mandell, *J. Pharm. Sci.*, 1970, **59**, 1001.
[50] S. Friberg and L. Mandell, *J. Amer. Oil Chemists' Soc.*, 1970, **27**, 149.
[51] S. Friberg, *J. Colloid Interface Sci.*, 1971, **37**, 291.
[52] S. Friberg and L. Rydhag, *Kolloid-Z.*, 1971, **244**, 233.
[53] S. Friberg, *Kolloid-Z.*, 1971, **244**, 333.

I

by polyelectrolytes, *e.g.* polyacrylates,[54] styrene–maleic acid copolymers,[55] and gelatin.[56]

The liquid paraffin + water emulsions prepared with poly(vinyl alcohol) mentioned earlier (p. 225)[22] were extremely stable, no visible change occurring over a period of months. Glass, Lundberg, and Bailey[57] have looked at the stability to coalescence at a water/organic liquid interface of liquid drops, having absorbed water-soluble polymers. Weakly adsorbed polymers, *e.g.* 99% hydrolysed poly(vinyl acetate) and polyvinylpyrrolidone, tended to increase the droplet stability with increasing molecular weight, whereas with strongly adsorbed polymers, *e.g.* poly(ethylene oxide), greater drop stability was observed with lower molecular weight samples. Some attempt was made to correlate the observed coalescence behaviour with dimensional changes of the adsorbed polymer molecules. As the authors state, the viscous and elastic properties of the adsorbed layers really need to be measured to make a real comparison with stability. They suggest this might be possible using a computerized Monte Carlo analysis of these properties, based on a knowledge of the chemical composition of the polymers, plus oscillating pendulum and canal viscometer measurements.

It is well known that non-ionic polymers complex with ionic surfactants above the c.m.c., and that the properties of the complex are different from those of either component. Polyelectrolyte complexes (with the exception of proteins) have been less studied. Addition of an ionic surfactant to a protein-stabilized emulsion, where they are of *opposite* charge, generally leads to aggregation. Davis[58,59] has looked at liquid paraffin + water emulsions stabilized by a potassium laurate (KL) + potassium arabate (KA) mixture, *i.e.* where surfactant and polymer are of the *same* charge. Viscometric and conductivity measurements on bulk aqueous solutions of the mixture show no complex formation and the c.m.c. of KL is unchanged by the addition of KA. Normally 1% KL (*i.e.* above the c.m.c.) will stabilize the o/w emulsions. Addition of KA up to a certain concentration leads to aggregation and pseudoplastic rheological behaviour, although there is a concurrent increase in zeta potential, so that changes in the electrical double layer are not controlling the system. The mechanism is not clear. Davis proposed that negatively charged arabate ions are able to adsorb, overcoming the negative charge on the surface owing to adsorbed laurate, because of the gain in entropy from released water molecules. One suspects that interparticle bridging by the polyelectrolyte may well play a role, particularly if salt is present, despite the fact the aggregates are apparently broken down reversibly under shear and

[54] B. V. Myasnikov, A. V. Ryabov, and D. N. Emelyanov, *Kolloid. Zhur.*, 1971, **33**, 253 [*Colloid J. (U.S.S.R.)*, 1971, **33**, 208].
[55] E. P. Shavrev, S. A. Nikitina, and N. M. Chernysleva, *Kolloid. Zhur.*, 1970, **32**, 916 [*Colloid J. (U.S.S.R.)*, 1970, **32**, 771].
[56] V. N. Izmailova, Z. D. Tuloskaya, A. F. El-Shimi, I. G. Nadel, and I. E. Alekseeva, *Doklady Akad. Nauk S.S.S.R.*, 1970, **191**, 1081 (*Doklady Phys. Chem.*, 1970, **191**, 300).
[57] J. E. Glass, R. D. Lundberg, and F. E. Bailey, *J. Colloid Interface Sci.*, 1970, **33**, 491.
[58] S. S. Davis, *J. Colloid Interface Sci.*, 1971, **35**, 665.
[59] S. S. Davis, *Kolloid-Z.*, 1971, **246**, 600.

by dilution. The emulsions were stabilized by 4% KA alone (except at very high paraffin phase volumes where interparticle bridging does lead to aggregation).[59] On adding KL (above the c.m.c.), aggregation again occurs.[58] However, the rate of aggregation and also the zeta potential both decrease with increased time of storage of the emulsions. The evidence points to penetration and (partial) displacement of the KA layer by laurate ions. Presumably with increased storage time the adsorbed polyelectrolyte has more time to pack down at the interface as well as to build up thicker layers through more adsorption. Penetration and displacement therefore become more difficult.

As mentioned above, the emulsions of high volume fraction (70.4%) are flocculated (*i.e.* aggregated but not coalesced).[59] They exhibit non-newtonian behaviour, the shear stress–shear rate curves showing hysteresis. High shear leads to redispersion but reflocculation does not occur on its removal, because the adsorbed macromolecules can rearrange at the interface to higher coverage levels and hence steric stabilization results rather than bridging. Addition of KL to the flocculated emulsions leads to redispersion,[59] *i.e.* the converse of the case with the low phase volume paraffin emulsions.[58] KL (anionic) seems to be more effective at this than either CTAB (cationic) or Tween-20 (non-ionic) when compared on the basis of added concentration relative to the c.m.c. The mechanism is again one of displacement of adsorbed segments by laurate ions, leading to uncoupling of the interparticle bridges.

8 Rheological Properties

Many of the factors which influence the viscosity of an emulsion may do so, at least in part, by altering the droplet size distribution. It is therefore necessary to know how droplet size distribution affects viscosity.

Parkinson, Matsumoto, and Sherman[60] have blended together monodisperse suspensions of poly(methyl methacrylate) in Nujol, over a size range 0.1 to 4 μm, to give different modal diameters. The relative viscosity, $\eta_{rel(\Sigma_i)}$, of the suspensions could be represented by the equation

$$\eta_{rel(\Sigma_i)} = \prod_i \eta_{rel(i)}, \tag{6}$$

where $\eta_{rel(i)}$ is the relative viscosity at the same rate of shear for the appropriate concentration of size fraction i, dispersed independently in Nujol. This equation proved valid also for w/o emulsions up to volume fraction 0.6, but only up to 0.4 for the o/w type. Deviations at higher volume fractions for the o/w emulsions were attributed to particle deformation on close-packing. As the emulsion did not exhibit permanent aggregation it is possible to transform the above equation into a Mooney-type equation, *i.e.*

$$\eta_{rel(\Sigma_i)} = \prod_i \exp\left(\frac{2.5 - \phi_i}{1 - k_i \phi_i}\right), \tag{7}$$

[60] C. Parkinson, S. Matsumoto, and P. Sherman, *J. Colloid Interface Sci.*, 1970, **33**, 150.

where ϕ_i is the volume fraction of phase i, and k_i is the hydrodynamic inter-action coefficient, which increases with increasing particle diameter (D), according to an empirical equation of the form,

$$k = 1.079 + \exp\left(\frac{0.01008}{D}\right) + \exp\left(\frac{0.00290}{D^2}\right). \qquad (8)$$

Direct photomicroscopic investigations of the state of dispersion of various dilute emulsion systems as a function of applied shear have been made by Suzuki and Watanabe[61] and the limiting conditions under which Einstein's equation is applicable determined. It was found, for example, somewhat surprisingly, that Einstein's equation could be applied, at a high rate of shear, for the o/w emulsions used at volume fractions below 0.02, even though some aggregation was observed.

Barry and Saunders[62-67] have continued to report on studies relating the structural and rheological properties of o/w emulsions, stabilized by surfactant mixtures with fatty alcohols. Their emulsions were prepared at 70°C, where the alcohol is largely present in the oil phase. On cooling, the alcohol is increasingly solubilized into the aqueous micelles, liquid crystalline phases arising which form network structures stabilizing the emulsion. These struc-tures are eventually weakened, however, by the crystallization of the alcohol and the formation of frozen smectic phases. Correlations between the emulsion behaviour and that of the corresponding ternary systems (minus hydrocarbon) could be made.[63-67] Rheological studies (creep and continuous shear experi-ments) of the effect of alcohol chain length, and mixed chain lengths, showed that sodium dodecyl sulphate (SDS), cetostearyl alcohol, or mixtures of the C_{16} and C_{18} alcohols in the same ratio as in cetostearyl alcohol*, gave the most stable emulsions. Stearyl alcohol alone gave poor emulsion stability owing to its long chain length (and therefore its difficulty in fitting into the sodium dodecyl sulphate micelles) and its high m.p. Cetostearyl alcohol and cetrimide mixtures gave even better rheological stability[63] than cetostearyl alcohol and SDS, owing to the lower c.m.c. of cetrimide and more nearly equal chain lengths of cetostearyl alcohol and cetrimide. It would seem, in fact, that the C_{14} and C_{16} alkyltrimethylammonium bromides give the strongest network structure and therefore highest stability.[65] Indeed, with the C_{12} and C_{18} cationics the gel structure is weak enough to allow close-packing from second-ary minimum flocculation on storage, and hence distortion of the droplet shape from spheres to hexagons.[65] Temperature studies[64,66] on these systems showed

---

* Cetostearyl alcohol is a commercially available mixture of long-chain alcohols (Loveridge Ltd., Southampton, U.K.).

[61] K. Suzuki and T. Watanabe, *Bull. Chem. Soc. Japan*, 1971, **44**, 2039.
[62] B. W. Barry, *J. Colloid Interface Sci.*, 1970, **32**, 551.
[63] B. W. Barry and G. M. Saunders, *J. Colloid Interface Sci.*, 1970, **34**, 300.
[64] B. W. Barry and G. M. Saunders, *J. Pharm. Pharmacol.*, (*Suppl.*), 1970, **22**, 139S.
[65] B. W. Barry and G. M. Saunders, *J. Colloid Interface Sci.*, 1971, **35**, 689.
[66] B. W. Barry and G. M. Saunders, *J. Colloid Interface Sci.*, 1971, **36**, 130.
[67] B. W. Barry, *Rheol. Acta*, 1971, **10**, 96.

that optimum stability was obtained at the temperature corresponding to the transition from the frozen smectic phase to liquid crystalline phase, and this corresponds closely to the 'penetration temperature',[63-66] that is, the minimum temperature at which a 1% aqueous soap solution penetrates the alcohol to form ternary liquid crystals. At higher temperatures the network structure weakens to the point where the ternary systems becomes isotropic.

9 Electrical Properties

Monodisperse emulsions may be prepared by the charging mechanism[68] developed in the 1950s. Abdullaev *et al.*[69] have investigated the effect of field intensity on droplet size and have given an equation which relates the two. Application of a uniform electric field to an emulsion where the droplets are stabilized by an adsorbed layer film can lead to coalescence.[70] The droplets are deformed in the field, as they are under shear, and beyond a critical eccentricity, coalescence results. Panchenkov and Tsabek[70] have developed an expression for the equivalent critical field strength in terms of the shear modulus and thickness of the stabilizing film, the radius of the droplets, and the dielectric constants of the droplets and continuous phase. Unfortunately no experimental test of their theory was carried out.

The dielectric properties of w/o emulsions depend markedly on the nature of the emulsifier.[71,72] Hanai[72] has shown that dielectric dispersion due to interfacial polarization could be expected for w/o emulsions. Agreement between theoretical and experimental values for both the high-frequency (ε_h) and low-frequency (ε_l) dielectric constants was poor. Hanai[72] could fit the ε_l values, however, if shear was applied. This implies that aggregation of water droplets may be significant. Chapman[71] has investigated water + liquid paraffin emulsions stabilized by (*a*) sorbitan mono-oleate (non-ionic) or (*b*) magnesium stearate (anionic). ε_l values with the non-ionic were greater than those for the anionic over the whole water volume fraction range studied (0.0—0.3). Microscope studies revealed that the emulsions with the non-ionic were partially aggregated. Good agreement with the classical Maxwell–Wagner equation for interfacial polarization relating ε_l to volume fraction could be obtained in the case of the anionically stabilized emulsions, on the other hand.

[68] See, for example, ref. 11, p. 55.
[69] R. Kh. Abdullaev, A. A. Agaev, T. G. Kurbanaliev, I. N. Rzabekov, and Kh. Bekmamedov, *Izvest. V. U. Z., Neft i Gaz*, 1971, **14**, 63 (*Chem. Abs.*, 1971, **75**, 25751).
[70] G. M. Panchenkov and L. K. Tsabek, *Kolloid Zhur.*, 1971, **32**, 710 [*Colloid J. (U.S.S.R.)*, 1971, **32**, 594.]
[71] I. D. Chapman, *J. Phys. Chem.*, 1971, **75**, 537.
[72] T. Hanai, ref. 11, ch. 5.

7
Non-aqueous Systems

BY B. VINCENT

1 Introduction

The most recent extensive review to appear dealing with dispersion stability in non-aqueous media is that by Lyklema (1968).[1] In the Introduction the author states: 'Reviewing the literature on this topic, it at once becomes clear that the majority of investigations, both theoretical and practical, are devoted to aqueous systems. As a result, non-aqueous systems, especially dispersions in solvents of low or zero polarity, are less well understood than their aqueous counterparts.' Since the middle 1960's, however, there has been considerably more effort put into understanding non-aqueous systems. On the theoretical side (Section 2), several detailed papers[2-4] have appeared on the theory of steric stabilization by adsorbed macromolecules. This work has largely resulted from the developments in polymer adsorption theory during the past ten years or so[5, 6] (see Chapter 3). During the period under review here (1970/1971), further advances have been made in this direction.[7-9] It could well be in fact that the current theories of steric stabilization are better tested with non-aqueous systems than with the more complex aqueous ones, since non-aqueous polymer solutions themselves are, on the whole, better understood. On the experimental side (Section 3), a much better appreciation of the structural requirements of polymeric stabilizers has evolved, and experiments on well-chosen, and well-characterized, systems have been made.[10, 11]

It is now generally accepted that in rigorously dry media of low polarity (*e.g.* aliphatic hydrocarbons), the stabilization of dispersions by surface-

[1] J. Lyklema, *Adv. Colloid. Interface Sci.*, 1968, 2, 65.
[2] E. J. Clayfield and E. C. Lumb, *J. Colloid Interface Sci.*, 1966, **22**, 269, 285.
[3] D. J. Meier, *J. Phys. Chem.*, 1967, **71**, 1861.
[4] E. J. Clayfield and E. C. Lumb, *Macromolecules*, 1968, **1**, 133.
[5] R. R. Stromberg, in 'Treatise on Adhesion and Adhesives', ed. R. L. Patrick, Marcel Dekker, New York, 1967, vol. 1.
[6] C. A. J. Hoeve, *J. Polymer Sci.*, *Part C, Polymer Symposia*, 1971, **34**, 1.
[7] F. Th. Hesselink, *J. Phys. Chem.*, 1969, **73**, 3488.
[8] F. Th. Hesselink, *J. Phys. Chem.*, 1971, **75**, 65.
[9] F. Th. Hesselink, A. Vrij, and J. Th. G. Overbeek, *J. Phys. Chem.*, 1971, **75**, 2094.
[10] D. J. Walbridge and J. A. Waters, *Discuss. Faraday Soc.*, 1966, No. 42, p. 247.
[11] D. H. Napper, *Trans. Faraday Soc.*, 1968, **64**, 1701.

charge mechanisms plays only a minor role, if any, compared to contributions from steric stabilization. Over recent years attention has been paid to evaluating more exactly the role played by water in these systems. In particular, spectroscopic and electrophoretic studies have been made. Studies have also been made with media of somewhat higher polarity (*e.g.* alcohols).[12,13] These and related topics are considered in Section 4.

In Section 5, some of the properties of association colloids in non-aqueous media are reviewed. Surfactants, *e.g.* Aerosol OT, are widely used in dry-cleaning fluids as solubilizing agents for water-soluble materials such as salt and sugar. Despite this, until comparatively recently, virtually no fundamental work on micelle structure, and the solubilization of water by non-aqueous surfactant systems, had been undertaken. A much clearer picture has now emerged from recent work, in particular that by the Japanese schools of Shinoda and Kitahara. Kitahara, in fact, has recently (1970) reviewed the field.[14] An earlier review had been compiled by Fowkes (1967).[15]

2 Dispersion Stability: Theoretical Aspects

Since the 1950's, three basic approaches to the theory of steric stabilization have evolved. One, originated by Mackor and van der Waals,[16] is based on evaluating the work required to desorb molecules from the region between two approaching interfaces. This interpretation is limited in the case of polymers to situations where the molecules are relatively weakly and, as a consequence, reversibly adsorbed and is probably restricted, therefore, to non-ionic surfactant species of low molecular weight. This approach has been recently revived by Ash and Findenegg.[17] They make use of a multilayer, hexagonal lattice model[18] of adsorption from solution, in which the solution is composed of monomer and r-mer ($r = 2$ or 4) species. The r-mers have either one active (preferentially adsorbed) end segment, or all segments equally active. Non-athermal conditions are allowed for by incorporating a segment–solvent interchange energy parameter, w ($w = 0$ corresponds to athermal conditions). The ratio of the interaction with the surface of active segments and monomer is included through a second parameter, K ($K = 1$ implies no preferential adsorption). The change in free energy of the solution between two parallel plates is calculated as the plates are brought from infinite separation to a separation of t lattice layers. On adding to this

[12] P. Jackson and G. D. Parfitt, *Kolloid-Z.*, 1971, **244**, 240.
[13] A. N. Zhukov, N. A. Kibirova, M. P. Sidorova, and D. A. Fridrikhsberg, *Doklady Akad. Nauk S.S.S.R.*, 1970, **194**, 130 (*Doklady. Phys. Chem.*, 1970, **194**, 666).
[14] A. Kitahara, in 'Cationic Surfactants', ed. E. Jungerman, Marcel Dekker, New York, 1970, p. 289.
[15] F. M. Fowkes, in 'Solvent Properties of Surfactant Solutions', ed. K. Shinoda, Marcel Dekker, New York, 1967, p. 65.
[16] E. L. Mackor and J. H. van der Waals, *J. Colloid Sci.*, 1952, **7**, 535.
[17] S. G. Ash and G. H. Findenegg, *Trans. Faraday Soc.*, 1971, **67**, 2122.
[18] S. G. Ash, D. H. Everett, and G. H. Findenegg, *Trans. Faraday Soc.*, 1968, **64**, 2645; 1970, **66**, 708.

term the change in the free energy of the bulk solution, one obtains the total change in free energy of the system. For r-mers with one terminal active segment this quantity is, for the most part, positive (implying repulsion between the plates), rising with decreasing t, increasing w, increasing K, increasing r, and increasing ϕ_r^b, where ϕ_r^b is the bulk solution concentration of r-mers. For homogeneous r-mers, on the other hand, the total free-energy change is generally *negative* (implying attraction). The bridging configurations that occur are the cause of this attraction between the plates. This approach is unfortunately restricted in that, being based on a lattice model, it is necessary to stipulate that the monomer species and segments have equal volume. At present, the upper limit on r is 4, and a complete analysis for interacting spherical particles appears mathematically difficult. It is also difficult to devise any experimental test of the theory.

The other two approaches to steric stabilization have been applied to *irreversibly* adsorbed polymers. One originated again with Mackor.[19] It has since been greatly improved by Clayfield and Lumb.[2] The other is due to Fischer.[20] The Mackor–Clayfield–Lumb approach considers the change in configurational entropy of *complete*, adsorbed macromolecules as a second impenetrable barrier is brought up. The Fischer analysis, on the other hand, considers the change in the free energy of mixing of the *segments* of the adsorbed macromolecules in the region of interaction. Meier[3] considered that these two effects are additive. This concept has been taken up in the work of Hesselink *et al.*[7–9] In the first paper,[7] the segment density distribution normal to the surface was calculated for terminally adsorbed tails and loops. The method[21, 22] used was first to calculate, using a six-choice cubic lattice, the number of possible conformations for the first k segments of a chain of size i, such that the kth segment terminates at a distance x normal to the surface. Having then determined the number of possible configurations for the remaining $(i-k)$ segments, it is possible to calculate the probability, $\rho(i,k,x)$, of finding the kth segment of the total chain at x. The total probability, $\rho(i,x)$, of finding a segment at x, *i.e.* the effective segment density, is found by integrating over all k.

In his second paper,[8] Hesselink made use of Hoeve's equation[23] for loop size distribution to calculate $\rho(i,x)$ for homopolymers.[22] The analysis seems to neglect contributions from the tails at the ends of molecules, which other authors[24, 25] suggest may be important, and also from trains of segments in the lattice layer adjacent to the interface which give rise to a discontinuity in the segment density distribution[22] for homopolymers. The normalized density

[19] E. L. Mackor, *J. Colloid Sci.*, 1951, **6**, 492.
[20] E. W. Fischer, *Kolloid-Z.*, 1958, **160**, 120.
[21] S. Chandrasekhar, *Rev. Mod. Phys.*, 1943, **15**, 1.
[22] C. A. J. Hoeve, *J. Chem. Phys.*, 1965, **43**, 3007.
[23] C. A. J. Hoeve, E. A. DiMarzio, and P. Peyser, *J. Chem. Phys.*, 1965, **42**, 2558.
[24] R. J. Roe, *J. Chem. Phys.*, 1965, **43**, 1591; 1966, **44**, 4204.
[25] K. Motomura and R. Matuura, *J. Chem. Phys.*, 1969, **50**, 1281.

distribution for copolymers (*i.e.* with 'anchoring' groups spaced randomly along the chain) is also given.[8]

The next problem[8] was to derive the total number of possible configurations, $z(i,d)$, for tails and loops, and hence, also the modified segment density distribution, $\rho(i,x,d)$, when a second, parallel interface is brought up to a distance d from the one on which the tails and loops are adsorbed. The loss in configurational entropy is given by the Boltzmann relationship,

$$\Delta S = -k \ln \frac{z(i,d)}{z(i,\infty)}. \tag{1}$$

In the third paper,[9] Hesselink, Vrij, and Overbeek calculate the total free-energy change,* ΔA, involved in bringing two parallel plates, both having adsorbed polymer molecules, from infinite separation to a separation d.

$$\Delta A = \Delta A_{\text{vr}} + \Delta A_{\text{m}} + \Delta A_{\text{a}}, \tag{2}$$

where ΔA_{vr} is the volume restriction (configurational) term, ΔA_{m} the segmental free energy of mixing term, and ΔA_{a} the London–van der Waals attraction between the plates. The model takes into account solvent effects by including the bulk solution expansion factor, α, that is, the ratio of the actual root mean square (r.m.s.) end-to-end distance of the free polymer molecules to their unperturbed r.m.s. dimensions (θ-state). One can, at present, only speculate as to the justification for using the same parameter for molecules in the adsorbed state.

ΔA_{vr} is derived from the configurational entropy term for loops and trains on a *single* plate given above,

$$\Delta A_{\text{vr}} = -2kT \sum_i n_i \ln \left[\frac{z(i,d)}{z(i,\infty)} \right], \tag{3}$$

where n_i is the number of loops (per unit area) containing i segments.

ΔA_{m} is calculated using the classical Flory–Huggins expression for the free energy of mixing of polymer solutions,

$$\Delta A_{\text{m}} = \frac{kTV_2^2}{V_1} (\tfrac{1}{2} - \chi) \left[\int_0^d (\rho_{\text{a}} + \rho_{\text{b}})_d^2 \, dx - \int_0^d (\rho_{\text{a}} + \rho_{\text{b}})_\infty^2 \, dx \right], \tag{4}$$

where V_1 and V_2 are the molecular volumes of solvent molecules and polymer segments, respectively, χ is the Flory–Huggins segment–solvent interaction parameter (related to α), and $\rho_{\text{a}}(x)$ and $\rho_{\text{b}}(x)$ are the derived segment density distributions for polymer on each plate.

The model[9] assumes direct additivity of the individual segment distributions, that is, the segment distribution for the polymer molecules adsorbed on one plate is modified only by the presence of the second plate itself and *not* by the

---

* The notation used by the authors for free energy has been changed from F to A to avoid confusion since F is later used here to represent force.

polymer molecules adsorbed on it. In this it differs in an important respect from the Clayfield–Lumb model[2] where the adsorbed layers themselves are regarded as impenetrable walls, and there is no segment overlap. Furthermore, Hesselink *et al.*[9] assume there is no change in the actual distribution of loop sizes, n_i, since they argue that there is no time during a typical Brownian collision for adsorption–desorption of segments to take place. One feels that this is still an open question, however. One feature where the Clayfield–Lumb model[2] is an improvement upon the Hesselink model[9] is that the former uses a four-choice cubic lattice (*i.e.* only 90° bond angles are allowed), rather than the six-choice cubic lattice, and it does go some way to incorporate excluded-volume effects in the ΔA_{vr} term.

The evaluation of ΔA_{vr} and ΔA_m both require numerical integration. ΔA has been computed[9] for the specific case: polystyrene (mol. wt. 10^3—10^5) under various solvency conditions (α 0.9—1.6); the amount adsorbed (ω) was varied from 10^{-10} to 5×10^{-7} g cm^{-2}, and the Hamaker constant for the system was taken to lie in the range 3×10^{-21}—3×10^{-19} J. In Figure 1, ΔA_{vr}, ΔA_m, ΔA_a, and ΔA are illustrated for one particular set of parameters in the case of terminally adsorbed polystyrene. It can be seen that ΔA_m exceeds ΔA_{vr}, both terms rising fairly steeply, with ΔA_m becoming effective at around $d = 18$ nm and ΔA_{vr} becoming effective at around $d = 15$ nm. These values of d lie between twice the calculated r.m.s. distance of segments from the interface (12 nm)* and twice the fully extended chain-length of the molecules (~ 26 nm). These figures are compared with experimental measurements of adsorbed layer thickness in the following section.

Hesselink *et al.*[9] have suggested that an approximate solution for the case of interacting spherical particles may be obtained by utilizing Derjaguin's approximation.[26]

3 Dispersion Stability: Experimental

One of the first direct measurements of the magnitude of steric repulsion forces has come from the work of Andrews, Manev, and Haydon.[27, 28] They studied, using a capacitance technique, the thickness of non-aqueous black films, formed from solutions of glycerol mono-oleate in hydrocarbon solvents of varying chain-length. The films were supported between bulk aqueous solutions of sodium chloride. The adsorption of the glycerol mono-oleate was calculated from interfacial tension measurements. A d.c. potential applied across the films resulted in compression and thinning. The major effects observed were as follows. The equilibrium thickness of the films decreased with increasing hydrocarbon chain-length of the solvent from C$_9$ to C$_{16}$,

---

* Twice the r.m.s. end-to-end distance in *bulk solution* would be about 6 nm.

[26] B. Derjaguin, *Kolloid-Z.*, 1934, **69**, 155.
[27] D. M. Andrews, E. D. Manev, and D. A. Haydon, *Special Discuss. Faraday Soc.*, 1970, No. 1, 46.
[28] D. M. Andrews, Ph.D. Thesis, Cambridge, 1970.

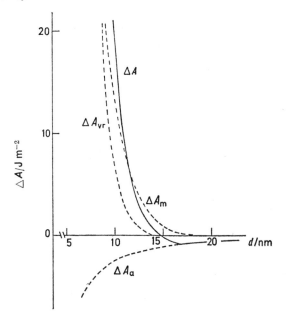

Figure 1 *The change in the total free energy of interaction, ΔA,* versus *the distance, d, between two flat plates covered by equal tails. ΔA_a is the London–van der Waals attraction; ΔA_{vr} the volume-restriction term; ΔA_m the segmental free energy of mixing term. The curves are calculated for $A = 10^{-20}$ J, $\alpha = 1.2$, $\omega = 2 \times 10^{-8}$ g cm^{-2}, mol. wt. = 6000. Area per chain = 50 nm^2*
(Adapted from a figure in *J. Phys. Chem.*, 1971, **75**, 2094)

i.e. as the chain-length of the stabilizing mono-oleate chains was approached. At zero applied potential the film thickness in n-decane was 4.4 nm; this decreased to 3.6 nm on applying a potential of 0.4 V. On increasing the potential, the film took times of the order of minutes to hours to re-equilibrate. This time was longer the larger the hydrocarbon chain-length of the solvent.

At equilibrium, the net force on the film is zero, so that:

$$F_a + F_s + F_e = 0, \tag{5}$$

where F_a is the London–van der Waals attraction, F_s the steric repulsion, and F_e the applied electrical compression force. From this it is readily shown that:[27,28]

$$F_s = \frac{A_H}{6\pi\delta_L{}^3} + \frac{cV^2}{2\delta_e}, \tag{6}$$

where A_H is the Hamaker constant for the system, V the applied potential, and

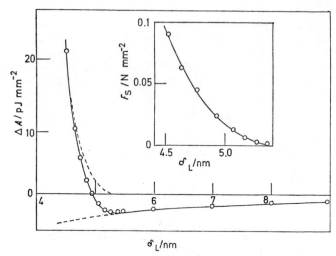

Figure 2 *The change in the total free energy of interaction, ΔA, and the steric repulsion force, F_s (see inset) as a function of equilibrium film thickness, δ_L, for glycerol mono-oleate plus n-decane films in saturated NaCl. Hamaker constant = 3.48×10^{-21} J*

(Reproduced from *Special Discuss. Faraday Soc.*, 1970, No. 1, 46)

c the specific capacitance. δ_e is the thickness of the hydrocarbon core, and is related to δ_L, the measured thickness, by

$$\delta_L = \delta_e + 2 \times 0.45 \text{ nm} \tag{7}$$

(0.45 nm is the thickness of the layer comprising the polar groups).

A plot of F_s *versus* δ_L is shown in Figure 2, for the case in which n-decane was the solvent. Also shown is the change in free energy, ΔA, as the film thins:

$$\Delta A = -\frac{A_H}{12\pi\delta_L^2} - \int_{\infty}^{\delta_L} F_s \, d\delta_L. \tag{8}$$

The thicker films, formed with the solvents of lower molecular weight in the absence of an externally applied potential, contain a greater volume fraction of solvent and are much more compressible. Calculations[28] of the steric repulsion were based on the Fischer model.[20] Comparing the computed and experimental values of F_s for these thicker films, it was concluded that the segment density in the overlap region must be very small compared with the average segment density in the film. Since, for the thick films, the measured values of δ_L are roughly twice the fully extended stabilizer chain dimensions, it must be that, at any given time, only a small fraction of the oleate chains are in fact in the fully extended state. This result implies that even for relatively small surfactant molecules, assumptions of uniform segment density in the adsorbed layer may be in error.

The chains in the thinner films, formed with the longer chain-length hydrocarbon solvents, seem to be in a compressed state even prior to any applied potential. As pointed out by Osmond[29] in a comment on the paper by Andrews *et al.*,[27] this point has a strong bearing on the possible conformations of high-molecular-weight polymers and the resultant effect on the level of steric stabilization, where the continuous phase is the polymer melt. It can be seen from Figure 2 that the free-energy change rises rapidly in the manner predicted by Hesselink[9] (Figure 1).

Doroszkowski and Lambourne[30] have attempted to measure directly the steric repulsion forces between latex particles having adsorbed polymer. They made compression studies on the particles spread at a heptane/water interface, using a surface-balance technique. The latex particles were stabilized with an AB-type graft copolymer where the A (anchoring) group was poly-(methyl methacrylate) (PMMA) and the B (soluble) group was poly(hydroxy-stearic acid) (PHS). It was established that the particles were on the heptane side of the interface. All the free, unadsorbed polymer in the continuous phase of the latex dispersion was washed out before spreading by a series of centrifugation–solvent dilution steps. A series of PMMA latices, covering the particle diameter range 0.14—0.49 μm, was studied. In each case the average area per adsorbed stabilizer molecule was about 5.4 nm², and the measured barrier thicknesses (calculated from the observed area where inter-particle repulsion started) were in the range 10—14.5 nm. These latter values may be compared with those (6—10 nm) obtained from viscometric measurements[31–33] with similar latices. The number-average molecular weight of the PHS chains was 1600, so that the fully extended chain-length would be of the order of 10 nm. The reason, presumably, why interactions at greater thicknesses were observed is that there was a distribution of chain lengths, some chains being appreciably longer than the average. Nevertheless, in the case of these PHS stabilizing moieties it is evident that all the chains are more or less in the fully extended state. This in turn would verify the assumption of uniform segment density in the adsorbed layer, in contradistinction to the conclusion from Andrew's experiments described above. This point will be taken up again later.

The work done in compressing the adsorbed layers on the latex particles could be calculated by integrating the experimental pressure–area curves, assuming that the adsorbed layer on each particle interacts with that from each of six nearest neighbours in the same plane. A comparison was made[30]

[29] D. W. J. Osmond, *Spec. Discuss. Faraday Soc.*, 1970, No. 1, 75.
[30] A. Doroszkowski and R. Lambourne, *J. Polymer. Sci., Part C, Polymer Symposia*, 1971, **34**, 253.
[31] A. Doroszkowski and R. Lambourne, *J. Colloid Interface Sci.*, 1968, **26**, 214.
[32] S. J. Barsted, L. J. Nowakowska, I. Wagstaff, and D. J. Walbridge, *Trans. Faraday Soc.*, 1971, **67**, 3598.
[33] D. W. J. Osmond and D. J. Walbridge, *J. Polymer. Sci., Part C, Polymer Symposia*, 1970, **30**, 381.

with theoretical computations. A remarkably good fit appears to be obtained with the simple Ottewill–Walker[34] extension, for spherical particles, of the Fischer approach for calculating the osmotic term. This also assumes a uniform segment density distribution in the adsorbed layer. Although it has to be remembered that a number of uncertain, adjustable parameters are involved, the fact that such close agreement is obtained is encouraging. An attempt by Doroszkowski and Lambourne to improve the Fischer theory further by assuming a more realistic volume for the overlap region did not in fact improve the fit with the experimental data.

Finally it is worth noting, as the authors point out,[30] that the osmotic term contributes even more strongly, relative to the volume-restriction term, in the case of spherical particles than it does in the case of flat plates. Hence, this is their justification for neglecting the volume-restriction term. This assumption would appear to be valid under 'good' solvent conditions, but under θ-conditions, of course, the volume-restriction term will be the only operative one.

Block or graft copolymers of the type used by Doroszkowski and Lambourne[30] are now widely used in industry as stabilizers for non-aqueous dispersions. They function most efficiently when the anchoring component is either chemisorbed at the interface, or, for example, in the case of latex dispersions, physically incorporated into the bulk of the particles. Methods for achieving these ends have been discussed by Waite,[35] and by Osmond and Walbridge.[33] In the latter paper a discussion is also presented of adsorption levels in systems where these types of stabilizers have been used. Two general types of stabilizer are discussed. In one, the soluble group poly(lauryl methacrylate) (PLMA) in heptane solvent was grafted *in situ* to various latex particles. This was achieved by allowing a 97/3 copolymer of lauryl methacrylate and glycidyl methacrylate to react with methacrylic acid, thus incorporating double bonds into the molecule (one or two on average).[35] This stabilizer 'precursor' was then copolymerized with the necessary monomer to form the (stabilized) latex particles. The second type of stabilizer is the preformed variety, for example, PMMA/PHS as discussed above. Since in this case there may be several PHS groups attached to one PMMA 'backbone', this type is often referred to as 'comb' stabilizer.[35]

Adsorption measurements[33] with such PMMA/PHS comb stabilizers indicate a reasonably constant area per molecule (3—5.5 nm^2) on a variety of latex surfaces, and for a variety of particle diameters (0.02—0.6 μm). (There was a small, but significant, decrease in area per molecule as the concentration of free stabilizer in solution increased. This is in line with the comments referred to earlier concerning the polymer melt situation.[29]) The upper limit, 5.5 nm^2, is close to that found by Doroszkowski and Lambourne[30] for their 'washed' PMMA latices. Again an extended conformation for the PHS chains is

[34] R. H. Ottewill and T. Walker, *Kolloid-Z.*, 1968, **222**, 108.
[35] F. A. Waite, *J. Oil. Colour Chemists' Assoc.*, 1971, **54**, 342.

implied. This must have to do with the special geometry of the PHS molecule, which has C_6 side-chains attached to the main backbone, giving a 'tree-like' configuration at the interface, and a resulting uniform segment density (~ 10—15%).

With the linear PLMA stabilizer systems, the surface coverage was again more or less independent of particle size or type,[33] but this time the lateral spacing of the adsorbed chains ('tails' on the Hesselink classification[7]) approximated better to the r.m.s. dimensions of PLMA molecules in *bulk solution*. These linear, higher molecular weight (3×10^4—4×10^5) PLMA chains would constitute, therefore, a better model for testing the Hesselink theory[7-9] than the PHS or oleate chains. Unfortunately, no experimental measurements of steric forces, or of adsorbed layer thicknesses, had been reported by the end of 1971. One is aware, however, of current experiments in this direction.

Another, potentially powerful technique for studying both the r.m.s. thickness and the average segment density of adsorbed layers is ellipsometry. Stromberg, Smith, and McCrackin[36] have discussed its applications to polymer adsorption and reviewed some of the data collected so far. The major limitations are the production of an optically smooth surface, and the necessity for having sufficiently high increments of refractive index between the substrate, the adsorbed layer, and the bulk solution.

Molau and Richardson[37] have studied the stability (sedimentation rate) of TiO_2 dispersions in toluene and *o*-dichlorobenzene using a 70/30 AB block copolymer of styrene and butadiene. Thioglycollic acid was used for partial carboxylation of the polybutadiene chains so as to increase their anchoring efficiency. 1% dispersions of TiO_2 in dichlorobenzene containing 1% stabilizer were quite stable, even with the uncarboxylated stabilizer. When, however, these primary dispersions were centrifuged, dried, and redispersed (at 1% solids) in toluene plus 1% stabilizer, the level of stability depended markedly on the degree of carboxylation of the stabilizer. 1—2% carboxylation of the butadiene groups was found to be optimum. A possible explanation that one could suggest for this is that over-efficient anchoring of the poly-butadiene chains may lead to a higher area per molecule for the soluble polystyrene chains (see Figure 3), and, as a consequence, weaker steric repulsion. Higher ratios of butadiene to styrene in the stabilizer led to even lower optimum values of the degree of carboxylation, whilst the opposite was true for lower ratios. These observations would fit the mechanism offered.

To illustrate the advantages that graft copolymers in general yield as stabilizers, in comparison with homopolymers, Molau and Richardson[37] have also looked at the stability of TiO_2 dispersions in toluene when either ungrafted,

[36] R. R. Stromberg, L. E. Smith, and F. L. McCrackin, *Symposia Faraday Soc.*, 1970, No. 4, 192.

[37] G. E. Molau and E. H. Richardson, in 'Multicomponent Polymer Systems', ed. R. F. Gould, Advances in Chemistry Series, American Chemical Society, 1971, No. 99, p. 379.

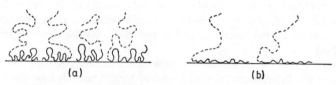

(a)   (b)

Figure 3 *Schematic representation of the effect of over-anchoring of the insoluble part of an AB block copolymer. In* (b) *(over-strong anchoring) the average segment density of the soluble chains (and hence the osmotic repulsion) is much less than in* (a) *(optimum anchoring)*

partially carboxylated polybutadiene or else polystyrene (copolymerized with 1—2% acrylic acid to introduce carboxy-groups) was added. In all cases the stability was very poor. One would have expected this in the case of the polybutadiene (poor solvent conditions), but one feels that if the number of carboxy-groups and the molecular weight of the polystyrene had been optimized, it might well have functioned as a reasonable stabilizer.

One must always consider the possibility, with polymers of high molecular weight, that inter-particle bridging may be the cause of decrease in stability. Molau and Richardson[37] suggest this as a possible cause of the observed instability at high degrees of carboxylation of the block copolymers. This would seem unlikely to be the case, however. The bridging entity is hardly likely to be a polybutadiene chain since this must, more or less, be adsorbed in a flat configuration (see Figure 3) at high degrees of carboxylation. Also the (unmodified) polystyrene chains would adsorb only weakly, if at all, on the surface of a second particle, even if there were unoccupied sites for them to do so. Bridging flocculation only occurs when the degree of surface coverage is low, and this is unlikely to be the case here. The mechanism offered above is surely the more likely one.

There is more positive evidence for bridging flocculation in the systems studied by Kitahara *et al.*[38] They showed that the rate of flocculation of TiO_2 dispersions in n-hexane went through a maximum with increasing concentrations of various chain-length tetra-alkyl titanates. No effect was observed with SiO_2 or copper phthalocyanine presumably since adsorption of the polymers is very much weaker on these solids.

If one attempts to mix solutions of two incompatible polymers, even in the same solvent, phase separation normally occurs. Molau[39, 40] has shown, however, that if a block copolymer of the two individual polymers concerned is added to the system a stable emulsion can be produced. In a recent paper, Molau[40] has reported further studies on the system (5% polybutadiene + 25% polystyrene + 70% benzene), to which is added polybutadiene–polystyrene

[37] A. Kitahara, M. Hoshino, T. Fujii, N. Yoshino, and T. Yoshino, *Kogyo Kagaku Zasshi*, 1970, **73**, 2081 (*Chem. Abs.*, 1971, **74**, 91 446).
[39] G. E. Molau, *J. Polymer Sci., Part A, General Papers*, 1965, **3**, 1267, 4235.
[40] G. E. Molau, *Kolloid-Z.*, 1970, **238**, 493.

block copolymer. He has shown that it is necessary for the molecular weight of each component of the block copolymer to be above a critical minimum value before adequate stabilization occurs.

The effect of the level of polymer adsorption on dispersion stability is also apparent from the work of Sato,[41] who showed that low-molecular-weight (500—3000) polyamides would stabilize α-Fe_2O_3 sols in cyclohexane but not in isopropyl alcohol. The polymer is preferentially adsorbed in the former case, and the solvent in the latter.

Even for species of low molecular weight (including surfactants) that are preferentially adsorbed at an interface, the configuration they adopt is all-important for stability. Bagchi and Vold[42] have looked at dispersions of Graphon (of specific surface area 85.2 m^2 g^{-1}) in heptane, containing varying concentrations of either 'Triton X-35' [*para*-(t-octyl)phenyl poly(ethylene oxide)$_{3.7}$] or dodecylbenzene. The former gave a stepped adsorption isotherm, but had no effect on the coagulation rate obtained from the observed changes in turbidity with time. The latter gave a Langmuir-type isotherm,[43] at low concentrations, with a steadily decreasing flocculation rate. In the case of Triton X-35, even at concentrations beyond the step in the isotherm, one suspects that the thickness of the adsorbed layer is still insufficient to bring about a noticeable change in stability. Only the short t-octyl group is likely to be protruding into the heptane solution since heptane is a non-solvent for poly(ethylene oxide). Bagchi and Vold[42] suggest, from viscosity measurements, that the second virial coefficient of Triton X-35 is close to zero in heptane solution, and that as a result there is no osmotic repulsion. However, it is only the protruding component(s) of the adsorbed molecule, and not the whole molecule, that contributes to the osmotic repulsions,[11, 44] so one doubts if this can be the explanation.

With regard to the dodecylbenzene results, Bagchi and Vold[42] have made theoretical calculations of the inter-particle forces involved, and hence have been able to derive theoretical stability ratios. However, the computed values are orders of magnitude greater than the experimental ones. This is hardly surprising considering the great number of assumptions and approximations which are made (for example, the steric component is calculated on the basis of a modification of the elementary Mackor model[19] for the volume-restriction term. This assumes that the adsorbed molecules are rigid rods, oriented perpendicular to the interface, and takes no account of solvency effects). Parfitt[45]* had previously studied this system, and had reached no firm

---

* Parfitt has written a review[46] of some of his earlier studies on stability in non-aqueous media.

[41] T. Sato, *J. Appl. Polymer Sci.*, 1971, **15**, 1053.
[42] P. Bagchi and R. D. Vold, *J. Colloid Interface Sci.*, 1970, **33**, 405.
[43] G. D. Parfitt and E. Willis, *J. Phys. Chem.*, 1964, **68**, 1780.
[44] D. H. Napper, *Ind. and Eng. Chem. (Product Res. and Development)*, 1970, **9**, 467.
[45] G. D. Parfitt and E. Willis, *J. Colloid Interface Sci.*, 1966, **22**, 100.
[46] G. D. Parfitt in 'Solid/Liquid Interfaces', ed. B. Tezak and V. Pravdic, Croatica Chemica Acta, Zagreb, 1971; or *Croat. Chem. Acta*, 1970, **42**, 215.

conclusions as to the apparent stability increase, but in a recent paper Lawrence and Parfitt[47] concluded, from particle-counting studies, that the change in flocculation rate could be completely accounted for in terms of changes in the viscosity of the bulk solution, and that the *adsorbed* dodecylbenzene molecules play no role. Lawrence and Parfitt[47] also question the use of simple turbidity measurements for following rates of coagulation. They suggest that only under special circumstances can one expect to obtain the same results as from direct particle-counting methods. In general, the wavelength-dependence of the turbidity has to be determined, and some idea of aggregate morphology is required.

Because, in the past, structural effects in solvent layers (particularly water)[48] adjacent to solid surfaces have sometimes been considered to play a role in controlling dispersion stability, it is worth mentioning in this section some interesting studies on Graphon in pure alkanes reported by Ash and Findenegg.[49,50] The heat of wetting per unit area of Graphon in n-alkanes is known[51-53] to increase with chain length (C_6 to C_{16}). To explain these results it had been predicted that the alkane layer(s) adjacent to the surface should have properties intermediate between those of the liquid and crystalline phases. The latest measurements show that there is indeed a density change for alkanes at the interface, compared with the bulk liquid, at temperatures approaching their freezing points. The surface excess mass per unit area, Γ, defined by locating the Gibbs' dividing surface at the Graphon/alkane physical boundary, is given by:

$$\Gamma = \int [\rho(x) - \rho_1^0] \, dx, \qquad (9)$$

where $\rho(x)$ is the actual density of the liquid as a function of the distance, x, normal to the interface and ρ_1^0 is the bulk liquid density. By assuming that the excess mass of the liquid is distributed *uniformly* over a zone of finite thickness, and that the properties of this zone are the same as those of the solid alkane at the freezing point, it was possible to obtain excellent correlation between these density measurements and the data for heat of wetting.[49,50] However, the effective thickness of the surface zone is very small ($= 1$ nm), so that one can hardly expect any significant contribution to particle interactions. For benzene (and water)[49] Γ actually turned out to be *negative* at, and close to, the freezing point itself, implying a *lower* density at the interface than in the bulk. Following the controversy over the anomalous properties of

[47] S. G. Lawrence and G. D. Parfitt, *J. Colloid Interface Sci.*, 1971, **35**, 675.
[48] For example, G. A. Johnson, J. Goldfarb, and B. A. Pethica, *Trans. Faraday Soc.*, 1965, **61**, 2321.
[49] S. G. Ash and G. H. Findenegg, *Special Discuss. Faraday Soc.*, 1970, No. 1, 105.
[50] G. H. Findenegg, *J. Colloid Interface Sci.*, 1971, **35**, 249.
[51] L. Robert, *Compt. rend.*, 1963, **256**, 655; *Bull. Soc. chim. France*, 1967, 2309.
[52] D. H. Everett and G. H. Findenegg, *J. Chem. Thermodynamics*, 1969, **1**, 573.
[53] J. H. Clint, J. S. Clunie, J. S. Goodman, and J. R. Tate, *Nature*, 1969, **223**, 51.

liquids[54,55] in narrow quartz capillaries, Churayev *et al.*[56] now report that carbon tetrachloride and benzene behave normally under these circumstances. Russell and Pattle[57] have tested the ability of fluids, close to their critical conditions, to disperse, or at least form stable slurries of, various solid powders in closed vessels (no added stabilizers present). It was noted that as the temperature was raised towards the critical temperature, T_c, this ability decreased. For example, liquid ethane ($T_c = 32 °C$) and propane ($T_c = 96 °C$) did not form stable slurries of ultramarine powder at $0 °C$, whereas liquid n-butane ($T_c = 153 °C$) and liquid n-pentane ($T_c = 197 °C$) did. In the latter case, the slurry was stable up to around $50 °C$. One suspects that viscosity is the controlling effect here.

Benitez *et al.*[58] have shown that methylated Aerosil (SiO_2) may be dispersed in a number of organic solvents, *e.g.* alcohols, without added stabilizers, but the addition of water, above a critical concentration, leads to flocculation. This critical concentration increases with alcohol chain-length. The authors interpret this phenomenon in terms of de-wetting of the solid surface, that is, the system has changed from an essentially lyophilic one to a lyophobic one. However, although wetting may be an important factor controlling the initial dispersibility of such particles, it is not, fundamentally, the controlling factor as far as their stability behaviour is concerned. It is known that silica particles in aqueous dispersions are covered with surface layers of poly(silicic acid).[59] Methylation of the particles would presumably lead to methylation of this material also. A possible explanation of the results, therefore, is that incipient flocculation of the type described by Napper[11,44] does occur, contrary to the authors' opinion, on adding water to the system, water being a non-solvent for methylated poly(silicic acid). The authors state that the observed changes were reversible on adding alcohol. According to Napper,[11,44] incipient flocculation is also reversible provided the systems are not allowed to stand in the flocculated state for too long.

The gelation of (unmethylated) silica dispersions in methanol on the addition of various additives (*e.g.*, water, electrolyte solutions, amines) has been investigated by Akabayashi *et al.*[60]

Finally, in this section on stability, one would mention some interesting work by Prigorodov[61] on the foaming of partially miscible liquid mixtures. The system studied was poly(ethylsiloxane) plus poly(methyl-γ-trifluoro-

[54] N. N. Fedyakin, *Kolloid. Zhur.*, 1962, **24**, 497 [*Colloid J.(U.S.S.R.)*, 1962, **24**, 425].
[55] D. H. Everett, J. M. Haynes, and P. J. McElroy, *Sci. Progr.*, 1971, **59**, 279.
[56] N. V. Churayev, V. D. Sobolev, and Z. M. Zurin, *Special Discuss. Faraday Soc.*, 1970, No. 1, 213.
[57] J. H. Russell and R. E. Pattle, *J. Appl. Chem. Biotechnol.*, 1971, **21**, 174.
[58] R. Benitez, S. Contreras, and J. Goldfarb, *J. Colloid Interface Sci.*, 1971, **36**, 146.
[59] W. Stöber in 'Equilibrium Concepts in Natural Water Systems', ed. R. F. Gould, Advances in Chemistry Series, American Chemical Society, 1967, No. 67, p. 161.
[60] H. Akabayashi, A. Yosmida, and Y. Otsubo, *Kogyo Kagaku Zasshi*, 1970, **73**, 866, 871 (*Chem. Abs.*, 1970, **73**, 81 002, 81 003).
[61] V. N. Progorodov, *Kolloid. Zhur.*, 1970, **32**, 793 [*Colloid J(U.S.S.R.)*, 1970, **32**, 662].

propylsiloxane). Stable foams were formed in the totally miscible regions at either end of the concentration scale, but not in the immiscible central region. The explanation, however, is not entirely clear.

4 Charge Effects

Surface-charging mechanisms, on the whole, are better understood for aqueous systems than non-aqueous ones. Particle electrophoresis* is one possible method for studying the nature of surface charge groups, but it is notoriously difficult to obtain consistent results in media of low polarity. For example, trace impurities of polar materials, especially water, can have considerable effects, not only on the magnitude, but also the sign, of the particle mobility. Indeed this effect has been utilized as a method for separating, *e.g.* mineral particles, by adding ionic surfactants in a suitable solvent.[63] Jackson and Parfitt[12] have carried out electrophoresis and i.r. studies on dispersions of rutile (TiO$_2$) in the C$_1$ to C$_{10}$ n-alcohols. Water, as usual, had a major effect. In rigorously dried alcohols the authors found that the mobility was negative for the lower members of the series ($< C_4$), but became positive for the higher members. It is suggested that two processes contribute: (*a*) exchange of surface protons with alcohol molecules: (*b*) the self-ionization of the alcohols. For the lower alcohols (*a*) probably dominates, whilst (*b*) is dominant in the case of the higher alcohols.

The effect of pre-treatment temperature, and the role of physi- and chemisorbed water on the zeta potential and i.r. spectra of silica dispersions in methanol and n-propanol have been studied by Maroto and Griot.[64] Furthermore, Zhukov *et al.*[13] have also illustrated the important effect that water has on the surface conductivity and zeta potential of quartz dispersions in dioxan. In the absence of water, cetyltrimethylammonium bromide (CTAB) had little effect on the zeta potential of the particles. With increasing water content, however, the zeta potential reverses sign. It must be that CTAB is adsorbed at the interface only when co-adsorbed with water. Water also produced a large increase in the surface to volume conductance ratio.

Arguments have been put forward against the use of classical electrical double-layer theory for media of low polarity, where the concentration of free ions present is normally very small.[65] Nevertheless, Sanfeld *et al.*[66, 67] have analysed the theoretical contribution of ionic molar volumes and ionic polarizabilities to the electrical double-layer repulsion between particles in a

---

* Tari[62] has reviewed electrokinetic techniques for measuring zeta potentials in non-aqueous media.

[62] I. Tari, *Shikizai Kyokaishi*, 1970, **43**, 510 (*Chem. Abs.*, 1971, **74**, 68 043).
[63] G. Todd and G. A. Wild, *Chem. and Ind.*, 1970, 715; *J. Colloid Interface Sci.*, 1970, **33**, 178.
[64] A. J. G. Moroto and O. Griot, *Anales Asoc. qum. Argentina*, 1970, **58**, 175.
[65] D. W. J. Osmond, *Discuss. Faraday Soc.*, 1966, No. 42, p. 247.
[66] A. Sanfeld, C. Devillez, and P. J. Terlinek, *J. Colloid Interface Sci.*, 1970, **32**, 33.
[67] A. Sanfeld, C. Devillez, and S. Wahrmann, *J. Colloid Interface Sci.*, 1971, **36**, 359.

non-aqueous medium. The particular system considered[67] was: 1 μm radius, spherical particles dispersed in benzene solutions of tetra-isoamylammonium picrate. In comparison with the classical DLVO theory the authors predict a large increase in the electrical double-layer repulsion from the strong polariz-ability of the cation, whereas its finite molar volume leads to a slight decrease. Both effects increase with increasing surface potential and solute concentra-tion.

Another charging effect that can arise (and can indeed be a hazard!) with media of low polarity occurs in flow through pipes, membranes, and porous media. Further work in this field has been reported.[68-70] The effect of such factors as liquid conductivity, flow rate, and flow time on the sign and magni-tude of the charge generated in the liquid were investigated. The charge appears to go through a maximum with increasing conductivity, the position of the maximum being dependent on the other factors such as flow rate.[68] The nature of the solid surface plays an important role,[68] as does the presence of ionic impurities in the liquid.[69]

5 Association Colloids

As indicated in the Introduction, although micelle formation in non-aqueous surfactant solutions has been recognized for some time, it is only in recent years that systematic studies have been undertaken. Most of the common techniques (*e.g.* light scattering) used for studying micellization in aqueous media can, in principle, be used with non-aqueous systems. Vipper *et al.*[71] have recently, for example, described spectrophotometric and refractometric methods, and Scibina *et al.*[72] have produced log (interfacial tension) *versus* log (surfactant concentration) plots for long-chain alkylammonium salts at the aqueous/benzene interfaces. They ascribe the discontinuity in these plots to micellization in the benzene phase.

Shinoda and Ogawa[73] have previously shown that for a fixed surfactant concentration in a given hydrocarbon solvent, there is a narrow, optimum temperature range over which solubilization is considerably enhanced. The surfactants studied were generally non-ionics of the alkyl (or alkylphenyl)–poly(ethylene oxide) type. A recent example,[74] which illustrates this effect, is shown in Figure 4, where the system is dodecyl hexa(ethylene oxide) ($C_{12}E_6$) in cyclohexane. In other publications, Shinoda[75,76] has shown how

[68] J. T. Leonard and H. W. Carhart, *J. Colloid Interface Sci.*, 1970, **32**, 383.
[69] J. L. Lauer and P. G. Antal, *J. Colloid Interface Sci.*, 1970, **32**, 407.
[70] J. Ginsburg, *J. Colloid Interface Sci.*, 1970, **32**, 424.
[71] A. B. Vipper, V. N. Baumann, E. J. Markova, and P. I. Sanin, *Neftekhimiya*, 1970, **10**, 748.
[72] G. Scibina, P. R. Danesi, A. Conte, and B. Scuppa, *J. Colloid Interface Sci.*, 1971, **35**, 631.
[73] K. Shinoda and T. Ogawa, *J. Colloid Interface Sci.*. 1967, **24**, 56.
[74] H. Saito and K. Shinoda, *J. Colloid Interface Sci.*, 1971, **35**, 359.
[75] K. Shinoda, *Chem. Ind. Japan*, 1966, **19**, 199.
[76] K. Shinoda in 'Solvent Properties of Surfactant Solutions', Marcel Dekker New York, 1967, p. 27.

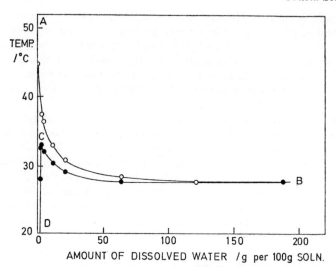

Figure 4 *The solubilization region (ABCD) of water in cyclohexane containing*
5.86 wt. % $C_{12}E_6$. ○, *solubilization end-point;* ●, *cloud (haze)*
point
(Reproduced by permission from *J. Colloid Interface Sci.*, 1971, **35**, 359)

the phase diagram for this type of non-ionic surfactant in water gradually
changes as increasing levels of hydrocarbon are added to the system.* The
region ABCD in Figure 4 seems to be a homogeneous, though ill-defined,
single phase, rich in surfactant and containing solubilized water as well as
hydrocarbon. It is in fact directly analogous to the 'surfactant' phase illustrated in Figure 1 in Chapter 6 (p. 222). Point B thus represents the maximum
level of solubilized water possible. The region above the line AB is a two-
phase region, excess water being emulsified in the continuous hydrocarbon
phase, which contains most of the surfactant. The region below the line DCB
is also a two-phase region, in which the hydrocarbon is largely dispersed in a
continuous water–surfactant phase. As discussed in Chapter 6, one can
probably best regard AB as the phase-separation curve for the surfactant in
the hydrocarbon media, and CB as the phase-separation curve for the surfactant in aqueous media. The single-phase region, in fact, not only corresponds to the region of maximum solubilization of water, but also to the region
for microemulsion formation, and to the phase-inversion region for normal
emulsions (again, see Chapter 6).

By determining the amount of water solubilized at a given temperature
(*i.e.* from the curve AB in Figure 4) as a function of surfactant concentration,
Saito and Shinoda[74] were able to determine the c.m.c. of $C_{12}E_6$ in cyclohexane

---

* The reader is strongly encouraged to refer to these articles[75, 76] for a fuller appreciation
  of the various facets involved.

and showed that it increased with increasing temperature. The c.m.c.'s obtained cannot be expected to be the same as in the absence of water, however.

A similar series of experiments has been carried out by Kon-No and Kitahara,[77-80] mainly to investigate the phenomenon of secondary solubilization of electrolytes by micelles in non-aqueous solutions. In their studies using various cationic surfactants in the dodecylammonium carboxylate series,[77] they used a water-soluble dye (Crystal Violet) to construct the phase diagrams for the systems. A fixed concentration (0.1 mol dm^{-3}) of surfactant in various hydrocarbon or chlorinated-hydrocarbon solvents was used. The general features of the phase diagram obtained are the same as those depicted in Figure 4. The major difference was that the single-phase region now splits into two homogeneous solubilization regions, one being colourless, the other being light blue and translucent in appearance in the presence of the dye. No explanation for this observation is offered. The transition between the two is sharp, however.

It was found that the level of maximum solubilization increased, and the optimum temperature for solubilization decreased, with increasing chain length of the carboxylate group. There is also a strong dependence on the nature of the solvent. The general trend with added electrolytes was also to increase the level of solubilization and to lower the optimum solubilization temperature. The orders of effectiveness for cations and anions, respectively, were:

$$Cs^+ > Na^+ \sim K^+ > Li^+ \sim NH_4^+ > \tfrac{1}{2}Ca^{2+} \sim \tfrac{1}{2}Mg^{2+} > \tfrac{1}{3}Al^{3+};$$

$$SCN^- > I^- > NO_3^- > Br^- > Cl^- > I^-.$$

A tentative explanation is offered in terms of the Lewis acidity–basicity of the ions and the nature of the ionization of the carboxylate groups in the interior of the micelles, along the lines proposed by Fowkes.[15]

With the non-ionic poly(ethylene oxide) nonylphenyl ethers,[78] cations had little effect on the water solubilization level, but the order for anions was essentially the reverse of that given above. There was a strong effect with NaOH but little effect with the protonic acids. The structural changes which salts induce in the poly(ethylene oxide)/water interiors of the micelles must reflect the complex changes that are known to occur in bulk aqueous solutions of this polymer when salts are added.[81]

The level of water solubilization in non-aqueous micelles may also be determined from measurements of water vapour pressure above the system.[79] From the variation of the vapour pressure with temperature, at a fixed water concentration, the enthalpy of solubilization, ΔH_s, may be determined by

[77] K. Kon-No and A. Kitahara, *J. Colloid Interface Sci.*, 1970, **33**, 124.
[78] K. Kon-No and A. Kitahara, *J. Colloid Interface Sci.*, 1970, **34**, 221.
[79] K. Kon-No and A. Kitahara, *J. Colloid Interface Sci.*, 1971, **35**, 409.
[80] K. Kon-No and A. Kitahara, *J. Colloid Interface Sci.*, 1971, **35**, 636.
[81] S. Erlander, *J. Colloid Interface Sci.*, 1970, **34**, 53.

use of the Clausius–Clapeyron relationship. With anionic surfactants, ΔH_s was found to decrease initially as the level of added water increased, and then levelled off at a value more or less corresponding to the enthalpy of condensation of pure water, ΔH_c. With cationics, ΔH_s increased initially to the ΔH_c value, whilst with non-ionics there was no obvious trend, ΔH_s tending to oscillate about the value of ΔH_c (probably reflecting experimental error) over the complete range of water concentration that was studied. The definite changes in ΔH_s with ionic surfactants at low water levels, however, must relate to the water–ionic group interactions in the micelle. One would have thought, intuitively, that this interaction would have been greater for cationic groups than anionic groups, whereas the results seem to imply the opposite. At high levels of added water, water–water interactions in the micelle must dominate.

The authors also investigated[79] the effect of: (*a*) varying the nature of the counter-ions in the case of the anionic surfactants; and (*b*) the size of the hydrocarbon group of the surfactant. ΔH_s was greater for Na^+ counter-ions than K^+, the former being more heavily solvated, of course. Less water seems to be associated with the micelles, the larger the hydrocarbon group. This reflects the decrease in micelle aggregation number that occurs with the bulkier hydrocarbon groups. In a further paper,[80] Kon-No and Kitahara have made a specific study of aggregation numbers, using a vapour-pressure-depression technique. For a given hydrocarbon chain-length, the aggregation number for various head groups lay in the order:

$$\text{anionic} \gg \text{cationic} > \text{non-ionic.}$$

The aggregation numbers obtained by this method are number averages. By making a comparison with the somewhat sparse data from light-scattering measurements, which give weight-average aggregation numbers, the authors were able to conclude that, as is the case with aqueous systems, micelles in non-aqueous systems appear to be monodisperse.

Finally, we would mention that molecules which micellize are good drag-reducing agents. Little *et al.*[82] have reported on some aspects of this effect for non-aqueous liquids.

[82] R. C. Little, R. H. Baker, R. N. Bolster, and P. B. Leach, *Ind. and Eng. Chem.* (*Product Res. and Development*), 1970, **9**, 541.

Author Index